DFG

**The MAK-Collection for
Occupational Health and Safety**
Part I: MAK Value Documentations

Forthcoming Volumes

Part II: BAT Value Documentations, Volume 4
Drexler, H. and Greim, H. (eds.)

2005. ISBN 3-527-27049-3

Part III: Air Monitoring Methods, Volume 9
Parlar, H. and Greim, H. (eds.)

2005. ISBN 3-527-31138-6

Part IV: Biomonitoring Methods, Volume 10
Angerer, J. and Greim, H. (eds.)

2006. ISBN 3-527-31137-8

The MAK-Collection online
www.mrw.interscience.wiley.com/makbat

DFG Deutsche Forschungsgemeinschaft

The MAK-Collection for Occupational Health and Safety

Part I: MAK Value Documentations

Volume 21

Edited by Helmut Greim

Commission for the Investigation of Health
Hazards of Chemical Compounds in the Work Area

WILEY-VCH Verlag GmbH & Co. KGaA

Prof. Dr. Helmut Greim
Commission
for the Investigation of Health Hazards
of Chemical Compounds in the Work Area
Hohenbachernstr. 15 -17
85354 Freising
Germany

All books published by Wiley-VCH are carefully produced. Nevertheless, authors, editors, and publisher do not warrant the information contained in these books, including this book, to be free of errors. Readers are advised to keep in mind that statements, data, illustrations, procedural details or other items may inadvertently be inaccurate.

We gratefully acknowledge the support of Brush Wellman Inc. Cleveland, Ohio, USA and the Scientific Committtee for Occupational Exposure Limits (SCOEL) of the European Union, Luxemburg.

Translators:
O. Bomhard
J. Handwerker-Sharman
Dr. A. E. Wild

Library of Congress Card No.: applied for
British Library Cataloging-in-Publication Data: A catalogue record for this book is available from the British Library.

Bibliographic information published by Die Deutsche Bibliothek
Die Deutsche Bibliothek lists this publication in the Deutsche Nationalbibliografie; detailed bibliographic data is available in the Internet at http://dnb.ddb.de.

ISSN: 1860-496X
ISBN-10: 3-527-31134-3
ISBN-13: 978-3-527-31134-7

© 2005 WILEY-VCH Verlag GmbH & Co. KGaA, Weinheim

Printed on acid-free paper

All rights reserved (including those of translation into other languages). No part of this book may be reproduced in any form – by photoprinting, microfilm, or any other means – nor transmitted or translated into a machine language without written permission from the publishers. Registered names, trademarks, etc. used in this book, even when not specifically marked as such, are not to be considered unprotected by law.

Layout: K. Schöpe
Printing: betz-druck GmbH, Darmstadt
Binding: Litges & Dopf Buchbinderei GmbH, Heppenheim

Printed in the Federal Republic of Germany

Note

In the present volume reference is frequently made to the various sections of the *List of MAK and BAT Values** (MAK List). The MAK List was introduced and its sections described in detail in the first volume of this series.

---

* Available from the publisher, WILEY-VCH, D-69451 Weinheim, both in the original German edition and as an English translation

Contents of Volume 21

General Papers
Monocyclic aromatic amino and nitro compounds 3

Chemical Substances
Arsenic and its inorganic compounds (except arsine) 49
Beryllium and its inorganic compounds 107
Butanethiol 161
Carbon disulfide 171
Diisopropyl ether 187
Ethanethiol 195
Nitrogen dioxide 205
Propargyl alcohol 261
Vinyl acetate 271

Authors of Documents in Volume 21 295
Index 297

General Papers

Monocyclic aromatic amino and nitro compounds:

toxicity, genotoxicity and carcinogenicity, classification in a carcinogen category

H.-G. Neumann, Institut für Toxikologie, Universität Würzburg

1 Introduction

It was long believed that monocyclic amino compounds (arylamines) are generally not carcinogenic and that for carcinogenicity arylamines had to have at least two benzene rings. Only after the hepatocarcinogenic properties of 2,4,6-trimethylaniline had been reported (Morris 1965) were the monocyclic aromatic amino and nitro compounds included in the NCI bioassay programme for testing substances for carcinogenicity (Weisburger and Fiala 1981). In the *List of MAK and BAT Values* numerous monocyclic aromatic amino and nitro compounds are classified in the carcinogen categories 1 to 3. Now that the two new categories, 4 and 5, have been created for the classification of carcinogens, it is necessary—especially for substances in category 3—to decide which compounds can be reclassified according to the new criteria in one of the new categories (Neumann *et al.* 1998, DFG 2004).

In a comparative review of the appropriate data, the present document aims to establish whether the monocyclic aromatic amino and nitro compounds which are currently classified in the *List of MAK and BAT Values* may be considered as a group for the purposes of classification in carcinogen categories and whether classification on the basis of analogy may be justified. Because it is considered that these substances share common mechanisms of action and that therefore any differences in effect are quantitative rather than qualitative, it seems unsatisfactory that individual substances of this group are found classified in the Carcinogen categories 1 to 3, mostly because of differences in the amount of data available (Table 1). And it has been suggested that the carcinogenic effects and the activity of these substances could be a result mainly of their acute and chronic toxicity and less of their genotoxicity, which could justify their classification in the new category 4.

Tables 1 to 4 present all the monocyclic aromatic amino and nitro compounds which are found to date in the *List of MAK and BAT Values*, both those which are classified in one of the carcinogen categories, those with a MAK value, and those for which insufficient data were available and which are therefore listed in Section IIb of the *List of MAK and BAT Values*. The substances are not presented in the order of their current classification in carcinogen categories but in six groups formed on the basis of structural considerations. Amino and nitro derivatives are considered together as *N*-substituted aryl compounds. The data have mostly been taken from the MAK documentation for the individual substances; the literature sources are cited there. These data have been supplemented with data published especially during the last 10 years and with information from the databases of the European Chemicals Bureau of the EU. An exhaustive review of the data has not been attempted and often only the most relevant data have been included. The six groups of substances:

mono-*N*-substituted benzene derivatives
(1) aniline and substances which can be metabolized to yield aniline derivatives
(2) *o*-toluidine and its derivatives
(3) *p*-toluidine and its derivatives
poly-*N*-substituted benzene derivatives
(4) phenylenediamine and substances which can be metabolized to yield phenylenediamine
(5) di-substituted and poly-substituted toluene derivatives and analogous benzene derivatives
(6) *N*-substituted aminophenols and nitrophenols and substances which can be metabolized to yield phenols

The structural formulae of the monocyclic aromatic amino and nitro compounds included in the *List of MAK and BAT Values* are shown in Figure 1.

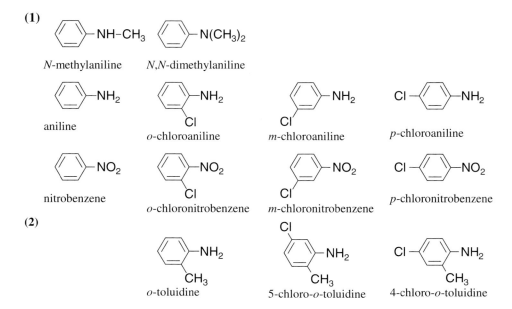

Figure 1. Structural formulae of the monocyclic aromatic amino and nitro compounds in the *List of MAK and BAT Values* 2004

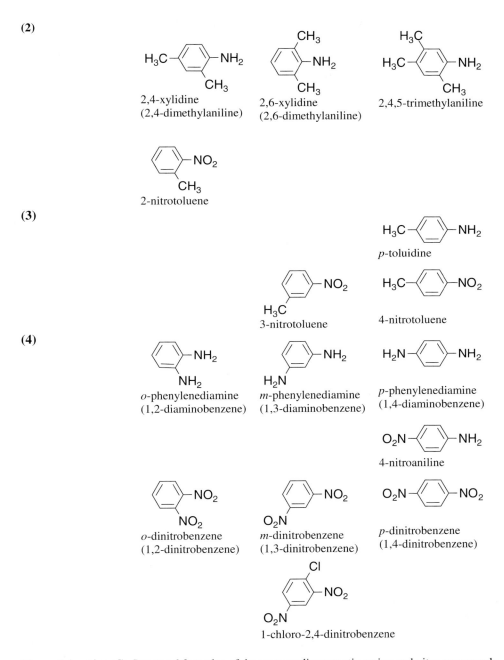

Figure 1 (continued). Structural formulae of the monocyclic aromatic amino and nitro compounds in the *List of MAK and BAT Values* 2004

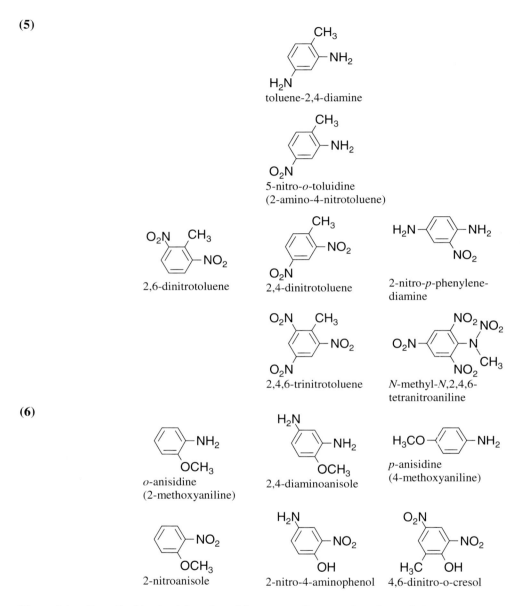

Figure 1 (continued). Structural formulae of the monocyclic aromatic amino and nitro compounds in the *List of MAK and BAT Values* 2004

2 Classification

The amount of data available for the individual substances is very varied and results in marked differences in classification (Table 1). When appropriate tests have been carried out, tumours have generally been observed; sometimes the incidences were in the range of the significance limits; the sex and species differences are mostly inexplicable.

3 Toxicity

Practically all the listed monocyclic aromatic amino and nitro compounds induce acute methaemoglobin (metHb) formation in experimental animals (Table 2). *p*-Phenylenediamine and 2,4-diaminoanisole seem to be exceptions; however, 2,4-diaminoanisole has been reported to cause anaemia. In some cases adaptation has been described after repeated uptake of the substance, that is, initial increases in the level of metHb disappear during the course of treatment. Usually the animals become anaemic. Heinz bodies are frequently reported but do not always seem to correlate with the metHb formation. Haemosiderosis is almost always observed and put down to the increased turnover of erythrocytes. The distribution of the deposits in the organism can differ widely.

Expressed in terms of the LD_{50} for the rat (Table 2), the acute toxicity of the listed monocyclic aromatic amino and nitro compounds ranges from moderate to low. The low values range from 300 to 400 mg/kg body weight, the high up to over 1000 mg/kg body weight. Exceptions are the dinitrobenzenes, *p*-phenylenediamine and 4,6-dinitro-*o*-cresol for which values below 100 mg/kg body weight have been reported. In the case of the dinitrobenzenes, the high toxicity is a result of the severe haematotoxicity: in tests of 24 monocyclic aromatic amino and nitro compounds, *p*-dinitrobenzene caused most metHb formation in the cat (Bodansky 1951). For *p*-phenylenediamine, the development of oedema and neuromyopathy is indicative of effects specific for the rat (Tainter *et al.* 1926). Likewise the highly toxic, not yet classified substance 4,6-dinitro-*o*-cresol (IIb in the *List of MAK and BAT Values*) operates by a different mechanism as a respiratory chain uncoupler (DFG 1976).

As a parameter for bioavailability of biologically active metabolites, the available haemoglobin binding indices (HBI) are listed in Table 2. With practically all the compounds such adduct formation can be demonstrated; given the simultaneous metHb formation, this is only to be expected (Sabbioni 1994). The lowest and highest values differ by a factor of 100. In summary,
- the comparable HBIs of aniline, *N*-methylaniline and *N*,*N*-dimethylaniline support the conclusion that the toxicity of these compounds is also comparable (see below);
- the values for *o*-chloroaniline and *p*-chloroaniline are very different;
- the HBIs for the three nitrotoluene isomers are the same;
- the value for dinitrobenzene is conspicuously high;
- the value for 2,4-diaminotoluene is conspicuously low.

Table 1. Monocyclic aromatic amino and nitro compounds: classification, MAK values and tumour locations

Substance [CAS number]	Category[1]	MAK value	Tumour location[2] rat	mouse	MAK documentation[3] References
(1) aniline and substances which can be metabolized to yield aniline derivatives					
aniline [62-53-3]	3B	2 ml/m^3	spleen sarcomas, haemangiosarcomas, fibrosarcomas, osteosarcomas	no tumours	Volume 6
N-methylaniline [100-61-8]	–	0.5 ml/m^3	no tumours (inadequately studied)	no tumours (inadequately studied)	Volume 6
N,N-dimethylaniline [121-69-7]	3B	5 ml/m^3	sarcomas in spleen and thymus, osteosarcomas	forestomach papillomas	Volume 3
o-chloroaniline [95-51-2]	–	–	not studied	not studied	Volume 3
m-chloroaniline [108-42-9]	–	–	not studied	not studied	Volume 3
p-chloroaniline [106-47-8]	2	–	spleen sarcomas	liver adenomas, carcinomas and haemangiosarcomas, spleen haemangiosarcomas	Volume 3, Chhabra et al. 1991
nitrobenzene [98-95-3]	3B	–	liver adenomas and carcinomas	lung adenomas	Volume 19, ECB 2000b
o-chloronitrobenzene [88-73-3]	3B	–	multiple tumours (inadequately studied)	liver carcinomas (inadequately studied)	Volume 4, Nair et al. 1986
m-chloronitrobenzene [121-73-3]	–	–	not studied	not studied	Volume 4
p-chloronitrobenzene [100-00-5]	3B	–	no tumours (inadequately studied)	vascular tumours (inadequately studied)	Volume 4
(2) o-toluidine and its derivatives					
o-toluidine [95-53-4]	2	–	fibrosarcomas, angiosarcomas, osteosarcomas, haemangiosarcomas mesotheliomas, multiple tumours e.g. in bladder, liver, mammary gland	haemangiosarcomas, liver adenomas and carcinomas	Volume 3, ECB 2000e

Table 1. continued

Substance [CAS number]	Category[1]	MAK value	Tumour location[2] rat	mouse	MAK documentation[3] References
5-chloro-o-toluidine [95-79-4]	3B	–	(adrenal glands)	haemangiosarcomas, liver	Volume 6
4-chloro-o-toluidine [95-69-2]	1	–	(hepatomas)	haemangiosarcomas, spleen	Volume 6
2,4-xylidine (2,4-dimethylaniline) [95-68-1]	2	–	no tumours (inadequately studied)	lung (inadequately studied)	Volume 19, BUA 1995
2,6-xylidine (2,6-dimethylaniline) [87-62-7]	2	–	adenomas, carcinomas and sarcomas of the nasal cavity, fibromas, and fibrosarcomas of the subcutis, liver nodules, hepatomas	not studied	Volume 19, BUA 1995, ECB 2000f
xylidine isomers: 2,3-xylidine, 2,5-xylidine, 3,4-xylidine, 3,5-xylidine) [87-59-2, 95-78-3, 95-64-7, 108-69-0]	3A	–	fibromas (2,5-xylidine; inadequately studied)	vascular tumours, hepatomas (2,5-xylidine; inadequately studied)	Volume 19, BUA 1995
2,4,5-trimethylaniline [137-17-7]	2	–	liver, lung, multiple tumours	liver, multiple tumours	Volume 4
2-nitrotoluene [88-72-2]	2	–	mesotheliomas, fibromas, fibrosarcomas, liver	haemangiosarcomas, fibromas, liver	Volume 8, Greim 2002
(3) *p*-**toluidine** and its derivatives					
***p*-toluidine** [106-49-0]	3B	–	no tumours (oral administration); liver, local subcutaneous sarcomas (s.c.)	liver adenomas and carcinomas (oral administration)	Volume 3, ECB 2000g
3-nitrotoluene [99-08-1]	–	5 ml/m^3	not studied	not studied	ECB 2000h
4-nitrotoluene [99-99-0]	–	5 ml/m^3	clitoris	(lung)	Bayer AG 2003, NTP 2002

Table 1. continued

Substance [CAS number]	Category [1]	MAK value	Tumour location [2] rat	Tumour location [2] mouse	MAK documentation [3] References
(4) phenylenediamine and substances which can be metabolized to yield phenylenediamine					
o-phenylenediamine (1,2-diaminobenzene) [95-54-5]	3B	–	liver carcinomas (oral administration)	liver carcinomas (oral administration); local subcutaneous sarcomas (s.c. injection)	Volume 13, BUA 1993, ECB 2000i
m-phenylenediamine (1,3-diaminobenzene) [108-45-2]	3B	–	no tumours (oral administration); local subcutaneous sarcomas (s.c. injection)	no tumours (oral, dermal administration)	Volume 6, BUA 1993, ECB 2000j
p-phenylenediamine (1,4-diaminobenzene) [106-50-3]	3B	0.1 mg/m^3	no tumours (oral administration)	no tumours (oral, dermal administration)	Volume 6, BUA 1993, ECB 2000k
4-nitroaniline [100-01-6]	3A	–	no tumours	(haemangiosarcomas)	Greim 1999, Nair et al. 1990
dinitrobenzene (all isomers) [25154-54-5]	3B	–	not studied	not studied	Volume 1
1-chloro-2,4-dinitrobenzene [97-00-7]	–	–	no tumours (inadequately studied)	no tumours (inadequately studied)	BG Chemie 1992; ECB 2000n
(5) di-substituted and poly-substituted toluene derivatives and analogous benzene derivatives					
toluene-2,4-diamine [95-80-7]	2	–	liver, multiple tumours	liver	Volume 6
5-nitro-*o*-toluidine (2-amino-4-nitrotoluene) [99-55-8]	2	–	(liver)	liver, haemangiosarcomas	Volume 6
2,6-dinitrotoluene [606-20-2]	2	–	liver, multiple tumours	not studied	Volume 6
2,4-dinitrotoluene [121-14-2]	2	–	fibromas, fibroadenomas, mammary gland, liver	kidney	Volume 6
2-nitro-*p*-phenylenediamine [5307-14-2]	3B	–	no tumours	liver	Volume 4
2,4,6-trinitrotoluene [118-96-7]	3B	0.011 ml/m^3	(urinary bladder)	not studied	Volume 1, Furedi et al. 1984

Table 1. continued

Substance [CAS number]	Category[1]	MAK value	Tumour location[2] rat	mouse	MAK documentation[3] References
N-methyl-N,2,4,6-tetra-nitroaniline [479-45-8]	3B	–	(stomach)	not studied	Volume 11
(6) N-substituted aminophenols and nitrophenols and substances which can be metabolized to yield phenols					
o-anisidine (2-methoxyaniline) [90-04-0]	2	–	urinary bladder, kidney, thyroid gland	urinary bladder	Volume 10
2,4-diaminoanisole [615-05-4]	2	–	thyroid gland, Zymbal gland, clitoris	thyroid gland	Volume 6
p-anisidine (4-methoxyaniline) [104-94-9]	3B	–	urinary bladder, preputial gland, liver	no tumours	Greim 2001
2-nitro-4-aminophenol [119-34-6]	3B	–	urinary bladder, fibroadenomas	lung (inadequately studied)	Volume 4
2-nitroanisole [91-23-6]	2	–	leukaemia, urinary bladder, kidney, colon, lower urinary tract	liver adenomas and carcinomas	Volume 9, ECB 2000p
4,6-dinitro-o-cresol [534-52-1]	–	–	no tumours	not studied	Volume 19

[1] Carcinogen categories
[2] especially the main target organs are listed; low incidences are indicated with brackets
[3] volume numbers refer to MAK documentation in present series, *Occupational Toxicants*, Volumes 1–20
s.c.: subcutaneous

Not all the results can be explained satisfactorily but the data suggest additional arguments for the comparative assessment of the substances.

In an evaluation of the acute and chronic toxicity of these substances, the neurotoxic effects should be included. These have often been described but, apart from a role of oxygen deficiency, practically no quantitative data or mechanistic suggestions have been published.

For assessment of the carcinogenic properties, the frequently observed but probably not always recorded development of fibrosis is of significance. Its appearance is regularly associated with damage in the typical target organs spleen, liver and kidneys and it could be causally involved as an important parameter in the development of tumours (NCI 1978). The mechanism of action is not fully understood. The study of the primary reactions which result in toxicity should be continued, for example, at the molecular level, in order to identify appropriate parameters for dose-dependency.

Table 2. Monocyclic aromatic amino and nitro compounds: toxicity

Substance	Effects of single exposures	Effects of repeated exposures	LD_{50} rat (mg/kg body weight, *p.o.*)	HBI[1]	MAK documentation,[2] References
(1) aniline and substances which can be metabolized to yield aniline derivatives					
aniline	spleen, haemosiderosis, metHb	anaemia, Heinz bodies liver, kidneys	440	22	Volume 6, Albrecht and Neumann 1985
***N*-methylaniline**	metHb, Heinz bodies	anaemia, spleen, liver, kidneys	no data	16	Volume 6, Birner and Neumann 1988
***N,N*-dimethylaniline**	metHb	anaemia, haemosiderosis, spleen, liver	1300	11	Volume 3, Birner and Neumann 1988
***o*-chloroaniline**	metHb, Heinz bodies	anaemia, kidney, haemosiderosis, spleen, liver	116	0.5	Volume 3, ECB 2000a, NTP 1998, Sabbioni 1994
***m*-chloroaniline**	metHb, Heinz bodies	anaemia, haemosiderosis, spleen, liver, kidney	1104	12.5	Volume 3, NTP 1998, Sabbioni 1994
***p*-chloroaniline**	metHb, liver, kidneys	anaemia, spleen, Heinz bodies, haemosiderosis, liver, kidney	300–400	569	Volume 3, Chhabra *et al.* 1990, 1991, NTP 1998, Birner and Neumann 1988
nitrobenzene	metHb, liver, kidneys	anaemia, haemosiderosis, spleen, liver (necrosis), kidney, testis	640	79	Volume 19, Albrecht and Neumann 1985
***o*-chloronitrobenzene**	metHb, Heinz bodies	anaemia, haemosiderosis, spleen, kidney, liver (necrosis)	180–500	no data	Volume 4, Travlos *et al.* 1996
***m*-chloronitrobenzene**	metHb	metHb, Heinz bodies, testis	470	no data	Volume 4
***p*-chloronitrobenzene**	metHb	anaemia, haemosiderosis, Heinz bodies, spleen, liver, kidney, testis	400–1000	"high"	Volume 4, Travlos *et al.* 1996

Table 2. continued

Substance	Effects of single exposures	Effects of repeated exposures	LD_{50} rat (mg/kg body weight, p.o.)	HBI[1]	MAK documentation,[2] References
(2) *o*-toluidine and its derivatives					
***o*-toluidine**	metHb	anaemia, haemosiderosis, spleen, urinary bladder, kidney	600–1300	4	Volume 3, Birner and Neumann 1988
5-chloro-*o*-toluidine	metHb	liver, kidney	about 1000	28	Volume 6, BIA 2003, Birner and Neumann
4-chloro-*o*-toluidine	metHb, Heinz bodies	anaemia, Heinz bodies, spleen, liver, urinary bladder	700–1000	28	Volume 6, Birner and Neumann 1988
2,4-xylidine (2,4-dimethylaniline)	metHb, liver	liver (increased endoplasmic reticulum, necrosis), spleen, kidney	470	2.3	Volume 19, Birner and Neumann 1988
2,6-xylidine (2,6-dimethylaniline)	metHb	spleen, haemosiderosis fibrosis, liver (enzyme induction), kidney	1230	1.1	Volume 19, ECB 2000f, Sabbioni 1994, Vernot et al. 1977
xylidine (2,3-xylidine, 2,5-xylidine, 3,4-xylidine, 3,5-xylidine)	metHb (3,5-xylidine)	fibromas, liver (2,5-xylidine; inadequately studied)	930 (2,3-) 1300 (2,5-) 810 (3,4-) 710 (3,5-)	7.3 (2,5-) 0.7 (3,4-) 14.0 (3,5-)	Volume 19, Sabbioni 1994, Vernot et al. 1977
2,4,5-trimethyl-aniline	metHb	spleen, liver, lung	1500	0.7 (0.2)	Volume 4, Birner and Neumann 1988
2-nitrotoluene	metHb, liver, olfactory system	haemosiderosis, fibrosis; spleen, liver, kidney, testis	890–2500	5	Volume 8, 2002, Dunnik et al. 1994
(3) *p*-toluidine and its derivatives					
***p*-toluidine**	metHb	anaemia, liver (enzyme induction)	650–1200	2	Volume 3, Birner and Neumann 1988

Table 2. continued

Substance	Effects of single exposures	Effects of repeated exposures	LD_{50} rat (mg/kg body weight, *p.o.*)	HBI[1]	MAK documentation,[2] References
3-nitrotoluene	(liver)	haemosiderosis, anaemia, fibrosis, spleen, liver, kidney, testis	1070–2200	4	Dunnik *et al.* 1994, ECB 2000 h
4-nitrotoluene	metHb, (liver)	haemosiderosis, fibrosis; spleen, liver, kidney, testis	2144–4700	4	Bayer AG 2003, Dunnik *et al.* 1994
(4) phenylenediamine and substances which can be metabolized to yield phenylenediamine					
o-**phenylenediamine** (1,2-diaminobenzene)	metHb (liver, kidney, spleen)	no data	500–1300	no data	Volume 13, BUA 1993
m-**phenylenediamine** (1,3-diaminobenzene)	metHb	liver, kidney	540–780	0.7	Volume 6, BUA 1993, Neumann *et al.* 1993
p-**phenylenediamine** (1,4-diaminobenzene)	(no metHb reported)	oedema, liver, kidney	56–114	no data	Volume 6, BUA 1993
4-nitroaniline	metHb, liver, kidney	spleen, haemosiderosis, liver, kidney	750–3000	no data	Greim 1999, Nair *et al.* 1986
dinitrobenzene (all isomers)	metHb, Heinz bodies (*p*>*m*>*o*)	anaemia, spleen, haemosiderosis, liver	30–250 (LD_{low})	69 *m*-DNB	Volume 1, Neumann *et al.* 1993
1-chloro-2,4-dinitrobenzene	metHb, Heinz bodies	(liver, spleen; inadequately studied)	640–1070	no data	BG Chemie 1992, ECB 2000n
(5) di-substituted and poly-substituted toluene derivatives and analogous benzene derivatives					
toluene-2,4-diamine	(metHb)	liver (cirrhosis)	170–300	< 0.02	Volume 6, Neumann *et al.* 1993
5-nitro-*o*-toluidine (2-amino-4-nitrotoluene)	metHb, Heinz bodies	anaemia, liver, kidney	574	1.0	Volume 6, Zwirner-Baier *et al.* 1994

Table 2. continued

Substance	Effects of single exposures	Effects of repeated exposures	LD_{50} rat (mg/kg body weight, p.o.)	HBI[1]	MAK documentation,[2] References
2,6-dinitrotoluene	metHb, Heinz bodies	anaemia, liver (necrosis)	535–800	1.2	Volume 6, Zwirner-Baier et al. 1994
2,4-dinitrotoluene	metHb (no Heinz bodies)	anaemia, Heinz bodies liver	500–1000	0.7	Volume 6, Zwirner-Baier et al. 1994
2-nitro-*p*-phenylenediamine	no data	liver	350–3000	no data	Volume 4
2,4,6-trinitrotoluene	metHb, Heinz bodies	spleen, liver, haemosiderosis	600–1800	6	Volume 1, Zwirner-Baier et al. 1994
N-methyl-N,2,4,6-tetranitroaniline	metHb	anaemia, spleen, liver, kidney	no data	no data	Volume 11

(6) *N*-substituted aminophenols and nitrophenols and substances which can be metabolized to yield phenols

***o*-anisidine** (2-methoxyaniline)	metHb, Heinz bodies	anaemia, spleen	2000	no data	Volume 10
2,4-diaminoanisole	(no metHb reported)	(anaemia), liver (hyperplasia)	500	no data	Volume 6
***p*-anisidine** (4-methoxyaniline)		metHb, Heinz bodies, anaemia, spleen, liver, urinary bladder	1400	no data	Greim 2001
2-nitro-4-aminophenol	no data	no data	3000–9000	no data	Volume 4
2-nitroanisole	gastrointestinal tract, kidney	anaemia, spleen, metHb, urinary bladder, liver, kidney	740–1000	no data	Volume 9
4,6-dinitro-*o*-cresol	(uncoupler)	metHb, Heinz bodies, liver, kidney, heart	25–85	no data	Volume 19

[1] HBI: haemoglobin binding index: binding [mmol/mol Hb] per dose [mmol/kg body weight]
[2] volume numbers refer to MAK documentation in present series, *Occupational Toxicants*, Volumes 1–20
metHb: methaemoglobin

4 Evidence of non-genotoxic effects

That the genotoxic effects of the aromatic amines cannot on their own account for the carcinogenicity of these substances and that the acute and chronic toxic effects could play an important role, especially in tumour promotion, has been suggested often in recent years (Bitsch *et al.* 1997). Therefore the non-genotoxic effects of these substances should be given more attention. There is considerable evidence of effects which are usually thought to be receptor-mediated, for example, the enzyme induction by many of these substances. The role of proliferation-promoting effects and fibrosis formation has already been mentioned. These develop as a result of the toxic effects in the typical target organs, spleen, liver and kidney. In the long-term studies with nitrobenzene, for example, all the tissues in which tumours developed had been altered by chronic toxic processes and showed increased proliferation rates or inflammation. As early as 1984, Goodman *et al.* (1984) demonstrated an association between the development of fibrosis and sarcomas. Aniline, *o*-toluidine and *p*-chloroaniline induced with similar potency fibrosis in the spleen parenchyma or capsule. The fibrosis-inducing properties of monocyclic aromatic amino and nitro compounds could be counted among the tumour-promoting properties of these substances.

It is not clear whether the fibrosis develops as a result of unspecific cytotoxicity—e.g. overloading of the spleen by the increased erythrocyte elimination—or whether a role is played by specific biochemical mechanisms—e.g. impairment of mitochondrial respiration and inhibition or acceleration of apoptosis.

With only few of the substances have initiation-promotion studies been carried out to test for tumour-promoting activity. In such initiation-promotion studies, 2,4-dinitrotoluene, for example, acted only as a promoter, 2,6-dinitrotoluene as initiator and promoter (Leonard *et al.* 1983, Popp and Leonard 1983).

There are also results available from *in vitro* tests for promoting properties. At a pH of 6.7 in a test for morphological transformation of Syrian hamster embryo (SHE) cells, toluene-2,4-diamine, 2,4-dinitrotoluene and *o*-anisidine had transforming activity, but toluene-2,6-diamine and *p*-phenylenediamine did not (Kerckaert *et al.* 1998). In another study, none of four dinitrotoluene isomers yielded positive results in a test for morphological cell transformation. Likewise the metabolites toluene-2,4-diamine, 5-nitro-*o*-toluidine and 3-nitro-*o*-toluidine yielded negative results in this study. 2,4-Dinitrotoluene and 2,6-dinitrotoluene inhibited intercellular communication, but at toxic concentrations (Holen *et al.* 1990).

5 Genotoxicity

5.1 *In vitro*

The *in vitro* genotoxicity of monocyclic aromatic amino and nitro compounds is not marked (Table 3). In the bacterial mutagenicity test (standard Ames test), especially aniline derivatives (group 1) and *o*-toluidine and 2-nitrotoluene (group 2) but also *p*-toluidine and 3-nitrotoluene (group 3) yield negative results even when metabolic activation is promoted by addition of enzymes (S9). Positive results can be obtained in most cases by the addition of norharman, a comutagen which forms mutagenic compounds from aromatic amines during metabolic activation. From this result and the observation that the *N*-hydroxylamines of *o*-toluidine and *p*-toluidine yield positive results in the Ames test, it may be concluded that the initial substances have weak genotoxic potential which is expressed when the metabolic pathway via the *N*-hydroxylamine and subsequent *O*-acetylation is sufficiently active. It has in the meantime been demonstrated that aniline is substituted by norharman in the *p*-position and that this leads to the formation in the test medium of a mutagenic compound which is responsible for the positive result (Totsuka *et al.* 1999).

Dimethylanilines (xylidines) and 2,4,5-trimethylaniline (group 2), 4-nitrotoluene (group 3) and the polyfunctional representatives (groups 4–6) yield positive results in the standard Ames test—generally with metabolic activation but in some cases also without—but their activity must be considered to be rather weak.

1-Chloro-2,4-dinitrobenzene (group 4), however, which yielded positive results in the Ames test both with and without metabolic activation, produces its mutagenic effects without reduction of its nitro group. This compound can probably react directly with DNA (Gupta *et al.* 1997). A second activation pathway could involve substitution with glutathione and subsequent reduction of the nitro group to form a more reactive compound (Kerklaan *et al.* 1987).

It may be seen that the substances of group 6 which possess a free or methylated phenolic hydroxyl group are more mutagenic in the Ames test than the corresponding compounds without the hydroxyl group. Thus 2,4-diaminoanisole is much more mutagenic than is toluene-2,4-diamine.

In a systematic comparison of the mutagenicity of 2,4-dinitrotoluene with that of 2,6-dinitrotoluene and of their partially reduced metabolites in the Ames test, all the substances were practically without effect in the strains TA98 and TA100. In strains with either higher nitroreductase activity (YG1021, YG1026) or higher *O*-acetyltransferase activity (YG1024, YG1029) positive results were obtained with some of the compounds. In strains in which both enzyme activities were increased (YG1041, YG1042) practically all the substances were mutagenic, and some of them highly mutagenic (Sayama *et al.* 1998). Thus the nitroreductase activity and especially the *O*-acetyltransferase activity must be high enough to reduce the nitro compound to the hydroxylamine; then *O*-acetylation can yield the ultimate mutagen. In the incubation media, two azoxy compounds, 4,4′-dimethyl-3,3′-dinitroazoxybenzene and 2,2-dimethyl-3,3′-dinitroazoxybenzene, are formed in a spontaneous reaction between a nitroso and a hydroxylamino derivative. These substances were tested in the Ames test and shown to be the most highly mutagenic of all the metabolites. The three nitroaminotoluenes were the least mutagenic of the substances and in any case were less mutagenic than the two dinitro compounds.

The genotoxic properties of the phenylenediamines are still being studied. Various results suggest that the three isomers have genotoxic potential. There is increasing evidence for an involvement of radical reactions. The role of peroxidases in the metabolic activation has received only little attention to date.

o-Phenylenediamine has been shown in the comet assay with human lymphocytes to cause the same DNA damage (DNA strand breaks) as gamma rays (Cebulska-Wasilewska *et al.* 1998). The mutagenicity of *m*-phenylenediamine in the Ames test is markedly increased by oxidation with ozone. The reaction product has been identified and shown to be a powerful mutagen (frame-shift mutations) (Kami *et al.* 2000). It is not known whether analogous oxidation reactions can also take place *in vivo*. Other studies of the genotoxicity of *m*-phenylenediamine also suggest that oxidative DNA damage predominates and not mutagenicity of the kind seen in bacterial tests (Chen *et al.* 1998).

Tobacco cells in suspension and an extract of the media from such suspensions both activated *m*-phenylenediamine to a mutagenic product (Gichner *et al.* 1995). *o*-Phenylenediamine, but not *m*-phenylenediamine or *p*-phenylenediamine, can be activated by light to a product which is mutagenic in the tobacco plant. This activation is inhibited by peroxidase inhibitors. With increasing light intensity, the peroxidase activity and the level of hydrogen peroxide—necessary for the metabolic activation via peroxidases—increase in the tobacco leaves (Gichner *et al.* 2001). These observations also suggest that active oxygen species are involved in the metabolic activation.

A weak mutagenicity of *p*-phenylenediamine has been confirmed with the Ames test (Chung *et al.* 1995).

5.2 *In vivo*

There is not much data available for *in vivo* genotoxicity of monocyclic aromatic amino and nitro compounds and the data which have been published are inconsistent. DNA binding is usually found (when studied) and clastogenic effects have been described (Table 3). Comparison of the results obtained for the first three groups of substances reveals that 4-chloro-*o*-toluidine (Carcinogen category 1) differs from the other compounds: it yields positive results in the coat colour spot test with the mouse. This result and a comparatively high *in vitro* mutagenicity are not in line with the toxicity and carcinogenicity shown by this compound in animal studies. The latter two parameters are like those found for the structurally related, weakly active substances.

The role of intestinal bacteria for the activation of nitroaromatics is made clear by the studies of Sayama *et al.* (1993) which demonstrate that in anaerobic incubations of 2,4-dinitrotoluene and 2,6-dinitrotoluene with intestinal bacteria, all the intermediate stages in the reduction of the nitro groups to amino groups are formed and also two nitroazoxy compounds. This too suggests that active secondary reaction products could play a role.

Table 3. Monocyclic aromatic amino and nitro compounds: genotoxicity, cell transformation [1]

Substance	in vitro		in vivo	MAK documentation,[2] References
	bacterial mutagenicity test (Ames test)	mammalian test systems	mammalian test systems, *Drosophila melanogaster*	
(1) aniline and substances which can be metabolized to yield aniline derivatives				
aniline	– (with/without MA) +(MA+norharman)	+ (SCE, CA) +/– (DNA-SB, CA, $TK^{+/-}$, transform.) – (UDS, MN)	+ (SCE, DNA-SB, DNA-binding, MN, mutagens in the urine) – (CA, DLM)	Volume 6, Bomhard 2003, EU 2002, Ohe et al. 2002
***N*-methylaniline**	– (with/without MA) +(MA+norharman)	not studied	not studied	Volume 6
***N,N*-dimethyl-aniline**	– (with/without MA) +(MA+norharman)	+ (SCE, CA, MN, $TK^{+/-}$) – (UDS)	+ (DNA-SB)	Volume 3, Taningher et al. 1993
***o*-chloroaniline**	– (with/without MA)	+ (HPRT, $TK^{+/-}$) – (UDS, CA, Na^+/K^+-ATPase, transform.)	+ (macromolecular binding) – (MN)	Volume 3, Dial et al. 1998, ECB 2000a, NTP 1998
***m*-chloroaniline**	– (with/without MA) +(MA+norharman)	+ (SCE, CA, HPRT, $TK^{+/-}$) – (UDS)	– (MN)	Volume 3, NTP 1998
***p*-chloroaniline**	+/– (with/without MA)	+ (SCE, CA, $TK^{+/-}$, transform., inhibition of mitotic activity) +/– (UDS) – (DNA-SB)	+ (macromolecular binding, MN)	Dial et al. 1998, Volume 3, NTP 1998
nitrobenzene	– (with/without MA) +(MA+norharman)	+/– (CA) – (SCE, UDS, transform.)	+ (SLRL; DNA binding) – (SCE, CA, MN, UDS)	Volume 19, ECB 2000b
***o*-chloronitro-benzene**	+/– (with/without MA) +(MA+norharman)	+ (SCE) – (UDS, CA, HPRT)	+ (DNA-SB) – (SLRL)	Volume 4, ECB 2000c

Table 3. continued

Substance	in vitro		in vivo	MAK documentation,[2] References
	bacterial mutagenicity test (Ames test)	mammalian test systems	mammalian test systems, *Drosophila melanogaster*	
***m*-chloronitro-benzene**	– (with/without MA, MA+norharman)	+/– (CA)– (SCE, HPRT)	not studied	Volume 4
***p*-chloronitro-benzene**	+/– (with/without MA) +(MA+norharman)	+ (SCE, TK$^{+/-}$) +/– (CA, UDS) – (HPRT)	+ (DNA-SB) +/– (MN) – (SLRL, SCE)	Volume 4, ECB 2000d
(2) *o*-toluidine and its derivatives				
***o*-toluidine**	– (with/without MA) +(MA+norharman)	+ (SCE, DNA-SB; CA, transform.) +/– (UDS, MN, TK$^{+/-}$) – (HPRT)	+ (DNA-SB, DNA binding, SMART) +/– (SCE) – (MN)	Volume 3, Batiste-Alentorn *et al.* 1995, ECB 2000e, Kerckaert *et al.* 1998, Ohkuma *et al.* 1999
5-chloro-*o*-toluidine	not studied	not studied	not studied	Volume 6
4-chloro-*o*-toluidine	+ (with MA) – (without MA)	+ (SCE, DNA-SB, UDS)	+ (DNA binding, coat colour spot test) – (MN)	Volume 6, Greim 2003
2,4-xylidine (2,4-dimethyl-aniline)	+ (with MA) – (without MA)	+ (DNA-SB, UDS, CA)	+ (DNA binding)	Volume 19, BUA 1995
2,6-xylidine (2,6-dimethyl-aniline)	+? (with MA) – (without MA)	+ (SCE, DNA binding, CA, TK$^{+/-}$) +/– (transform.)	+ (DNA-SB, DNA binding) – (SLRL, UDS, MN; host-mediated assay)	Volume 19, BUA 1995, ECB 2000f, Goncalves *et al.* 2001, Jeffrey *et al.* 2002
xylidine (2,3-xylidine, 2,5-xylidine, 3,4-xylidine, 3,5-xylidine)	+ (with MA) – (without MA)	+ (UDS: 2,5-xylidine)	not studied	Volume 19, BUA 1995

Table 3. continued

Substance	*in vitro*		*in vivo*	MAK documentation,[2] References
	bacterial mutagenicity test (Ames test)	mammalian test systems	mammalian test systems, *Drosophila melanogaster*	
2,4,5-trimethyl-aniline	+ (with MA) – (without MA)	+ (UDS)	– (SLRL)	Volume 4
2-nitrotoluene	– (with/without MA) +(MA+norharman)	+ (transform.) +/– (SCE, CA) – (UDS)	+ (DNA binding, UDS) – (MN)	Volume 8, Greim 2002
(3) *p*-toluidine and its derivatives				
***p*-toluidine**	– (with/without MA, MA + norharman)	+ (UDS) – (DNA-SB)	+ (DNA-SB, DNA binding)	Volume 3, ECB 2000g
3-nitrotoluene	– (with/without MA)	+ (SCE) – (CA)	+ (macromolecular binding in the liver) – (DNA binding in the liver, UDS)	ECB 2000h, Rickert *et al.* 1984
4-nitrotoluene	+/– (with/without MA)	+ (SCE, UDS, CA, TK$^{+/-}$) – (UDS, MN)	+ (macromolecular binding in the liver) – (DNA binding in the liver, UDS, CA, MN)	Bayer AG 2003, Rickert *et al.* 1984
(4) phenylenediamine and substances which can be metabolized to yield phenylenediamine				
***o*-phenylene-diamine** (1,2-diamino-benzene)	+ (with MA) – (without MA)	+ (DNA-SB) +/– (SCE, UDS, CA, TK$^{+/-}$)	+ (MN, CA) – (SLRL, SCE, coat colour spot test, DLM)	Volume 13, BUA 1993, Cebulska-Wasilewska *et al.* 1998, ECB 2000i
***m*-phenylene-diamine** (1,3-diamino-benzene)	+ (with MA) – (without MA)	+ (CA, TK$^{+/-}$, transform.) +/– (SCE, HPRT) – (DNA-SB, UDS)	– (SLRL, UDS, MN, DLM)	Volume 6, BUA 1993, Chen *et al.* 1998, ECB 2000j, Gichner *et al.* 1995

Table 3. continued

Substance	*in vitro*		*in vivo*	MAK documentation,[2] references
	bacterial mutagenicity test (Ames test)	mammalian test systems	mammalian test systems, *Drosophila melanogaster*	
***p*-phenylene-diamine** (1,4-diamino-benzene)	+ (with MA; less potent than the isomers) – (without MA)	+ (SCE, CA, MN, TK$^{+/-}$) +/– (transform.) – (UDS)	+/– (SLRL) – (MN, DLM)	Volume 6, BUA 1993, Chung *et al.* 1995, Kerckaert *et al.* 1998
4-nitroaniline	+/– (with/without MA)	+ (SCE, CA, TK$^{+/-}$) – (UDS)	– (SLRL, MN, UDS)	Greim 1999, ECB 2000*l*
dinitrobenzene (all isomers)	+ (*m*-isomer > *p*-isomer: with/without MA) – (*o*-isomer: with/without MA)	+ (CA: *m*-isomer) – (UDS: *m*-isomer)	not studied	Volume 1, ECB 2000m
1-chloro-2,4-dinitrobenzene	+ (with/without MA)	+ (DNA-SB, UDS, CA, transform.)	+ (DNA-SB) – (DLM)	BG Chemie 1992, ECB 2000n
(5) di-substituted and poly-substituted toluene derivatives and analogous benzene derivatives				
toluene-2,4-diamine	+ (with MA: TA98, TA1538) – (without MA: TA98)	+ (DNA-SB, UDS, TK$^{+/-}$, transform.)	+ (SLRL, SCE, DNA binding, UDS) – (MN, DLM)	Volume 6, Kerckaert *et al.* 1998, La and Froines 1993, 1994
5-nitro-*o*-toluidine (2-amino-4-nitrotoluene)	+ (with/without MA; more mutagenic than 2,4-DNT)	+ (transform.) – (UDS)	– (MN)	Volume 6
dinitrotoluene (mixture of isomers)	+/– (without MA)	+ (HPRT: 2,4-isomer with MA and reduced O$_2$-concentration) – (UDS, HPRT) +/– (transform.: 2,4-isomer) – (UDS, HPRT)	+ (macromolecular binding, DNA binding, UDS, mutagens in the urine) – (MN, DLM: 2,4-isomer)	Ashby *et al.* 1985, Volume 6, La and Froines 1993

Table 3. continued

Substance	in vitro		in vivo	MAK documentation,[2] references
	bacterial mutagenicity test (Ames test)	mammalian test systems	mammalian test systems, *Drosophila melanogaster*	
2,6-dinitrotoluene	+/– (without MA)	– (UDS, HPRT, transform.)	+ (macromolecular binding 2 to 5 times that with 2,4-isomer, DNA binding, UDS, mutagens in the urine)	Volume 6, George *et al.* 1996, 2001, Holen *et al.* 1990
2-nitro-*p*-phenylenediamine	+ (with/without MA)	+ (SCE, UDS, CA, TK$^{+/-}$, transform.)	– (SCE, MN, DLM)	Volume 4
2,4,6-trinitro-toluene	+ (with/without MA; more mutagenic than the dinitro-toluenes)	not studied	+ (mutagens in the urine of workers) – (UDS, MN)	Ashby *et al.* 1985, Volume 1, George *et al.* 2001
N-methyl-N,2,4,6-tetranitroaniline	+ (with/without MA)	not studied	not studied	Volume 11, George *et al.* 2001

(6) N-substituted aminophenols and nitrophenols and substances which can be metabolized to yield phenols

***o*-anisidine** (2-methoxyaniline)	+ (with MA, with norharman) – (without MA)	+ (SCE, DNA-SB; DNA binding, CA, TK$^{+/-}$, transform.) – (UDS)	+ (SMART, DNA-SB in bladder, lac I$^+$-mutation in the bladder of transgenic mice) +/– (host-mediated assay) – (SLRL, DNA-SB, DNA binding, UDS, MN)	Ashby 1995, Volume 10, ECB 2000o, Kerckaert *et al.* 1998, Ohkuma and Kawanishi 2001, Rodriguez-Arnaiz and Téllez 2002, Sasaki *et al.* 1998, Stiborova *et al.* 2002
2,4-diamino-anisole	+ (with MA; more mutagenic than the diamino-toluenes)	+ (DNA-SB, UDS, CA, TK$^{+/-}$)	+ (SCE, DNA-SB) – (SLRL, MN, DLM)	Volume 6

Table 3. continued

Substance	in vitro		in vivo	MAK documentation;[2] references
	bacterial mutagenicity test (Ames test)	mammalian test systems	mammalian test systems, *Drosophila melanogaster*	
***p*-anisidine** (4-methoxyaniline)	+ (with MA) − (without MA)	+ (SCE, DNA-SB, DNA binding, CA, TK$^{+/-}$)	+ (host-mediated assay) − (SMART)	Greim 2001, Rodriguez-Arnaiz and Téllez 2002
2-nitro-4-aminophenol	+ (with/without MA) − (highly purified substance)	+ (transform.) − (UDS)	− (DLM)	Volume 4
2-nitroanisole	+ (with/without MA)	+ (SCE, CA, TK$^{+/-}$) − (HPRT)	not studied	Volume 9, ECB 2000p
4,6-dinitro-*o*-cresol	+ (with MA)	+/− (CA, HPRT)	+ (herbicide: CA, DLM) − (pure substance: CA, MN)	Volume 19

[1] +: results mainly positive; +/−: results both positive and negative; − results mainly negative
[2] volume numbers refer to MAK documentations in present series, *Occupational Toxicants*, Volumes 1–20

Abbreviations: CA: chromosomal aberrations, DLM: dominant lethal mutations; DNA-SB: DNA strand breaks; HPRT: hypoxanthine phosphoribosyl transferase mutations; MN: micronuclei; MA: metabolic activation; SCE: sister chromatid exchange; SMART: somatic mutations and recombinations in *Drosophila* (eye spot test or wing spot test); SLRL: X-chromosomal recessive lethal mutations in *Drosophila*; transform.: morphological cell transformation; TK: thymidine kinase; UDS: unscheduled DNA repair synthesis

The data shown in Table 3 suggest that *in vitro* bacterial mutagenicity tests (Ames test) are not able to determine reliably the genotoxic potential of monocyclic aromatic amino and nitro compounds. The few results of *in vivo* studies and, in particular, the DNA binding detected with most of the substances support the hypothesis that some genotoxic potential must be expected. Which role this genotoxic potential plays in initiation of cells and tumour development remains to be established.

6 Tumour formation

It seems that, given appropriate test conditions, a carcinogenic potential may be demonstrated for the monocyclic aromatic amino and nitro compounds listed here. With most of these substances, the results of animal studies indicate at least a suspicion of carcinogenic activity (Table 1). These substances were classified in Carcinogen category 3 mostly because tumours were found in only one species or only one sex, or because the increases in tumour incidences were not convincing. The interpretation of the results of animal studies is made more difficult by the fact that the "right" dose plays an important role in these tests because the tumours develop mostly at exposure levels at which also toxic effects are seen.

The acute toxicity expressed as the LD_{50} for the rat rarely permits conclusions as to the chronic toxicity of the substance (Table 2).

In spite of all the differences in detail, a comparison of the target organs in which tumours develop reveals a common pattern (Table 1). In the rat and mouse vascular tumours predominate. The haemangiosarcomas can be concentrated in the spleen but are also found at other sites. The substances in group 6 are different from the rest; with these, tumours of the urinary bladder predominate. In addition, fibromas and fibrosarcomas are often observed. The liver is often affected, more rarely the kidney as well. Together with toxic liver and kidney damage, adenomas and carcinomas develop in these organs (nitrobenzene, dinitrotoluenes).

7 Potency

In Table 4 are listed the lowest doses at which tumours have been observed in carcinogenicity studies in the rat and mouse and the TD_{50} values (carcinogenic potency) in mg/kg body weight and day calculated by Gold *et al.* (1993, 1999). Comparative evaluation of the results is made more difficult by differences in the administration routes used. In addition, it seems that the lowest tumour-inducing doses correlate poorly with the calculated TD_{50} values so that from these data only the order of magnitude of the TD_{25} values can be estimated. According to the criteria for classification in the Carcinogen categories 4 and 5 of the *List of MAK and BAT Values* which are based on the definition of low potency suggested by the EU ($TD_{25} > 100$ mg/kg body weight and day), the listed monocyclic aromatic amino and nitro compounds do not fulfill the criteria for low potency but, in the rat, are more potent.

Although in a study with *o*-anisidine, for example, doses of 6000 mg/kg diet (300 mg/kg body weight and day) were necessary for tumour production, the TD_{50} calculated for the rat was as little as 27.8 mg/kg body weight and day. For several substances, concentrations of 3000 to 5000 mg/kg diet are effective; only for *o*-phenylenediamine are the two available TD_{50} values markedly over 100 mg/kg body weight and day. In comparison, 4-chloro-*o*-toluidine, *p*-chloroaniline, toluene-2,4-diamine and 2,6-dinitrotoluene are relatively highly potent members of this group of substances. As often seen with arylamines, the rat is more sensitive than the mouse. For substances for which TD_{50} values are available for both species, the TD_{50} for the mouse is about 10 times that for the rat (exceptions are 2,4,5-trimethylaniline, *o*-phenylenediamine).

Table 4. Monocyclic aromatic amino and nitro compounds: lowest doses producing tumours, TD_{50} values

Substance	Lowest doses producing tumours [1]		TD_{50} (mg/kg body weight and day) [2]	
	rat	mouse	rat	mouse
(1) aniline and substances which can be metabolized to yield aniline derivatives				
aniline	100 mg/kg body weight and day, gavage	–	160	
***N*-methylaniline**	(–)	(–)		
***N,N*-dimethylaniline**	30 mg/kg body weight and day, gavage	30 mg/kg body weight and day, gavage	125	99.3
***o*-chloroaniline**	not studied	not studied		
***m*-chloroaniline**	not studied	not studied		
***p*-chloroaniline**	2 mg/kg body weight and day, gavage	3 mg/kg body weight and day, gavage	7.6	33.8
nitrobenzene	25 ml/m^3, inhalation	5–25 ml/m^3, inhalation (lung model)		
***o*-chloronitrobenzene**	500–1000 mg/kg diet	1500–3000 mg/kg diet		108
***m*-chloronitrobenzene**	not studied	not studied		
***p*-chloronitrobenzene**	not studied	3000 mg/kg diet		430
(2) *o*-toluidine and its derivatives				
***o*-toluidine**	3000 mg/kg diet	1000 mg/kg diet	23.3	646
5-chloro-*o*-toluidine	(2500 mg/kg diet)	2000 mg/kg diet	714	134
4-chloro-*o*-toluidine	(20 mg/kg diet)	20, 2000 mg/kg diet		15.4
2,4-xylidine (2,4-dimethylaniline)	(–)	(250 mg/kg diet)		12.4

Table 4. continued

Substance	Lowest doses producing tumours [1]		TD_{50} (mg/kg body weight and day) [2]	
	rat	mouse	rat	mouse
2,6-xylidine (2,6-dimethylaniline)	3000 mg/kg diet (about 150 mg/kg body weight and day)	not studied	20.4	
xylidine (2,5-xylidine)	12000 + 6000 mg/kg diet	6000 mg/kg diet		723
2,4,5-trimethylaniline	200 mg/kg diet	50 mg/kg diet	27.2	6.13
2-nitrotoluene	625 mg/kg diet	1250 mg/kg diet		
(3) *p*-toluidine and its derivatives				
***p*-toluidine**	25–75 mg/kg body weight and week, subcutaneous injection	500–1000 mg/kg diet		49.1
3-nitrotoluene	not studied	not studied		
4-nitrotoluene	2500 mg/kg diet	(1250 mg/kg diet)		
(4) phenylenediamine and substances which can be metabolized to yield phenylenediamine				
***o*-phenylenediamine** (1,2-diaminobenzene)	4000 mg/kg diet (about 240 mg/kg body weight and day)	4000/8000 mg/kg diet (about 600 mg/kg body weight and day)	248	611
***m*-phenylenediamine** (1,3-diaminobenzene)	– (oral) (24 mg/kg body weight, subcutaneous injection)	– (oral, dermal)		
***p*-phenylenediamine** (1,4-diaminobenzene)	– (oral)	– (oral)		
4-nitroaniline	(–)	100 mg/kg body weight and day		
dinitrobenzene (all isomers)	not studied	not studied		
1-chloro-2,4-dinitro-benzene	(–)	(–)		
(5) di-substituted and poly-substituted toluene derivatives and analogous benzene derivatives				
toluene-2,4-diamine	50–125, 500 mg/kg diet (about 5.9 mg/kg body weight and day)	100 mg/kg diet (about 15 mg/kg body weight and day)	1.43	26.7
5-nitro-*o*-toluidine (2-amino-4-nitro-toluene)	(100 mg/kg diet)	1500–3000 mg/kg diet		

Table 4. continued

Substance	Lowest doses producing tumours [1]		TD_{50} (mg/kg body weight and day)[2]	
	rat	mouse	rat	mouse
2,6-dinitrotoluene	7 mg/kg body weight and day (substance administered with the diet)	not studied	0.574	–
2,4-dinitrotoluene	700 mg/kg diet (about 40 mg/kg body weight and day)	100 mg/kg diet (about 13.5 mg/kg body weight and day)		
2-nitro-*p*-phenylene-diamine	–	2200 mg/kg diet		614
2,4,6-trinitrotoluene	(50 mg/kg body weight and day)	not studied		
***N*-methyl-*N*,2,4,6-tetranitroaniline**	(10 × 40 mg, gavage)	not studied		

(6) *N*-substituted aminophenols and nitrophenols and substances which can be metabolized to yield phenols

***o*-anisidine** (2-methoxyaniline)	6000 mg/kg diet (about 300 mg/kg body weight and day)	2500–5000 mg/kg diet	27.8	935
2,4-diaminoanisole	1200 mg/kg diet	1200 mg/kg diet	72.6	791
***p*-anisidine** (4-methoxyaniline)	3000 mg/kg diet (about 150 mg/kg body weight and day)	–		
2-nitro-4-amino-phenol	1250–2500 mg/kg diet	–	309	
2-nitroanisole	6000 mg/kg diet (about 300 mg/kg body weight and day)	666 mg/kg diet (about 100 mg/kg body weight and day)	7.48	177
4,6-dinitro-*o*-cresol	–	not studied		

[1] for references see Table 1
[2] Gold *et al.* (1993, 1999) and current database entry (http://potency.berkely.edu)

8 Discussion

8.1 Mechanism of Action

The discussion of the causes of the toxicity of aniline and other simple monocyclic aromatic amino compounds in the spleen and its significance in tumour development is still in progress. The toxic effects of aniline are caused by the *N*-oxidation products phenylhydroxylamine and nitrosobenzene. After administration by gavage, both compounds cause more damage in the spleen than in other tissues (Khan *et al.* 1998, 2000). The total amount of iron in the organ is increased and parameters for lipid peroxidation increase with the dose; oxidative stress is concluded to be the main precondition for spleen damage. The formation of free radicals has been described by several authors. It has been confirmed by Brennan and Schiestl (1999) for aniline, *o*-toluidine, and *o*-anisidine.

During long-term administration of aniline, bound iron is continually released so that the levels of free iron in the spleen and the Kupffer cells of the liver are increased (Ciccoli *et al.* 1999). It has been suggested that the formation of haemoglobin thiyl free radicals as a result of the reaction of phenylhydroxylamine with oxyhaemoglobin could be the cause of haemolysis or premature aging of erythrocytes (Bradshaw *et al.* 1995, Maples *et al.* 1990). According to this hypothesis, the release of iron in erythrocytes is the cause of the damage and not the result of the increased elimination of damaged erythrocytes, a suggestion which fits with the observed occurrence of haemosiderosis in various parts of the organism. It is still difficult to understand the chain of events from metHb formation to toxic effects in the spleen and the role of oxidative stress (Khan *et al.* 1997).

Whereas it is generally the *N*-oxidation products, especially hydroxylamines, which are thought to be the biologically active metabolites, in the urine of animals given toluene derivatives the products of *C*-oxidation predominate. Little is known of the role played by benzyl and benzoic acid derivatives (Mori *et al.* 1996). The corresponding aldehydes are highly mutagenic and have been detected, for example, in the bile. They can be formed in the liver, perhaps by oxidation of benzyl alcohol absorbed from the intestine (Mori *et al.* 1997).

The idea that the spleen is "overloaded" by the erythrocyte elimination is not supported by the fact that the tumour development does not correlate with the level of metHb formation. For example, nitrobenzene causes much more metHb formation than does aniline and the corresponding effects in the spleen are also observed. But the tumours induced by nitroaniline do not develop in the spleen, as they do with aniline, but in various other tissues. Of the three nitroaniline isomers, *p*-nitroaniline is the most potent metHb inducer and much more potent than aniline (French *et al.* 1995). Repeated exposure to *p*-nitroaniline causes spleen damage and haemosiderosis but no tumours. Which role is played in the toxicity of these substances by the release of iron in the reaction with oxyhaemoglobin and during increased erythrocyte elimination is not known. It has been suggested that the formation of reactive oxygen species is increased by redox cycling (Dybing *et al.* 1979, Lum and Roebuck 2001) but it is not known whether this leads mainly to genotoxic effects or to tumour promotion. In any case, one consequence of the release of iron is the haemosiderosis.

The mechanism of action will only then be clear when we are in a position to explain the inconsistencies such as those described here. Until then we must attempt to interpret the information which is available and to use it to estimate the risks associated with this group of substances (see also Butterworth 1990).

8.2 Structure-effect relationships

There have been numerous attempts to account for the effects of the variously substituted compounds in terms of simple physicochemical or biological principles and so to explain differences in the metabolism or the reactions with macromolecules. For example, the effects of substitution on the oxidation potential of alkylanilines, the mutagenicity of *N*-hydroxyalkylanilines and the conformations of the DNA adducts have been studied (Marques *et al.* 1997). Differences in the oxidation potentials which corresponded to the electron-donating properties of the substituents were detected but the differences were too small to account for the wide range of genotoxic effects of the alkylanilines. The data suggest that the conformation of the DNA adducts, and especially the ability to adopt *syn*-conformations, determines the genotoxic response.

In a series of monocyclic aromatic amino and nitro compounds, the haemoglobin binding of the *N*-hydroxylamine and nitroso derivatives and the correlation of this dosimeter with various physicochemical parameters was determined (Sabbioni 1994). The mutagenic potency of the amines increased with their oxidizability but decreased with the level of haemoglobin binding, except for those amines with halogen substituents in *ortho* or *meta* position. For the latter the haemoglobin binding is directly proportional to the pKa. For the nitroaromatics, the haemoglobin binding increases with the reducibility but does not correlate with the genotoxicity. The genotoxicity and carcinogenicity of the amines are less well correlated with the electronic properties of the compounds than is the haemoglobin binding. Thus the biological properties of the compounds cannot be entirely explained in terms of their electronic properties. Particularly the sex-specific and species-specific differences show that the results of an animal study are determined by a much more complex interaction of activation, reactivity and the properties of the target organ and that simple predictions are difficult. One of the reasons for the poor correlation of genotoxicity and carcinogenicity is the fact that the genotoxicity of most of the substances in these groups probably plays only a minor role.

8.3 The effect of position isomerism

The groups of substances listed here include several examples which permit the comparison of position isomers. It has been suggested that metabolic elimination, which reduces the effective dose of the substance, is favoured by a free position *para* to the functional group (Aßmann *et al.* 1997). For the three chloroaniline isomers (group 1) the increase in effect resulting from *p*-substitution seems to be confirmed. The three isomers have comparable toxicity patterns, they are all haematotoxic, and toxic for the spleen, liver and kidney, but there are quantitative differences. *p*-Chloroaniline causes more metHb formation than does *m*-chloroaniline and this more than *o*-chloroaniline (Table 2). This order also applies for other parameters. Marked quantitative differences are also seen in the genotoxic properties. Whereas *p*-chloroaniline is mutagenic in all test systems, the results obtained with the other two isomers are inconsistent and they are considered to be non-mutagenic or only weakly mutagenic (NTP 1998). Thus, the isomer with the substituent in *para* position has the highest biological activity.

The hepatotoxic and nephrotoxic properties of the three acetylated isomers, that is, the chloroacetanilides, have been studied by Rankin *et al.* (1993). Conclusions for the amines may be drawn from results obtained with the chloroacetanilides because an equilibrium is rapidly established *in vivo* between amine and acetamide, the relative levels of which depend on the acetylator status. The amine is acetylated, the acetamide deacetylated. The bioavailability of the critical metabolite (*N*-hydroxylamine) is comparable in the two cases (Zwirner-Baier and Neumann 1998). In the study of the chloroacetanilides, the order of severity of liver damage caused by the three isomers (*para* > *meta* = *ortho*) was similar to that of kidney damage (*para* > or = *meta* > *ortho*). In contrast, in an earlier study (Rankin *et al.* 1986) these authors found the *ortho*-substituted 2-chloroaniline to be most nephrotoxic.

The different potencies of *o*-chloroaniline and *p*-chloroaniline may be explained in terms of toxicokinetics. *o*-Chloroaniline—with no substituent in *para* position—is more rapidly excreted and binds less to proteins in the liver and kidney than the more potent *para*-isomer (Dial *et al.* 1998).

Analogous differences between the three chloronitro isomers (group 1) are not seen. The *para*-isomer of this trio, 1-chloro-4-nitrobenzene, does not produce tumours in the rat and only in the mouse vascular tumours (haemangiomas and haemangiosarcomas). The *ortho* isomer, 1-chloro-2-nitrobenzene, on the other hand, increases the incidence of multiple tumours in the rat but has only slight effects on the liver of the mouse (Table 1). Focal necrosis and inflammation in the mouse liver are seen with the *ortho*-isomer but not with the *para*-isomer (Table 2). Thus, in this series the *o*-isomer—with a free *para* position—is most potent (Travlos *et al.* 1996).

The situation seems to be similar with 2,6-xylidine (2,6-dimethylaniline) which, although less mutagenic than 2,4-xylidine (2,4-dimethylaniline) with its substituted *para* position (Table 3), binds, after repeated administration, to the DNA of the target tissue (nasal cavity) in the rat (Short *et al.* 1989). Unlike the 2,4-isomer which does not produce tumours in the rat, the 2,6-isomer induces tumours in the nasal cavity, the subcutis and the liver (Table 1).

Of the three nitrotoluene isomers (groups 2 and 3), *o*-nitrotoluene was most potent in the rat. Even in a 13-week study, it produced mesotheliomas in the *tunica vaginalis*. Nephrotoxic effects were seen with all three isomers. Effects on the liver were most pronounced with *o*-nitrotoluene but were seen in a less severe form also with *m*-nitrotoluene and *p*-nitrotoluene (3-nitrotoluene and 4-nitrotoluene). Changes in the spleen were slightly more marked with *o*-nitrotoluene and *p*-nitrotoluene. And all the isomers had adverse effects on testis function and caused lengthening of the oestrous cycle. The spectra of adverse effects produced by the three isomers were very similar (Table 2) but the *ortho*-isomer was generally most toxic (Dunnik *et al.* 1994). *o*-Nitrotoluene produced not only mesotheliomas but also fibromas and fibrosarcomas, the *p*-isomer, on the other hand, only clitoris tumours (Table 1).

4-Chloro-*o*-toluidine, which has caused bladder tumours in exposed persons (Carcinogen category 1), is another example of an amine with a substituent in *para* position. In the rat it caused only a slight and not significant increase in the incidence of hepatomas and would hardly have been classified as a carcinogen on this basis. In the mouse, however, the substance produces haemangiosarcomas and haemangiomas even at relatively low doses (20 mg/kg diet) (Table 1).

The results available for the three phenylenediamines are not consistent. Because they are readily oxidized and because of the experimental difficulties this causes, they have often been administered subcutaneously. In the Ames test with metabolic activation, *p*-phenylenediamine was less potent than the other two isomers (Table 3). Of the three isomers, *o*-phenylenediamine was the most potent carcinogen, causing liver tumours after oral administration which were not seen with the other isomers (Table 1).

Most of the studies with dinitrobenzene have been carried out with mixtures of isomers and so the individual isomers cannot be compared.

In the comparative study of 17 monocyclic nitroaromatics mentioned above (Aßmann *et al.* 1997) all the compounds with a nitro group in *meta* position or *para* position to an amino or nitro group were mutagenic in the Ames test (TA98 or TA100), in most cases only in the presence of S9, whereas nitrobenzene and aniline themselves and their derivatives with *ortho* substituents were not mutagenic. The authors conclude from these results that derivatives with *para* substituents, with few exceptions, are mutagenic, unlike those with *ortho* or *meta* substituents.

8.4 Analogy between arylamines and nitro compounds

In this review, analogous monocyclic aromatic amino and nitro compounds have been placed in the same group. Now the discussion focuses on how far this analogy goes and whether or to what extent effect and potency of an arylamine differ from that of the corresponding nitro compound, for example, for aniline and nitrobenzene. For both of these two compounds, the biologically active metabolites are phenylhydroxylamine and nitrosobenzene. The location of the tumours induced by these substances and the species sensitivity are, however, very different. Aniline caused sarcomas at various locations only in the rat; nitrobenzene caused liver tumours only in the rat (Table 1) although the chronic toxicity is similar in the rat and mouse. With both substances, anaemia is seen in the rat and mouse and also liver and kidney damage (Table 2).

The data for *o*-chloroaniline and *o*-chloronitrobenzene are inadequate (Table 1). But comparison of the substances in group 1 suggests that *o*-chloroaniline could have carcinogenic potential.

o-Toluidine and 2-nitrotoluene (group 2) differ very little. Both substances cause mesotheliomas and haemangiosarcomas (Table 1), both yield negative results in the Ames test but positive in the presence of norharman, both bind *in vivo* to DNA (Table 3).

It is of interest to compare substances with two amino or two nitro groups when data for the nitroamine are also available. *p*-Phenylenediamine, 4-nitroaniline and dinitrobenzene (*p*-isomer) (group 4) all yield positive results in the Ames test, although it must be admitted that the results obtained with 4-nitroaniline are inconsistent and seem to be negative for *in vivo* mutagenicity (Table 3). The acute and chronic toxicity of 4-nitroaniline and dinitrobenzene are very similar: with both compounds spleen damage and haemosiderosis are the main effects and the liver is also affected (Table 2). Although the database for *p*-phenylenediamine is inadequate, similarities are evident.

Another such trio is formed by toluene-2,4-diamine, 5-nitro-*o*-toluidine (2-amino-4-nitrotoluene) and 2,4-dinitrotoluene (group 5). Here too the amount of data available for the three substances differs but, even so, marked differences may be seen. Toluene-2,4-diamine produces tumours in the rat not only in the liver but also in several other tissues. 2-Amino-4-nitrotoluene causes a slight increase in the incidence of liver tumours in the rat and liver tumours and haemangiosarcomas in the mouse. With the dinitrotoluenes, fibromas and fibroadenomas predominate (Table 1). Toluene-2,4-diamine causes less metHb formation than do the other two compounds (Table 2) but its hepatotoxic effects are more pronounced and this can favour the development of liver tumours. 2-Amino-4-nitrotoluene is more mutagenic than the corresponding dinitro compound (Table 3). The results for *in vivo* genotoxicity cannot be compared. However, mutagenic substances have been found in the urine of persons exposed to dinitrotoluene at work.

In the corresponding group of 2,6-substituted toluene derivatives (group 5), the substances differ in their ability to cause tumours. Toluene-2,6-diamine (NTP 1980) has been shown not to be carcinogenic but 2,6-dinitrotoluene is a relatively powerful carcinogen (Table 1). In agreement with these conclusions, DNA adducts were found in the livers of animals treated with 2,6-dinitrotoluene but not with toluene-2,6-diamine. The adduct pattern produced by the dinitro compound was qualitatively like that obtained with 2-amino-6-nitrotoluene. This substance was also highly hepatotoxic and produced haemorrhagic necrosis which was not seen with the amine (La and Froines 1993).

A corresponding pattern was seen when *o*-anisidine (2-methoxyaniline) and 2-nitroanisole were compared (group 6). Both substances are mutagenic in the Ames test (Table 3), cause anaemia when administered long-term (Table 2) and produce tumours in the urinary bladder in the rat (Table 1).

Thus there are examples of very unexpectedly similar patterns of effects and others in which the results obtained with the amine and nitro derivatives are very different. The latter seems to apply when the nitro compound is eliminated much more slowly than the amine, as with the example aniline and nitrobenzene. For the present assessment, however, it is helpful and appropriate to recognize the common pattern of haematotoxic effects and kinds of tumours: haemangiosarcomas, fibromas and mesotheliomas.

8.5 Similarities

That the monocyclic aromatic amino and nitro compounds may be considered as a single group is suggested by the fact that their acute and chronic toxic effects are similar. It is characterized by the induction of metHb formation. Practically all the substances discussed here induce metHb formation. This requires that all the substances react with oxyhaemoglobin via a common intermediate, the amines after oxidation, the nitro compounds after reduction to the reactive N-hydroxylamine. The divalent iron in haemoglobin is converted in a co-oxidation reaction to trivalent iron in metHb and the hydroxylamine is oxidized to a nitroso compound. Both oxidation products are then reduced back to the original compounds and a cyclic process is set in motion. The course of metHb formation and the release of iron in the oxidation reaction depend on the supply of reactive metabolites and the reduction capacity of the erythrocytes. The metHb levels can increase steeply and reach high peak values, as with aniline, or can increase after a delay to lower values, as with nitrobenzene. Thus it is plausible that the effects can be different even for the same AUC (the area under the curve obtained by plotting blood haemoglobin concentration against time). The release of iron and the denaturation of proteins could be a function of the slope of the curve and of the levels of the peak concentrations. These phenomena have mostly been studied in experimental animals; their role in persons exposed to low levels of the substances remains unclear. But in man too the substances are known to cause metHb formation and even lethal cyanosis. Thus it has been demonstrated that the substances can be metabolized to yield comparable amounts of the reactive N-hydroxylamines in man as well.

Another phenomenon which speaks for a common mechanism of action is the haemosiderosis produced by most of the substances. That the haemosiderosis can develop in very different locations is not understood.

The formation of N-hydroxylamines, mostly after activation by esterification, is also decisive for the genotoxicity. The monocyclic aromatic amino and nitro compounds have in common a low genotoxic potency which in many cases (groups 1 to 3) is not detectable in routine tests. The often inconsistent or questionable results obtained in the usual tests for DNA damage are presumably a consequence of this low potency. Mutagenic secondary products are formed with norharman but their relevance is not clear. The mutagenicity of many N-hydroxylamines and the fact that formation of DNA adducts has been observed with the N-hydroxylamines of many of these substances is evidence for the genotoxic properties of the monocyclic aromatic amino and nitro compounds. A certain, although weak genotoxic potential must be assumed for the substances discussed here; to date it is not possible to establish criteria for the exclusion of a contribution of genotoxic effects to tumour development.

Finally, a very similar range of tumours is produced by these substances. It is characterized by tumours of the blood vessels and connective tissue. The location of the tumours can, however, be very different for the different substances. When they develop in the spleen, the liver or the kidney, they are mostly preceded by recognizable toxic tissue damage and fibrosis.

The currently available data for these substances suggest that they have in common those properties which determine the acute toxic and carcinogenic effects. Mechanisms of action which do not involve the N-hydroxyl derivatives and end points which are not described above have not been proposed to date.

8.6 The differences

However, some of the properties of the substances differ in detail. This raises questions and reveals contradictions which make it doubtful whether it is possible to draw conclusions on the basis of analogy.

A general problem in comparative assessment is that, apart from the Ames test, very different methods have been used for the *in vitro* and *in vivo* genotoxicity tests, different doses have been administered and the results have been obtained in very different laboratories. This could result in apparent differences between substances which are actually caused by differences in methods. Relatively often, the binding of metabolites to DNA or the levels of oxidative DNA damage have been determined and the micronucleus test and tests for morphological transformation (e.g. with SHE cells) have been used. Even when the many methodical differences are taken into account, it is conspicuous how often the parameters determined in these tests do not correlate with each other or with the tumour incidence.

That the potency of *o*-chloronitrobenzene is unexpectedly higher than that of the *p*-isomer was described above in the section on position isomerism.

It is difficult to understand why 2-amino-4-nitrotoluene is only weakly carcinogenic in the rat even after high doses when its oxidation state lies between those of toluene-2,4-diamine and 2,4-dinitrotoluene, both of which produce tumours in this species. All three compounds are mutagenic in the Ames test and cause liver damage after long-term administration. They differ, however, in their haematotoxic properties (Table 2). Toluene-2,4-diamine causes practically no metHb formation, 2-amino-4-nitrotoluene causes most haemoglobin formation (with Heinz bodies) and 2,4-dinitrotoluene causes acute metHb formation without Heinz bodies. The HBI values correspond (< 0.02, 2.4 and 0.7). The three substances differ also in their target organs. Toluene-2,4-diamine causes liver tumours in the rat, together with liver cirrhosis; 2-amino-4-nitrotoluene causes low levels of liver tumours, 2,4-dinitrotoluene mainly fibromas and fibroadenomas. In the mouse the tumours develop in the liver (toluene-2,4-diamine), liver (2-amino-4-nitrotoluene) and kidney (2,4-dinitrotoluene). The marked discrepancy between the high mutagenicity and haematotoxicity produced by 4-chloro-*o*-toluidine and its minimal or lack of carcinogenicity in the rat has already been mentioned.

Why do some of the substances produce tumours in only one of the two species tested or why do they affect different tissues in the two species? *p*-Chloronitrobenzene and 2,4-xylidine have effects only in the mouse, aniline, *m*-phenylenediamine and 2-nitro-*p*-phenylenediamine only in the rat. *m*-Phenylenediamine and *p*-phenylenediamine produced no tumours after oral administration either in the mouse or in the rat; after subcutaneous injection of *m*-phenylenediamine, however, local subcutaneous sarcomas developed in the rat. A similar experiment has not been carried out with mice.

8.7 Conclusions for the classification

After the above discussion of the results and problems, here the current classifications of the substances are to be discussed (Table 1) and, as examples, some conclusions drawn.

- *N*,*N*-dimethylamine derivatives are rapidly converted to the amines via the monomethyl derivatives (Metzler and Neumann 1971). With stilbene derivatives, for example, it has been demonstrated that comparable amounts and patterns of DNA adducts are formed in the liver after oral administration of the *N*,*N*-dimethyl and the *N*-acetyl derivatives (Gaugler and Neumann 1979). This can be explained by the fact that an equilibrium is established very rapidly between a little of the initial substance (*N*,*N*-dimethylamino-stilbene), the monomethyl derivative, the free amine and its acetylation product. The acetamide accounts for most of the non-conjugated metabolites (Metzler and Neumann 1971). *N*,*N*-Dimethylaniline and aniline are both classified in the same carcinogen category 3B.
- It is extremely unlikely that the carcinogenicity of *N*-methylaniline can be different from that of aniline and of *N*,*N*-dimethylaniline.
- Similarly, *o*-chloroaniline and *m*-chloroaniline and the three chloronitrobenzenes are so similar in their acute and chronic toxicity that it does not make sense that they have different carcinogen classifications.
- The different classification of *p*-chloroaniline should be documented in more detail.
- The difference between 4-chloro-*o*-toluidine which is classified as a human carcinogen (Carcinogen category 1) and 5-chloro-*o*-toluidine which is merely a suspected carcinogen (category 3B) should be re-examined on the basis of new data for toxicity, species differences between man and animals and genotoxicity.
- The data for the toxicity and genotoxicity of 3-nitrotoluene and 4-nitrotoluene do not suggest that these substances are different and do not require classification. A recent NTP carcinogenicity study of 4-nitrotoluene makes a re-examination necessary.

8.8 Conclusions

Which conclusions are to be drawn for the classification of the substances into categories for carcinogenic substances? Although it seems at first glance that the comparatively weakly carcinogenic monocyclic aromatic amino and nitro compounds should be classified in Carcinogen category 5 on the basis of their genotoxic properties, detailed examination of the data suggests that category 4 might be more appropriate. But before the substances could be classified in category 4, it would be necessary to understand the reasons for the tissue damage they cause; that is, a better understanding of the mechanisms leading to toxicity and the dose-dependence of the relevant biochemical and biological end points is required. There is every reason to believe that the dose-response relationship for such end points is very flat in the low dose range and in any case that it is not linear. An effect threshold cannot be determined at present, but it should be possible, on the basis of NOELs for biochemical effects, to define occupational exposure levels (MAK or BAT values) at which the contribution to cancer risk at the workplace is no longer significant. In this process it must be taken into account that especially for this group of substances the acute toxic effects are expected to be additive.

The data comparison for these substances has supported the hypothesis that all the *N*-substituted aromatics discussed here may have acute toxic and genotoxic effects. The potency in both cases is a function of the biologically active dose and the reactivity of certain metabolites, especially the hydroxylamine derivatives. The bioavailability of these ultimately active metabolites varies widely in this group of compounds, a fact which is reflected in the wide range of HBI values. The relationship of chemical structure to the kinetics and mechanism of action of a compound is not sufficiently understood to predict with any certainty the potency of a substance for which insufficient data are available. But conclusions on the basis of analogy can still be justified. Whether or not it is justified must be decided on a case to case basis depending on the information which is available for the substance to be classified and its structural analogues. In any case, the use of such comparisons should play an increasingly important role in the evaluation of structurally related chemicals. In this way, numerous animal studies could be avoided.

Growing evidence suggests that the haematotoxic effects of this group of substances are the main prerequisite for the development of tissue damage and tumours. If this hypothesis is correct, then exposure to all monocyclic aromatic amino and nitro compounds which cause metHb formation and thus erythrocyte damage or premature aging must be seen as a cancer risk factor. This aspect should be taken into account in the evaluation and classification of the substances. If data are available for cytotoxicity and a NOEL, and if the genotoxicity data are in the range of those for weakly mutagenic structural analogues, the substance could be classified in Carcinogen category 4.

The establishment of a limit value could be based on a reference value determined by human biomonitoring for the substance in question (Lewalter and Neumann 1998). Here it is of relevance that persons may often be exposed environmentally to numerous monocyclic aromatic amino and nitro compounds. As metHb formation in the erythrocytes is the common effect of these substances, evaluation of the effects of the individual substances is anyway inappropriate.

The external and internal exposure levels of monocyclic aromatic amino and nitro compounds for persons in the general population cover a wide range and the 95th percentiles of the exposure parameters which have been determined are correspondingly far above the median values. In the case of the class of substances discussed here, the internal exposure is best determined by biochemical effect monitoring, for example, by the determination of haemoglobin adducts in blood. At the workplace, biomonitoring is anyway recommended for substances with a BAT value, as part of the secondary preventive measures defined in TRGS 710. If a worker's background values are below the reference value, that is, in the range of the median value or lower, the reference value could be retained as target value and ceiling value during additional occupational exposure during work with amines. But if the background value for a worker is found to be at this ceiling value already, appropriate measures must be taken. These could include determination of the level of exposure to structurally related substances, for example, by analysis of the cleavage products from the haemoglobin adducts. There is always background exposure to numerous amino and nitro aromatics and present-day methods readily permit analysis of several substances in parallel. Generally high values are suggestive of a metabolic status which favours metabolic activation and reveals potential sensitivity to this class of compounds. Additional exposure should then be avoided. Less stringent interpretations are also conceivable and should be discussed. A MAK or BAT value should at least guarantee that the normal erythrocyte and iron turnover in the exposed persons is not affected adversely.

The question remains as to whether under these conditions the total of genotoxic effects is increased in the occupationally exposed persons. In the environment there are not only numerous monocyclic aromatic amino and nitro compounds but—taking only this class of substances into account—also polycyclic aromatic amino and nitro compounds. Given this background exposure and the low mutagenic potency of the monocyclic compounds, exposures in line with threshold values would not make a significant contribution to the total effects on the DNA caused by this group of substances, and especially by the much more potent polycyclic amino and nitro aromatics.

8.9 Prospects

The immediate question of classification requires further consideration. We need to know which consequences for future developments result from the presently available data and which further studies are required.

At present, the assessment and classification of carcinogenic substances is mostly based on the results of animal studies and tests designed on the basis of an initiation-promotion-progression multistage model of carcinogenesis. In this model, genotoxicity plays a predominant role. For the first time with the introduction of the Carcinogen categories 4 and 5, the reversible tumour-promoting properties of a substance were put on a level with the genotoxic properties and were recognized as a risk factor. There are, however, still practically no validated test systems. The current developments in our understanding of the process of carcinogenesis make it necessary, with new test systems, to put much more emphasis on non-genotoxic effects and to distinguish between biochemical and biological parameters.

Many of the described effects, for example, differences in sensitivity of species, sexes and organs, cannot be explained at present. Only hypotheses can be formulated, and they must be tested experimentally. It seems increasingly likely that the substances discussed here must be seen as genotoxic—although in many cases the genotoxicity is weak—but that this property is of much less importance than the organotropic toxic effects. The complex interaction of activating and inactivating metabolic reactions is modulated, on the one hand, by the effects of the substituents on reaction rates and, on the other, by the distribution of the compound in the organism and its interactions with host structures which are determined by its solubility. It is conceivable that the concentrations of critical metabolites can be locally very different because structurally determined differences in kinetic parameters can affect local bioavailability and result in differences in cell exposure. It is, however, not very likely that all differences can be explained by toxicokinetics. Therefore, particular attention should be given to the study of the mechanism of action of these compounds. The primary reactions should be studied at the molecular and cell biological level.

Recent results of molecular biological studies have revealed complex networks of feedback systems which regulate cell proliferation and determine the course of tumour development. An example is the signal sent by the mitochondria for the induction of apoptosis (Klöhn *et al.* 1998). The internal exposure to carcinogens (and other xenobiotics) has effects on such signal transduction pathways and it is to be expected that disturbance at this level can determine organotropy and other specificities. Differentiation of the much-used term "oxidative stress" is required in the context of mitochondrial function.

The question of whether or not a threshold value can be defined becomes increasingly the question of the capacity of physiological regulatory reactions to cope with the stress, that is, of the exposure level which overloads the adaptation mechanisms of the cell. The hypothesis described below was developed from the results of our studies of the role of toxicity of 2-acetylaminofluorene, a polycyclic aromatic amine, in the tumour development induced, in this case, by a highly mutagenic substance (Bitsch *et al.* 1999, 2000).

One of the sensitive parameters determining the survival of a cell is the ATP supply provided by the mitochondrial respiratory chain. If the flow of electrons is reduced, for example, when electrons are withdrawn during redox cycling of a xenobiotic, the cell reacts. If sufficient ATP for the reaction is not available, the cell induces its own elimination by apoptosis. But apoptosis is also subject to regulation which seems to prevent the process happening too quickly or too often. Apoptosis is inhibited and its rate is then constant and independent of the dose for a certain time interval before it increases further. Subsequently the proliferation rate increases. Because the respiratory chain is affected by numerous substances and in various ways, it is hardly possible to determine an ineffective dose of a single substance unless the range in which adaptation can take place is known (Preston *et al.* 2001).

For the much less mutagenic monocyclic members of this group it is even more important to clarify the significance of genotoxic and cytotoxic effects for the process of tumour development. It has been suggested that the cytotoxicity alone is responsible for the tumour development caused by these substances. But it is not yet known which role is played by the reactive oxygen species arising in the redox cycling induced by the iron released from haemoglobin. It can have both genotoxic and promoting effects.

Finally, progress in the answering of these questions could be obtained in hypothesis-based initiation-promotion studies in which biochemical parameters are determined. The monocyclic aromatic amino and nitro compounds are a class of substances particularly well suited for the study of the mechanisms of toxicity and their contribution to tumour development.

The question remains as to whether the differences can be explained in terms of a common mechanism of action or whether quite different previously ignored substance-specific effects could be responsible. This is shown by the signals sent by the mitochondria for the induction of apoptosis (Klöhn *et al.* 1998).

Detailed comparative studies should search for the causes of the differences in species-susceptibility and organotropy with the aim of identifying biochemical parameters at the beginning of the chain of events. The objective is, on the basis of appropriate examples, to establish the real role of toxicokinetics and determine whether the observed differences in cytotoxicity and DNA damage can be accounted for in terms of the bioavailability of the reactive metabolites known to date.

9 References

Albrecht W, Neumann HG (1985) Biomonitoring of aniline and nitrobenzene: Hemoglobin binding in rats and analysis of adducts. *Arch Toxicol 57*: 1–5

Ashby J (1995) Genetic toxicity in relation to receptor-mediated carcinogenesis. *Mutat Res 333*: 209–213

Ashby J, Burlinson B, Lefevre PA, Topham L (1985) Non-genotoxicity of 2,4,6-trinitrotoluene (TNT) to the mouse bone marrow and the rat liver: implications for its carcinogenicity. *Arch Toxicol 58*: 14–19

Aßmann N, Emmrich M, Kampf G, Kaiser M (1997) Genotoxic activity of important nitrobenzenes and nitroanilines in the Ames test and their structure-activity relationship. *Mutat Res 395*: 139–144

Batiste-Alentorn M, Xamena N, Creus A, Marcos R (1995) Genotoxic evaluation of ten carcinogens in the *Drosophila melanogaster* wing spot test. *Experientia 51*: 73–76

Bayer AG (2003) 4-Nitrotoluene. IUCLID Data Set, 03.07.2003

BG Chemie (Berufsgenossenschaft der chemischen Industrie) (1992) *1-Chlor-2,4-dinitrobenzol. Toxikologische Bewertung Nr. 43,* BG Chemie, Heidelberg and in English translation: Toxicological Evaluations 5, Springer-Verlag, Berlin, Heidelberg, 1993, 1–28

BIA (2003) 5-Chlor-o-toluidin. GESTIS-Stoffdatenbank, Gefahrstoffinformationssystem der gewerblichen Berufsgenossenschaften, http://www.hvbg.de/d/bia/fac/zesp/zesp.htm

Birner G, Neumann HG (1988) Biomonitoring of aromatic amines II: Hemoglobin binding of some monocyclic aromatic amines. *Arch Toxicol 62*: 110–115

Bitsch A, Fecher J, Jost M, Klöhn PC, Neumann HG (1997) Genotoxic and chronic toxic effects in the carcinogenicity of aromatic amines. *Recent Results Cancer Res 143*: 209–223

Bitsch A, Klöhn PC, Hadjiolov N, Bergmann O, Neumann HG (1999) New insights into carcinogenesis of the classical model arylamine 2-acetylaminofluorene. *Cancer Lett 143*: 223–227

Bitsch A, Hadjiolov N, Klöhn PC, Bergmann O, Zwirner-Baier I, Neumann HG (2000) Dose response of early effects related to tumor promotion of 2-acetylaminofluorene. *Toxicol Sci 55*: 44–51

Bodansky O (1951) Methemoglobinemia and methemoglobin-producing compounds. *Pharmacol Rev 3*: 144–196

Bomhard EM (2003) High-dose clastogenic activity of aniline in the rat bone marrow and its relationship to the carcinogenicity in the spleen of rats. *Arch Toxicol 77*: 291–297

Bradshaw TP, McMillan DC, Crouch RK, Jollow DJ (1995) Identification of free radicals produced in rat erythrocytes exposed to hemolytic concentrations of phenylhydroxylamine. *Free Radical Biol Med 18*: 279–285

Brennan RJ, Schiestl RH (1999) The aromatic amine carcinogens *o*-toluidine and *o*-anisidine induce free radicals and intrachromosomal recombination in *Saccharomyces cerevisiae*. *Mutat Res 430*: 37–45

BUA (Beratergremium für umweltrelevante Altstoffe) (1987) *p-Nitroanilin (4-Nitrobenzolamin). BUA-Stoffbericht 19*, VCH, Weinheim and in English translation: *p*-Nitroaniline, BUA Report 19, S. Hirzel, Stuttgart, 1995

BUA (Beratergremium für umweltrelevante Altstoffe) (1993) *Phenylendiamine (1,2-Diaminobenzol, 1,3-Diaminobenzol, 1,4-Diaminobenzol). BUA-Stoffbericht 97*, S. Hirzel, Stuttgart and in English translation: Phenylenediamines, BUA Report 97, S. Hirzel, Stuttgart, 1995

BUA (Beratergremium für umweltrelevante Altstoffe) (1995) *Xylidine (Aminodimethylbenzole). BUA-Stoffbericht 161*, S. Hirzel, Stuttgart and in English translation: Xylidines, BUA Report 161, S. Hirzel, Stuttgart, 1997

Butterworth BE (1990) Consideration of both genotoxic and non-genotoxic mechanisms in predicting carcinogenic potential. *Mutat Res 239*: 117–132

Cebulska-Wasilewska A, Nowak D, Niedzwiedz W, Anderson D (1998) Correlation between DNA and cytogenetic damage induced after chemical treatment and radiation. *Mutat Res 421*: 83–91

Chen F, Murata M, Hiraku Y, Yamashita N, Oikawa S, Kawanishi S (1998) DNA damage induced by *m*-phenylenediamine and its derivatives in the presence of copper ion. *Free Radical Res 29*: 197–205

Chhabra RS, Thompson M, Elwell MR, Gerken DK (1990) Toxicity of *p*-chloroaniline in rats and mice. *Food Chem Toxicol 28*: 717–722

Chhabra RS, Huff JE, Haseman JK, Elwell MR, Peters AC (1991) Carcinogenicity of *p*-chloroaniline in rats and mice. *Food Chem Toxicol 29*: 119–124

Chung KT, Murdock CA, Stevens SE Jr, Li YS, Wei CI, Huang TS, Chou MW (1995) Mutagenicity and toxicity studies of *p*-phenylenediamine and its derivatives. *Toxicol Lett 81*: 23–32

Ciccoli I, Ferrali M, Rossi V, Signorini C, Alessandrini C, Comporti M (1999) Hemolytic drugs aniline and dapsone induce iron release in erythrocytes and increase the free iron pool in spleen and liver. *Toxicol Lett 110*: 57–66

DFG (Deutsche Forschungsgemeinschaft) (1976) *Datensammlung zur Toxikologie der Herbizide: DNOC*, (Data collection on the toxicology of the herbicide DNOC) (German) Arbeitsgruppe Toxikologie der Kommission für Pflanzenschutz, Pflanzenbehandlungs- und Vorratsschutzmittel. Verlag Chemie, Weinheim

DFG (2004) Deutsche Forschungsgemeinschaft, *List of MAK and BAT Values*, Report 40, Wiley-VCH Verlag, Weinheim

Dial LD, Anestis DK, Kennedy SR, Rankin GO (1998) Tissue distribution, subcellular localization and covalent binding of 2-chloroaniline and 4-chloroaniline in Fischer 344 rats. *Toxicology 131*: 109–119

Dunnik JK, Elwell MR, Bucher JR (1994) Comparative toxicities of *o*-, *m*-, and *p*-nitrotoluene in 13-week feed studies in F344 rats and B6C3F$_1$ mice. *Fundam Appl Toxicol 22*: 411–421

Dybing E, Aune T, Nelson SD (1979) Metabolic activation of 2,4-diaminoanisole, a hair-dye component. II. Role of cytochrome P-450 metabolism in irreversible binding *in vitro*. *Biochem Pharmacol 28*: 43–50

EU (2002) *Risk assessment aniline*. Draft 13.02.2002, Bundesanstalt für Arbeitsschutz und Arbeitsmedizin, Anmeldestelle Chemikaliengesetz (BAuA), Dortmund

ECB (European Chemicals Bureau) (2000a) *o-Chloroaniline. Substance ID 95-51-2*, Ispra, Italy, 19.02.2000

ECB (European Chemicals Bureau) (2000b) *Nitrobenzene. Substance ID 98-95-3*, Ispra, Italy, 19.02.2000

ECB (European Chemicals Bureau) (2000c) *1-Chloro-2-nitrobenzene. Substance ID 88-73-3*, Ispra, Italy, 19.02.2000

ECB (European Chemicals Bureau) (2000d) *1-Chloro-4-nitrobenzene. Substance ID 100-00-5*, Ispra, Italy, 22.02.2000

ECB (European Chemicals Bureau) (2000e) *o-Toluidine. Substance ID 95-53-4*, Ispra, Italy, 19.02.2000

ECB (European Chemicals Bureau) (2000f) *2,6-Xylidine. Substance ID 87-62-7*, Ispra, Italy, 19.02.2000

ECB (European Chemicals Bureau) (2000g) *p-Toluidine. Substance ID 106-49-0*, Ispra, Italy, 18.02.2000

ECB (European Chemicals Bureau) (2000h) *3-Nitrotoluene. Substance ID 99-08-1*, Ispra, Italy, 19.02.2000

ECB (European Chemicals Bureau) (2000i) *o-Phenylenediamine. Substance ID 95-54-5*, Ispra, Italy, 19.02.2000

ECB (European Chemicals Bureau) (2000j) *m-Phenylenediamine. Substance ID 108-45-2*, Ispra, Italy, 18.02.2000

ECB (European Chemicals Bureau) (2000k) *p-Phenylenediamine. Substance ID 106-50-3*, Ispra, Italy, 19.02.2000

ECB (European Chemicals Bureau) (2000l) *4-Nitroaniline. Substance ID 100-01-6*, Ispra, Italy, 22.02.2000

ECB (European Chemicals Bureau) (2000m) *1,3-Dinitrobenzene. Substance ID 99-65-0*, Ispra, Italy, 19.02.2000

ECB (European Chemicals Bureau) (2000n) *1-Chloro-2,4-dinitrobenzene. Substance ID 97-00-7*, Ispra, Italy, 19.02.2000

ECB (European Chemicals Bureau) (2000o) *o-Anisidine. Substance ID 90-04-0*, Ispra, Italy, 19.02.2000

ECB (European Chemicals Bureau) (2000p) *2-Nitroanisole. Substance ID 91-23-6*, Ispra, Italy, 19.02.2000

French CI, Yaun SS, Baldwin LA, Leonard DA, Zhao XQ, Calabrese EJ (1995) Potency ranking of methemoglobin-forming agents. *J Appl Toxicol 15*: 167–174

Furedi E, Levine BS, Gordon DE, Rac VS, Lish PM (1984) A determination of the chronic mammalian toxicological effects of TNT [twenty-four months chronic toxicity carcinogenicity study of Trinitrotoluene (TNT) in the Fischer 344 rat]. IITRJ Project No L6116, study 9, ADA168637, ITT Research Institute, Chicago, USA. cited from EPA (1993) IRIS, integrated Risk Information System 6/2 US Environmental Protection Agency, Washington DC

Gaugler JM, Neumann HG (1979) The binding of metabolites formed from aminostilbene derivatives to nucleic acids in the liver of rats. *Chem Biol Interact 24*: 355–372

George SE, Kohan MJ, Warren SH (1996) Hepatic DNA adducts and production of mutagenic urine in 2,6-dinitrotoluene-treated B6C3F$_1$ male mice. *Cancer Lett 102*: 107–111

George SE, Huggins-Clark G, Brooks LR (2001) Use of *Salmonella* microsuspension bioassay to detect the mutagenicity of munitions compounds at low concentrations. *Mutat Res 490*: 45–56

Gichner T, Stavreva DA, Cerovska N, Wagner ED, Plewa MJ (1995) Metabolic activation of *m*-phenylenediamine to products mutagenic in *Salmonella typhimurium* by medium isolated from tobacco suspension cell cultures. *Mutat Res 331*: 127–132

Gichner T, Stavreva DA, Van Breusegem F (2001) *o*-Phenylenediamine-induced DNA damage and mutagenicity in tobacco seedlings is light-dependent. *Mutat Res 495*: 117–125

Gold LS, Manley NB, Slone TH, Garfinkel GB, Rohrbach L, Ames BN (1993) The fifth plot of Carcinogenic Potency Database: results of animal bioassays published in the general literature through 1988 and by the National Toxicology Program through 1989. *Environ Health Perspect 100*: 65–135

Gold LS, Manley NB, Slone TH, Rohrbach L (1999) Supplement to the Carcinogenic Potency Database (CPDB): results of animal bioassays published in the general literature in 1993 to 1994 and by the National Toxicology Program in 1995 to 1996. *Environ Health Perspect 107, Suppl 4*: 527–600

Goncalves LL, Beland FA, Marques MM (2001) Synthesis, characterization, and comparative ^{32}P-postlabeling of 2,6-dimethylaniline-DNA adducts. *Chem Res Toxicol 14*: 165–174

Goodman DG, Ward JM, Reichardt WD (1984) Splenic fibrosis and sarcomas in F344 rats fed diets containing aniline hydrochloride, *p*-chloroaniline, azobenzene, *o*-toluidine hydrochloride, 4,4'-sulfonyldianiline, or D & C red No. 9. *J Natl Cancer Inst 73*: 265–273

Greim H (Ed.) (1999) 4-Nitroanilin. in *Toxikologisch-arbeitsmedizinische Begründungen von MAK-Werten*, 28th issue, Wiley-VCH Verlag GmbH, Weinheim

Greim H (Ed.) (2001) 4-Methoxyanilin. in *Toxikologisch-arbeitsmedizinische Begründungen von MAK-Werten*, 33rd issue, Wiley-VCH Verlag GmbH, Weinheim

Greim H (Ed.) (2002) 2-Nitrotoluol. in *Toxikologisch-arbeitsmedizinische Begründungen von MAK-Werten*, 34th issue, Wiley-VCH Verlag GmbH, Weinheim

Greim H (Ed.) (2003) 4-Chlor-*o*-toluidin. in *Toxikologisch-arbeitsmedizinische Begründungen von MAK-Werten*, 36th issue, Wiley-VCH Verlag GmbH, Weinheim

Gupta RL, Saini BH, Juneja TR (1997) Nitroreductase independent mutagenicity of 1-halogenated-2,4-dinitrobenzenes. *Mutat Res 381*: 41–47

Holen I, Mikalsen SO, Sanner T (1990) Effects of dinitrotoluenes and morphological transformation and intercellular communication in Syrian hamster embryo cells. *Toxicol Environ Health 29*: 89–98

Jeffrey AM, Luo FQ, Amin S, Krzeminski J, Zech K, Williams GM (2002) Lack of DNA binding in the rat nasal mucosa and other tissues of the nasal toxicants roflumilast, a phosphodiesterase 4 inhibitor, and a metabolite, 4-amino-3,5-dichloropyridine, in contrast to the nasal carcinogen 2,6-dimethylaniline. *Drug Chem Toxicol 25*: 93–107

Kami H, Watanabe T, Takemura S, Kameda Y, Hirayama T (2000) Isolation and chemical-structural identification of a novel aromatic amine mutagen in an ozonized solution of *m*-phenylenediamine. *Chem Res Toxicol 13*: 165–169

Kerckaert GA, LeBoeuf RA, Isfort RJ (1998) Assessing the predictiveness of the Syrian hamster embryo cell transforming assay for determining the rodent carcinogenic potential of single ring aromatic/nitroaromatic amine compounds. *Toxicol Sci 41*: 189–197

Kerklaan PR, Bouter S, te Koppele JM, Vermeulen NP, van Bladeren PJ, Mohn GR (1987) Mutagenicity of halogenated and other substituted dinitrobenzenes in *Salmonella typhimurium* TA 100 and derivatives deficient in glutathione (TA 100/GSH-) and nitroreductase (TA 100NR). *Mutat Res 176*: 171–178

Khan MF, Boor PJ, Gu Y, Alcock NW, Ansari GAS (1997) Oxidative stress in the splenotoxicity of aniline. *Fundam Appl Toxicol 35*: 22–30

Khan MF, Green SM, Ansari GA, Boor PJ (1998) Phenylhydroxylamine: role in aniline associated splenic oxidative stress and induction of subendocardial necrosis. *Toxicol Sci 42*: 64–71

Khan MF, Wu X, Ansari GA (2000) Contribution of nitrosobenzene to splenic toxicity of aniline. *J Toxicol Environ Health 60*: 263–273

Klöhn PC, Bitsch A, Neumann HG (1998) Mitochondrial permeability transition is altered in early stages of carcinogenesis of 2-acetylaminofluorene. *Carcinogenesis 19*: 1185–1190

Kuroda Y (1986) Genetic and chemical factors affecting chemical mutagenesis in cultured mammalian cells. *Basic Life Sci 39*: 359–375

La DK, Froines JR (1993) Comparison of DNA binding between the carcinogen 2,6-dinitrotoluene and its noncarcinogenic analog 2,6-diaminotoluene. *Mutat Res 301*: 79–85

La DK, Froines JR (1994) Formation and removal of DNA adducts in Fischer-344 rats exposed to 2,4-diaminotoluene. *Arch Toxicol 69*: 8–13

Leonard TB, Lyght O, Popp JA (1983) Dinitrotoluene structure-dependent initiation of hepatocytes *in vivo*. *Carcinogenesis 4*: 1059–1061

Lewalter J, Neumann HG (1998) Biologische Arbeitsstoff-Toleranzwerte (Biomonitoring). Teil XIII. Die Bedeutung von Referenzwerten für die Bewertung von Fremdstoffbelastungen. (Biological tolerance values (biomonitoring). Part XIII. The significance of reference values for the evaluation of exposures to foreign substances.) (German) *Arbeitsmed Sozialmed Umweltmed 33*: 388–393

Lum H, Roebuck KA (2001) Oxidant stress and endothelial cell dysfunction. *Am J Physiol Cell Physiol 280*: C719–C741

Maples KR, Eyer P, Mason RP (1990) Aniline-, phenylhydroxylamine-, nitrosobenzene-, and nitrobenzene-induced hemoglobin thyil free radical formation *in vivo* and *in vitro*. *Mol Pharmacol 37*: 311–318

Marques MM, Mourato LL, Amorim MT, Santos MA, Melchior WB, Beland FA (1997) Effect of substitution site upon the oxidation potentials of alkylanilines, the mutagenicities of *N*-hydroxylamines, and the conformations of alkylaniline-DNA adducts. *Chem Res Toxicol 10*: 1266–1274

Metzler M, Neumann HG (1971) Zur Bedeutung chemisch-biologischer Wechselwirkungen für die toxische und krebserzeugende Wirkung aromatische Amine. IV. Stoffwechselmuster von trans-4-Dimethylaminostilben, cis-4-Dimethylaminostilben und 4-Dimethylaminobibenzyl in Leber, Niere und den Ausscheidungsprodukten der Ratte (Importance of chemico-biological interactions for the toxic and carcinogenic effect of aromatic amines. IV. Metabolic patterns of trans-4-dimethylaminostilbene, cis-4-dimethylaminostilbene and 4-dimethylaminobibenzyl in liver, kidney and excretion products of the rat) (German). *Z Krebsforsch 76*: 16–39

Mori MA, Shoji M, Dohrin M, Kawagoshi T, Honda T, Kozuka H (1996) Further studies on the urinary metabolites of 2,4-dinitrotoluene and 2,6-dinitrotoluene in the male Wistar rat. *Xenobiotica 26*: 79–88

Mori MA, Sayama M, Shoji M, Inoue M, Kawagoshi T, Maeda M, Honda T (1997) Biliary excretion and microfloral transformation of major conjugated metabolites of 2,4-dinitrotoluene and 2,6-dinitrotoluene in the male Wistar rat. *Xenobiotica 27*: 1225–1236

Morris HP (1965) Studies on the development, biochemistry, and biology of experimental hepatomas. *Adv Cancer Res 9*: 228–303

Nair RS, Johannsen FR, Levinskas GJ, Terril JB (1986) Subchronic inhalation toxicity of *p*-nitroaniline and *p*-nitrochlorobenzene in rats. *Fundam Appl Toxicol 6*: 618–627

Nair RS, Auletta CS, Schroeder RE, Johannsen FR (1990) Chronic toxicity, oncogenic potential, and reproductive toxicity of *p*-nitroaniline in rats. *Fundam Appl Toxicol 15*: 607–621

NCI (National Cancer Institute) (1978) *Bioassay of aniline hydrochloride for possible carcinogenicity*. NCI Bioassay Report No 130, USDHEW, Bethesda, MD, USA

Neumann HG, Birner G, Kowallik P, Schütze D, Zwirner-Baier I (1993) Hemoglobin adducts of N-substituted aryl compounds in exposure control and risk assessment. *Environ Health Perspect 99*: 65–70

Neumann HG, Thielmann HW, Filser JG, Gelbke HP, Greim H, Kappus H, Norpoth KH, Reuter U, Vamvakas S, Wardenbach P, Wichmann HE (1998) Changes in the classification of carcinogenic chemicals in the work area (section III of the German list of MAK and BAT values). *J Cancer Res Clin Oncol 124*: 661–669

NTP (1980) *Bioassay of 2,6-toluenediamine dihydrochloride for possible carcinogenicity (CAS No. 15481-70-6)*. NTP Technical Report 200, National Toxicology Program, Research Triangle Park, NC, USA

NTP (1998) *NTP comparative toxicity studies of o-, m-, and p-chloroanilines (CAS Nos. 95-51-2; 108-42-9; 106-47-8) administered by gavage to F344/N rats and B6C3F$_1$ mice*. NTP Toxicity Report Series 43, National Toxicology Program, Research Triangle Park, NC, USA

NTP (2002) *Toxicology and carcinogenesis studies of p-nitrotoluene (CAS No 99-99-0) in F344/N rats and B6C3F$_1$ mice (feed studies)*. NTP Technical Report 498, National Toxicology Program, Research Triangle Park, NC, USA

Ohe T, Takata T, Maeda Y, Totsuka Y, Hada N, Matsuoka A, Tanaka N, Wakabayashi K (2002) Induction of sister chromatid exchanges and chromosome aberrations in cultured mammalian cells treated with aminophenylnorharman formed by norharman with aniline. *Mutat Res 515*: 181–188

Ohkuma Y, Hiraku Y, Oikawa S, Yamashita N, Murata M, Kawanishi S (1999) Distinct mechanisms of oxidative DNA damage by two metabolites of carcinogenic *o*-toluidine. *Arch Biochem Biophys 372*: 97–106

Ohkuma Y, Kawanishi S (2001) Oxidative DNA damage induced by a metabolite of carcinogenic *o*-anisidine; enhancement of DNA damage and alteration in its sequence specificity by superoxide dismutase. *Arch Biochem Biophys 389*: 49–56

Popp JA, Leonard TB (1983) Hepatocarcinogenesis of 2,6-dinitrotoluene (DNT). *Proc Am Assoc Cancer Res 24*: 91 (Abstract 361)

Preston TJ, Abadi A, Wilson L, Singh G (2001) Mitochondrial contributions to cancer cell physiology: potential for drug development. *Adv Drug Delivery Rev 49*: 45–61

Rankin GO, Yang DJ, Cressey-Veneziano K, Casto S, Wang RT, Brown PI (1986) *In vivo* and *in vitro* nephrotoxicity of aniline and its monochlorophenyl derivatives in the Fischer 344 rat. *Toxicology 38*: 269–283

Rankin GO, Valentovic MA, Beers KW, Nicoll DW, Ball JG, Anestis DK, Brown PI, Hubbard JL (1993) Renal and hepatic toxicity of monochloroacetanilides in the Fischer 344 rat. *Toxicology 79*: 181–193

Rickert DE, Long RM, Dyroff MC, Kedderis GL (1984) Hepatic macromolecular covalent binding of mononitrotoluenes in Fischer-344 rats. *Chem Biol Interact 52*: 131–139

Rodriguez-Arnaiz R, Tellez GO (2002) Structure-activity relationships of several anisidine and dibenzanthracene isomers in the w/w+ somatic assay of *Drosophila melanogaster*. *Mutat Res 514*: 193–200

Sabbioni G (1994) Hemoglobin binding of arylamines and nitroarenes: molecular dosimetry and quantitative structure-activity relationships. *Environ Health Perspect 102, Suppl 6*: 61–67

Sasaki YF, Nishidate E, Su YQ, Matsusaka N, Tsuda S, Susa N, Furukawa Y, Ueno S (1998) Organ-specific genotoxicity of the potent bladder carcinogens *o*-anisidine and *p*-cresidine. *Mutat Res 412*: 155–160

Sayama M, Mori MA, Maruyama Y, Inoue M, Kozuka H (1993) Intestinal transformation of 2,6-dinitrotoluene in male Wistar rats. *Xenobiotica 23*: 123–131

Sayama M, Mori M, Shoji M, Uda S, Kakikawa M, Kondo T, Kodaira KI (1998) Mutagenicity of 2,4- and 2,6-dinitrotoluenes and their reduction products in *Salmonella typhimurium* nitroreductase- and O-acetyltransferase-overproducing Ames test strains. *Mutat Res 420*: 27–32

Short CR, Joseph M, Hardy ML (1989) Covalent binding of [^{14}C]-2,6-dimethylaniline to DNA of rat liver and ethmoid turbinate. *Toxicol Environ Health 27*: 85–94

Stiborova M, Miksanova M, Havlicek V, Schmeisser HH, Frei E (2002) Mechanism of peroxidase-mediated oxidation of carcinogenic *o*-anisidine and its binding to DNA. *Mutat Res 500*: 49–66

Tainter ML, James M, Vandeventer W (1926) Comparative edemic actions of *ortho*-, *meta*, and *para*-phenylenediamines in different species. *Arch Int Pharmacodyn Ther 36*: 152–162

Taningher M, Pasquini R, Bonnati S (1993) Genotoxicity analysis of *N,N*-dimethylaniline and *N,N*-dimethyl-*p*-toluidine. *Environ Mol Mutagen 21*: 349–356

Totsuka Y, Ushiyama H, Ishihara J, Sinha R, Goto S, Sugimura T, Wakabayashi K (1999) Quantification of the co-mutagenic β-carbolines, norharman and harman, in cigarette smoke condensates and cooked food. *Cancer Let 143*: 139–143

Travlos GS, Mahler J, Ragan HA, Chou BJ, Bucher JR (1996) Thirteen-week inhalation toxicity of 2- and 4-chloronitrobenzene in F344/N rats and B6C3F$_1$ mice. *Fundam Appl Toxicol 30*:75–92

Vernot EH, MacEwen JD, Haun CC, Kinkead ER (1977) Acute toxicity and skin corrosion data for some organic and inorganic compounds and aqueous solutions. *Toxicol Appl Pharmacol 42*: 417–423

Weisburger JH, Fiala ES (1981) Mechanisms of species, strain, and dose effects in arylamine carcinogenesis. *Natl Cancer Inst Monograph 58*: 41–48

Zwirner-Baier I, Kordowich FJ, Neumann HG (1994) Hydrolyzable hemoglobin adducts of polyfunctional monocyclic *N*-substituted arenes: dosimeters of exposure and markers of metabolism. *Environ Health Perspect 102, Suppl 6*: 43–45

Zwirner-Baier I, Neumann HG (1998) Biomonitoring of aromatic amines. V: Acetylation and deacetylation in the metabolic activation of aromatic amines as determined by haemoglobin binding. *Arch Toxicol 72*: 499–504

completed 12.08.2003

Chemical Substances

Arsenic and its inorganic compounds

(with the exception of arsine)

MAK value	–
Peak limitation	–
Absorption through the skin	–
Sensitization	–
Carcinogenicity (1971)	Category 1
Prenatal toxicity	–
Germ cell mutagenicity (2002)	3A

Substance	CAS No.	Formula	Molecular weight
arsenic	7440-38-2	As	74.92
trivalent arsenic compounds:			
arsenic trioxide, arsenic(III) oxide	1327-53-3	As_2O_3	197.82
arsenous acid, arsenic(III) acid	13464-58-9 36465-76-6	H_3AsO_3 (ortho form) $HAsO_2$ (meta form)	125.94
salts of arsenic(III) acid, e.g. sodium arsenite	13464-37-4	Na_3AsO_3	191.89
pentavalent arsenic compounds:			
arsenic pentoxide, arsenic(V) oxide	1303-28-2	As_2O_5	229.82
arsenic acid, arsenic(V) acid	1327-52-2 7778-39-4	H_3AsO_4	141.94
salts of arsenic(V) acid, e.g. sodium arsenate	13464-38-5	Na_3AsO_4	207.89

This documentation is based mainly on a few reviews, such as WHO 2001, ATSDR 2000 and IARC 1980. More recent literature and original studies are cited only as far as it is necessary for the evaluation.

In this documentation, with the exception of monomethylarsonate (MMA) and dimethylarsinate (DMA), only arsenic and inorganic arsenic compounds are discussed. As, however, MMA and DMA always occur during the metabolism of arsenic, they are taken into consideration for relevant end points, e.g. genotoxicity and carcinogenicity.

1 Toxic Effects and Mode of Action

Arsenic and inorganic arsenic compounds can be absorbed via inhalation and ingestion. They are distributed rapidly in all organs. Accumulation is observed particularly in the liver, kidneys and lungs. The metabolism of the inorganic arsenic compounds is independent of the route of absorption. Reduction and oxidation reactions lead to mutual transformation, with the preferred reduction of arsenate to arsenite. Only arsenite can be methylated to form first MMA and then DMA. DMA, the main metabolite of the inorganic arsenic compounds, is excreted with the urine.

In case studies of the acute toxicity of the arsenic compounds, above all after ingestion, the first symptoms of intoxication were reported after 30 to 60 minutes. Depending on the severity of the clinical symptoms, intoxication with arsenic is described by the terms acute paralytical syndrome and acute gastrointestinal syndrome. The acute paralytical syndrome is characterized by cardiovascular collapse, central nervous weakness and death within hours. Characteristic of the gastrointestinal syndrome is a metallic or garlic-like taste, a dry mouth, burning lips, difficulties swallowing, headaches, dizziness and vomiting. This can then lead to multi-organ failure.

Long-term exposure to arsenic in man causes fever, sleep disorders, weight loss, swelling of the liver, dark discoloration of the skin, sensory and motor neuropathy, and encephalopathy with symptoms such as headaches, poor concentration, deficits in the learning of new information, difficulties in remembering, mental confusion, anxiety and depression. Long-term exposure to arsenic can also cause peripheral vascular effects such as acrocyanosis, Raynaud's disease and tissue necrosis on the extremities (blackfoot disease). Also cardiovascular diseases and the occurrence of *diabetes mellitus* are described.

Long-term exposure to dust containing arsenic causes irritation of the conjunctiva and mucous membranes of the nose and throat. The pustular or follicular skin reactions observed after contact with inorganic arsenic compounds are usually attributed to irritative effects and not to sensitizing effects of the arsenic compounds.

In animal experiments, repeated ingestion of MMA induced fertility disorders in mice.

In man there is evidence that long-term exposure to inorganic arsenic compounds significantly increases the incidence of spontaneous abortions, single or multiple malformations and stillbirths. The results of animal experiments confirm these toxic effects on development.

In persons exposed long-term to arsenic, an increase in the incidence of micronuclei was found in epithelial cells of the cheeks and bladder, urothelial cells and lymphocytes, and additionally in lymphocytes an increase in the incidence of chromosomal aberrations and sister chromatid exchange.

Arsenic and inorganic arsenic compounds are carcinogenic in man. The target organs of the carcinogenic effects after inhalation are the lungs and after ingestion the bladder, kidneys, skin and lungs.

The prerequisite for the formation of reactive, systemically effective, genotoxic and carcinogenic arsenic species is the reduction of the arsenic compound to the arsenite and possibly its subsequent methylation. The relevant enzymes for these reactions are subject to genetic polymorphism.

2 Mechanism of Action

As inorganic arsenic compounds, above all arsenic(III) compounds, react readily with SH groups, the reaction with proteins is regarded as a mechanism for the toxicity of arsenic and inorganic arsenic compounds. It has only become clear over recent years that the mechanism of arsenic toxicity is much more complex than the unspecific reaction with proteins. Arsenic(III) compounds inactivate, possibly as a result of the reaction with SH groups, the phosphatases responsible for the inactivation of the *jun-N*-terminal kinases (JNKs). This results in an increase in the activity of the JNKs and of proto-oncogenes of the *c-jun* family. This activation of JNKs was also discussed in the context of the carcinogenicity of arsenic(III) compounds (Cavigelli *et al.* 1996).

The metabolism of inorganic arsenic compounds, above all their methylation, was thought for a long time to be a route of detoxification. In recent years it has become clear, however, that in particular the trivalent methylated metabolites are of greater toxicity in various systems. For DMA there is, in addition, clear evidence of co-carcinogenic and possibly also carcinogenic effects in animal experiments (Kenyon and Hughes 2001, Kitchin 2001, Thomas *et al.* 2001; see Section 5.7.2). This question has not, however, been systematically investigated. At least for the genotoxic effects of inorganic arsenic compounds there is an increasing body of evidence from experiments that DNA-damaging metabolites are also formed as a result of the methylation. While in earlier investigations methylated arsenic metabolites were less mutagenic and clastogenic (Moore 1997b), in the comet assay (*in vitro*) practically only the methylated arsenic(III) compounds were found to be DNA-damaging (Mass *et al.* 2001). The different composition of oxidoreductases in the cell systems used *in vitro* for the reduction of arsenic(V) to arsenic(III) and of methyltransferases for the subsequent methylation may explain the many contradictory results. As these enzymes are also subject to genetic polymorphism (Vahter 2000), this could explain the species and interindividual differences in the carcinogenesis of the arsenic compounds (Kenyon and Hughes 2001). The animal experiments with inorganic arsenic compounds to date, however, were not carried out under optimum conditions (Huff *et al.* 2000; see Section 5.7.2).

A suggested mechanism for the genotoxicity of the substance is the generation of reactive oxygen species (Liu *et al.* 2001a, Nordenson and Beckman 1991, Wang and Huang 1994). It was concluded from various studies that DNA damage caused by arsenic(III) compounds is the result of the calcium-mediated formation of peroxynitrite, hypochlorous acid and hydroxyl radicals (Wang *et al.* 2001). DMA forms a radical, possibly intracellularly, and is transformed in the presence of molecular oxygen into a peroxy radical (Kitchin 2001), which could be responsible for the observed oxidative DNA damage. Other studies show inorganic arsenic compounds to inhibit DNA repair processes (Bau *et al.* 2001, Hartwig *et al.* 1997, Li and Rossman 1989, Yager and Wiencke 1997). To date, however, no isolated DNA repair enzyme could be identified which is inhibited by arsenic(III) compounds in low concentrations. Most likely the regulation of repair is affected (Hu *et al.* 1998).

Also the interference of inorganic arsenic compounds with DNA methylation is described (Mass and Wang 1997, Goering *et al.* 1999, Zhao *et al.* 1997). Whether this occurs as a result of the depletion of the methyl pool in the cell is unclear (Kitchin 2001). The result is changes in gene expression, as determined for the tumor supressor gene *p53* and the oncogene *c-myc* (Hsu *et al.* 1999, Mass and Wang 1997, Menéndez *et al.* 2001, Salnikow and Cohen 2002, Vogt and Rossman 2001).

In addition it was shown that inorganic arsenic compounds make certain cells more sensitive to mitogenic stimulus and that cell proliferation is increased, but cell differentiation inhibited (Germolec *et al.* 1998, Trouba *et al.* 2000, Wauson *et al.* 2002). This imbalance in cellular control is thought to be the reason for the carcinogenicity of inorganic arsenic compounds (Salnikow and Cohen 2002).

Also the changes in cell cycle regulation are of importance. In various cells exposure to arsenic causes the induction of stress proteins, the stimulation of growth factors and hormone receptors, changes in cytokine production and signal transduction, and the expression of various oncogenes (Bernstam and Nriagu 2000, Chen *et al.* 2001, Liu and Huang 1997, Liu *et al.* 2001b). It was only recently, however, that a correlation between the toxicity of the substance and its carcinogenicity was investigated (Chen *et al.* 2001). On the other hand, in recent years arsenic trioxide has been used successfully in chemotherapy for promyelocytic leukaemia (Chen *et al.* 1996, Huang *et al.* 1998, Shen *et al.* 1997, Soignet *et al.* 1998). The anti-carcinogenic effects are thought to be the result e.g. of the induction of apoptosis and the inhibition of the formation of microtubules. The induction of apoptosis was detected in leukaemia cells (Chen *et al.* 1996, Dai *et al.* 1999, Ishitsuka *et al.* 1998) and in lymphoid cell lines (Bazarbachi *et al.* 1999, Zhang *et al.* 1998), and the inhibition of microtubule formation was demonstrated for arsenic(III) compounds (Huang and Lee 1998, Li and Brome 1999).

The false regulation of certain interleukins is also held responsible for impairment of the immune system in patients exposed to arsenic (Salnikow and Cohen 2002).

3 Toxicokinetics and Metabolism

3.1 Absorption, distribution, elimination

Absorption

Inhalation: In air, inorganic arsenic compounds are mainly found particle-bound. These are usually arsenites [As(III)] and arsenates [As(V)]. The levels in the outdoor ambient air are about 0.001 µg/m^3, and near the source of emission even > 0.01 µg/m^3 (EC 2000). Data for the levels of arsenic in workplace air vary considerably (WHO 2001). In studies from the 1950s, values of >> 50 µg/m^3 were found. Arsenic concentrations in air of > 10 µg/m^3 are still found today. Cigarette smoking is thought to cause additional inhalation exposure to arsenic (ATSDR 2000). Smoking, however, was not found to have an influence on the levels of arsenic excreted with the urine (Seiwert et al. 1999).
    The absorption of arsenic after inhalation takes place in two steps: deposition of the particles and their subsequent absorption. The extent of absorption varies considerably depending on the solubility of the arsenic compounds and the size of the particles. It is estimated to be in the range of 30 % to 90 % (ATSDR 2000). A more exact quantification of absorption is not possible (WHO 2001).
Ingestion: The main sources of exposure of the general population not occupationally exposed to arsenic are food and drinks. Around 25 % is taken up in the form of inorganic arsenic compounds. In regions with high levels of arsenic in the drinking water (> 100 µg/l), drinking water is the main source of inorganic arsenic compounds. Gastrointestinal absorption is rapid and the amounts absorbed are in the range of 45 % to 75 %. Of environmental-medical relevance is also the ingestion of arsenic in young children as a result of contaminated soils (WHO 2001).
Dermal absorption: Dermal absorption of individual arsenic compounds at the workplace is possible. Arsenic acid penetrates the skin *in vivo* and *in vitro* in amounts of less than 10 % and sodium arsenate from aqueous solution or as the solid *in vitro* in amounts of about 30 % to 60 % (WHO 2001). Several studies have shown that the internal exposure to arsenic correlates well with the external exposure (ACGIH 1999, Greim and Lehnert 1994).

Distribution

The half-life for the elimination of inorganic arsenic compounds from the blood is short. The arsenic concentration in blood is therefore not a suitable parameter for biological monitoring (WHO 2001).
    Arsenic compounds can pass the blood-brain barrier and the placental barrier. Studies with animals showed that the substance is distributed in all organs, with accumulation particularly in the liver, kidneys, bladder and tissues rich in keratin and sulfhydryl

groups, such as hair, nails and skin (WHO 2001). Inorganic arsenic compounds have a tendency to accumulate in the epididymis, thyroid gland and lenses of the eyes (Lindgren *et al.* 1982). With increasing age an increase in the accumulation of arsenic in the various tissues is observed (WHO 2001).

Elimination

Inorganic arsenic compounds are mainly eliminated as DMA via the kidneys (WHO 2001). In 101 male test persons the amounts of the following four arsenic species were determined in urine by means of anion exchange chromatography with atomic absorption spectroscopy: DMA 88.1 %, As(III) 11.9 %; the levels of MMA and As(V) were below the detection limit (Heinrich-Ramm *et al.* 2001).

Small amounts of absorbed arsenic are eliminated via the skin, hair, nails and breast milk. Even with low-level internal exposure the substance enters the placenta (WHO 2001).

3.2 Metabolism

Metabolism takes place via reduction and oxidation reactions with the mutual transformation of arsenite and arsenate—preferred is the reduction of As(V) to As(III). Via methylation arsenite is transformed into MMA(V), DMA(V) and DMA(III) (see Figure 1). These steps take place independent of the route of absorption. As the capacity of this enzymatic reaction is limited, the methylation capacity must be exceeded after a certain arsenic dose. Exact data are, however, not available. Habituation resulting from long-term exposure to arsenic can lead to more efficient metabolism (WHO 2001).

Arsenate can be reduced to arsenite by glutathione. Methylation takes place only with As(III). As(V) is first of all reduced to As(III) and then methylated. Methylation takes place mainly in the liver with the involvement of enzymes which use *S*-adenosylmethionine (SAM) as a methyl donor and form *S*-adenosylhomocysteine (SAH), and use GSH as an essential cofactor and form oxidized glutathione (GSSG) (WHO 2001).

$$As^{V}O_4^{3-} \xrightarrow[-GSSG]{+2GSH} As^{III}O_3^{3-} \xrightarrow[-SAH]{+SAM} H_3C-As^{V}O_3^{2-} \xrightarrow[-GSSG]{+2GSH} H_3C-As^{III}O_2^{2-}$$

Arsenate   Arsenite   Methylarsenate   Methylarsenite

$$H_3C-As^{III}O_2^{2-} \xrightarrow[-SAH]{+SAM} (H_3C)_2As^{V}O_2^{-} \xrightarrow[-GSSG]{+2GSH} (H_3C)_2As^{III}O^{-}$$

Methylarsenite   Dimethylarsenate   Dimethylarsenite

Figure 1. Metabolism of inorganic arsenic compounds

The methylation capacity is reduced by a reduction in the methyl donors provided in the diet. The quantity and quality of the methylation of inorganic arsenic compounds differs considerably in different species (see Section 2).

4 Effects in Man

4.1 Single exposures

4.1.1 Inhalation

There are various reports that short-term inhalation of arsenic led to nausea, vomiting and diarrhoea. As these symptoms, however, are not typical of intoxication, it was assumed that the arsenic particles were transported to the larynx by mucociliary clearance and then reached the gastrointestinal tract as a result of swallowing. Arsenic was not found to have effects on the blood count, skeletal muscles, liver or kidneys (ATSDR 2000). In other case studies, with no data for the levels of exposure, peripheral neuropathy (loss of muscle reflexes, muscle weakness, tremor) and encephalopathy (hallucinations, increased excitability, emotional lability, memory loss, difficulties in learning new information) were reported after inhalation exposure to arsenic (ATSDR 2000, Beckett *et al.* 1986, Bolla-Wilson and Bleecker 1987). This peripheral neuropathy and encephalopathy were attributed, however, to ingestion of arsenic (ATSDR 2000).

4.1.2 Ingestion

Ingestion of arsenic compounds led to the first symptoms usually within 30 to 60 minutes. Depending on the severity of the clinical symptoms, intoxication with arsenic is described by the terms acute paralytical syndrome and acute gastrointestinal syndrome. The acute paralytical syndrome is characterized by cardiovascular collapse, central nervous weakness and death within hours. The most frequent form of acute arsenic intoxication was, however, the gastrointestinal syndrome, which began with a metallic or garlic-like taste, a dry mouth, burning lips, difficulties in swallowing, headaches, dizziness, vomiting and diarrhoea. In some cases multi-organ failure occurred (ATSDR 2000, IARC 1980).

In 4 case studies in which the person died arsenic doses of between 22 and 121 mg/kg body weight were reported (ATSDR 2000).

Severe damage to the respiratory tract, such as shortness of breath, haemorrhagic bronchitis and pulmonary oedema, were described after acute intoxication with arsenic doses of 8 mg/kg body weight and above. These symptoms are probably the result of damage to pulmonary vessels (ATSDR 2000).

Characteristic arsenic-induced effects on the cardiovascular system were changes in the ECG (delayed Q-T interval, unspecific S-T segment) and cardiac irregularity. After a lethal dose of arsenic (93 mg/kg body weight), hypertrophy of the ventricle walls was reported (ATSDR 2000).

There are numerous case reports available that show that a single, high oral dose of arsenic (2 mg/kg body weight and above) causes damage to the nervous system. Usually encephalopathy occurs first with symptoms such as headaches, confusion, lethargy, hallucinations, convulsions and coma, and later peripheral neuropathy. Peripheral neuropathy, however, has also been observed without previous encephalopathy (ATSDR 2000).

In a man who swallowed 8 to 9 g arsenic with suicidal intent, vomiting and diarrhoea were observed after a few hours. Haemodialysis was carried out when he was taken to hospital. 10 days later the patient developed symmetrical polyneuropathy. After 7 weeks Wallerian degeneration of the myelinated nerve fibres was diagnosed and chelate therapy was commenced. Nerve biopsy revealed that arsenic was detectable in the fibular nerve, although it could not be clarified whether the arsenic was situated in the axon or the myelin sheath. In an examination 3 years later, arsenic could no longer be detected in the fibular nerve. The condition of the patient had improved, apart from slight polyneuropathy. The biopsy showed that regeneration and proliferation of the axons had taken place and there was no longer evidence of Wallerian degeneration (Goebel *et al*. 1990).

In another study the electrophysiological profiles of 13 persons with arsenic-induced neuropathy were drawn up; 12 of the persons were exposed to arsenic only once. The conductivity of sensory nerves was impaired as a result of a lack of or low action potentials, the conductivity of motor nerves, however, was only slightly impaired. Biopsy of the fibular nerve revealed axon degeneration (Oh 1991).

Various studies reported that single oral doses of arsenic cause anaemia and leukopenia; this was attributed to direct haemolytic effects or the inhibition of erythropoiesis. There are also, however, other case studies in which no arsenic-induced effects on the blood system were found (ATSDR 2000).

4.2 Repeated exposures

4.2.1 Inhalation

Various studies indicate that long-term exposure to dust containing arsenic can cause laryngitis, bronchitis and rhinitis (ATSDR 2000). Copper smelters in Anaconda, Montana and Tacoma, Washington, who were exposed to arsenic, died of respiratory diseases such as pulmonary emphysema significantly more frequently than persons not exposed. A clear dependence on the level of arsenic exposure could not, however, be deduced and smoking habits were not taken into account (ATSDR 2000).

In various epidemiological studies, peripheral vascular effects such as acrocyanosis, Raynaud's disease (episodes of ischaemia resulting from spasms in vessels, usually in the arteries of the fingers) and tissue necrosis on the extremities (blackfoot disease) were described after long-term inhalation exposure to arsenic (ATSDR 2000. Lagerkvist *et al.* 1988).

The risk of dying from cardiovascular diseases, such as ischaemia of the heart, or cerebrovascular diseases, was increased in many cohort studies. A clear dose–response relationship was not observed, however. In addition there was exposure to a mixture of substances, including copper, lead and radon (ATSDR 2000).

In a copper smelting plant, peripheral neuropathy was investigated in 70 employees exposed to arsenic trioxide and 41 control persons who were not exposed. The results show that the level of arsenic, which was determined by analysing urine, hair and finger nails, was associated with a higher number of cases of sensory and motor neuropathy and electrophysiological changes (Feldman *et al.* 1979).

In another study, the occurrence of chronic encephalopathy resulting from occupational exposure to arsenic was reported. After repeated exposure to arsenic during wood impregnation, one person developed encephalopathy with such symptoms as concentration difficulties, deficits in learning new information and in short-term memory, mental confusion, anxiety and depression. Another man suffered from irritation, headaches and memory difficulties. The condition of both persons improved after the end of the arsenic exposure (Morton and Caron 1989).

Evidence was found in a Swedish study that long-term inhalation exposure to arsenic can play a role in the development of *diabetes mellitus* (Rahman and Axelson 1995).

4.2.2 Ingestion

Case reports showed daily uptake of arsenic with the drinking water of 0.05 to 0.1 mg/kg body weight to be lethal in young children and of 0.014 to 3 mg/kg body weight to be lethal in adults (ATSDR 2000).

In one of two men exposed short-term to arsenic (no other details), indisposition, fever, coughing and a headache developed, and in the other man vomiting and diarrhoea. In both patients a scaly, papular skin rash developed. Blood analyses yielded increased haemoglobin values (1120 g/l; normal value: 140–180 g/l) in one of the exposed men and a reduced haematocrit value (32 %; normal value: 40–50 %) in the other. Both persons developed severe sensory deficits (tremor, proprioception) and motor deficits within 4 weeks. In the first examination in both patients action potentials of various sensory nerves were no longer present. Motor performance decreased in the first three months and then improved. Two years later, however, sensory and motor neuropathy was still detected, which indicates destruction of the sensory nerves (Murphy *et al.* 1981).

As a result of contaminated milk powder, around 12000 Japanese children ingested up to 3.5 mg arsenic daily for 33 days. The most frequent symptoms were fever, sleep disorders, weight loss, swelling of the liver and dark coloured skin. 130 deaths occurred (IARC 1980).

In a family which were exposed to arsenic long-term (no other details), only a 16-year-old girl was found to have extremely delayed thought processes, severe hypotonic paraparesis, generalized areflexia and white finger and toe nails. The girl's symptoms of intoxication were explained at the time, among other things, by a genetic defect in methylene tetrahydrofolate reductase, with the result that methylation of the arsenic compounds, and thus detoxification, could not take place (Brouwer *et al.* 1992).

Long-term uptake of arsenic doses of about 0.03–0.1 mg/kg body weight and day resulted in symmetrical peripheral neuropathy, which began with a loss of feeling in the hands and feet and developed into painful paresthesia. Sensory and motor nerves were affected. In some cases muscle weakness developed, which led to paralysis of the wrist and ankle, and changes in reflexes were observed. Histological examinations revealed the death of axons in conjunction with demyelination. After the end of the exposure to arsenic the damage regressed only very slowly and also not completely (ATSDR 2000).

Other studies reported that neurotoxic effects could not yet be observed with daily doses of arsenic of 0.006 mg/kg body weight. Another study from China reported, however, that exposed persons described fatigue, headaches, dizziness, sleep disorders, nightmares or a loss of feeling in the extremities after daily arsenic doses of 0.005 mg/kg body weight and more (ATSDR 2000). Various studies reported a relationship between the repeated oral uptake of arsenic with the drinking water and the increased occurrence of cerebrovascular diseases and circulatory disturbances of the heart (Taiwan), high blood pressure (Bangladesh), Raynaud's disease and cyanosis of the fingers and feet (with daily arsenic doses of 0.02 to 0.06 mg/kg body weight; Chile) (ATSDR 2000). Long-term oral uptake of arsenic caused damage to the peripheral vascular system. The characteristic example of this is blackfoot disease, which is endemic above all in Taiwan in those areas in which the level of arsenic in the drinking water varies between 0.17 and 0.8 mg/l, which corresponds to a daily dose of about 0.014 to 0.065 mg/kg body weight. The disease is characterized by a progressive loss of blood circulation in the hands and feet, and finally leads to necrosis and gangrene (ATSDR 2000). In a retrospective cohort study of 789 persons affected by blackfoot disease, mortality from peripheral vascular or cardiovascular diseases was significantly increased after an observation period of 15 years (Chen *et al.* 1988).

The uptake of arsenic (about 0.02 mg/kg body weight) with the drinking water over many years led also to a reduction in body weight (ATSDR 2000). Other studies showed a relationship between repeated oral uptake of arsenic with the drinking water and the increased occurrence of *diabetes mellitus* (Lai *et al.* 1994, Rahman *et al.* 1998).

4.3 Local effects on skin and mucous membranes

Skin

In an employee of a glass factory, who had contact with arsenic trioxide while mixing the raw materials, papulo-vesicular skin reactions developed on the hands, the bends of the arms, the hollows of the knees, and the neck (Paschoud 1964).

In a very extensive investigation of employees of a Swedish company for refining arsenic, toxic and irritative effects on the skin, above all caused by arsenic trioxide, were found in 71 employees. The skin changes were observed particularly on the face, while the hands, which were protected by gloves, were affected less often. Although the course of the skin diseases was acute in most cases, in some cases it was chronic. A latency period for the development of skin diseases of 1 week to 10 years was given. In view of the variable clinical picture with erythema, oedema, the formation of papula or vesicles, or even purely follicular reactions, it was often not possible to differentiate between an irritative and an allergic genesis (Holmqvist 1951; see Section 4.4).

Other pustular or follicular skin reactions which occurred after contact with inorganic arsenic compounds were attributed to the irritative effects of the arsenic compounds (Birmingham *et al.* 1965, Eberhartinger *et al.* 1969, Goncalo *et al.* 1989, Mohamed 1998).

Eyes and mucous membranes

Long-term exposure to dust containing arsenic caused irritation of the mucous membranes of the nose and throat and the conjunctiva (ATSDR 2000).

4.4 Allergenic effects

In several earlier reports sensitization resulting from skin contact with inorganic (but above all organic) arsenic compounds was described.

One case was an employee of a glass factory who had contact with arsenic trioxide while mixing the raw materials and as a result developed papulo-vesicular skin reactions on the hands, the bends of the arms, the hollows of the knees, and the neck. Patch tests were carried out with arsenic trioxide powder and with 1 % potassium arsenite in water and yielded positive results. No details were given, however, about the test conditions and the reading procedure (Paschoud 1964).

After long-term occupational contact with sodium arsenite, 3 persons developed suspected allergic contact dermatitis. In the patch test 2 of the 3 patients produced a weak reaction and 1 patient a marked reaction to 1 % sodium arsenite in water, yet no reaction to 5 % sodium arsenate in water (Schulz 1962).

Patch tests with arsenic trioxide produced papulous reactions in 71 of 163 exposed employees of a Swedish company that refined arsenic. Only 12 of 145 new employees and 17 of 105 employees with low-level exposure to arsenic produced such reactions, however. In the three groups skin reactions to arsenic trioxide were observed in 131 of 163 employees (80.4 %), 44 of 145 employees (30.3 %) and 34 of 105 employees (32.4 %), respectively. A 1 % sodium arsenite preparation caused a marked reaction in one of the affected employees, which persisted for at least 14 days and was regarded as allergic. Of 23 persons tested who previously did not react to arsenic trioxide, none reacted to 1 % sodium arsenite, but all of them to 10 % sodium arsenite. In patch tests with 1 % to 40 % sodium arsenate 4 of 10 persons not exposed also produced (weak) irritative reactions, in 2 cases even with 1 % sodium arsenate (Holmqvist 1951). Nevertheless, the severity and course of the reactions described for several employees indicate an allergic genesis rather than an irritative one. Further evidence for an immunological mechanism is the flaring up of the original skin reactions during the patch test or oral provocation. In some cases reactions that occurred only after several tests indicate the possibility of sensitization as a result of the testing. All in all, however, despite extensive documentation of the test results, no decisive statement can be made at present about the sensitizing effects of inorganic arsenic compounds on the basis of these investigations, as no standardized preparations were used for testing and no details are available about the purity of the test substances. In addition, it is not stated whether the tests were carried out during a period in which the skin had been free of symptoms for a longer period. These limitations are very probably true also for earlier findings listed by the author in a historical review and are therefore not taken into consideration here.

Later studies seem to indicate that even the lowest test concentrations can yield false positive results, so that 0.1 % sodium arsenate and 0.05 % sodium arsenite were used in preliminary comparative tests in exposed employees and for the consecutive testing of patients of a dermatological clinic. Two of 379 clinic patients produced marked reactions to both preparations. Previous exposure to arsenic was not evident from the persons' working history (Wahlberg and Boman 1986). Test results from the investigation of the employees were evidently not published.

In another earlier study the increased occurrence of eczematous or follicular skin reactions was reported in employees of a smelting plant and their families in connection with an inadequately functioning filter system. Only in 14 persons (12 children, 2 employees) with eczematous or follicular skin reactions were patch tests carried out 6 months later with 5 % arsenic trioxide in starch. This produced oedematous erythema in one case and oedematous-vesicular reactions in 2 cases. Only one of the persons who produced a reaction previously had eczematous skin changes but tolerated further contact with small amounts of arsenic without recurrence of the symptoms. The persons who reacted to arsenic trioxide also reacted to 1 % sodium arsenite and 10 % sodium arsenate in water, but some control persons (no other details) produced weak irritative reactions with these preparations (Birmingham *et al.* 1965).

Also in an arsenic-processing works erythematous, eczematous and bullous skin changes were observed, as was a specific manifestation of pustulous reactions ("arsenic scabies"). Allergological investigations, however, were not carried out (Rossberg 1967). Such pustulous or follicular reactions occur also after contact or testing with other metal salts, are not always reproducible and their genesis is usually difficult to evaluate. Between 1966 and 1968, 1638 patients were tested with standard substances, which also included an arsenic compound (it is unclear whether this was 1 % sodium arsenite or 1 % sodium arsenate). Pustulous or follicular reactions were observed in 109 patients, in 88 of these persons to the arsenic compound, in 33 to 10 % nickel sulfate and in 17 to 2 % formalin solution. In none of the 33 patients who reacted in such a way to nickel sulfate was evidence of a nickel allergy found in subsequent investigations (Eberhartinger *et al.* 1969).

In an employee who had previously never suffered from skin complaints, skin changes were observed on the hands, arms and legs after employment for 3 months in a crystal factory. In a patch test with various substances with which the employee had contact at work, 1 % sodium arsenate in water was the only substance which yielded a positive result, with reactions after 48 and 96 hours. 22 control persons did not react to the test preparation (Barbaud *et al.* 1995).

In a tin smelting works, in which 11 employees suffered from dry, itching, hyperpigmented skin with folliculitis, the concentration of arsenic trioxide in the air was between 5.2 and 14.4 µg/m^3. The authors attributed the complaints to probably irritative contact dermatitis resulting from dust containing arsenic. Patch tests were not carried out (Mohamed 1998).

Three workers in the glass industry suffered among other things from pustular and follicular skin changes on exposed areas of skin. In patch tests with the arsenic trioxide powder to which the workers were exposed, a clear pustular-follicular reaction was observed, while the workers produced only weak reactions to a 5 % arsenic trioxide preparation. The authors attributed the skin changes to irritative effects of the arsenic trioxide (Goncalo et al. 1989).

In the more recent literature there is only a report about a 26-year-old female worker who was occupationally exposed to DMA and developed dermatitis on the face. In patch tests with 0.1 % and 1 % DMA in water, reactions were observed after 3 days at both concentrations (Bourrain et al. 1998).

One study investigated the immunological effects after inhalation of inorganic arsenic compounds. In workers exposed to arsenic in a coal-firing power plant, the serum levels of the immunoglobulins were not unusual (Bencko et al. 1988).

4.5 Reproductive and developmental toxicity

4.5.1 Fertility

There are no studies available of the influence of arsenic and inorganic arsenic compounds on the fertility of man after inhalation or oral exposure (ATSDR 2000).

4.5.2 Developmental toxicity

Single exposures

A 17-year-old, who swallowed arsenic trioxide by drinking rat and mouse poison (about 0.39 mg/kg body weight) in week 30 of pregnancy, was admitted to hospital 24 hours later with acute kidney failure. On day 3 her condition worsened and she became progressively disoriented. She was delivered of a 1.1 kg baby girl and underwent haemodialysis (no other details). After the first minute of life the baby had an APGAR score of 4 (a measure of heart rate, respiratory effort, muscle tone, reflex irritability and colour; optimum score: 9–10), had increasing difficulties breathing, a reduced heart rate, hypoxia, hypercabia and acidosis and died after 11 hours. Autopsy revealed in addition to the symptoms of premature birth generalized skin haemorrhage and severe intraalveolar bleeding (Lugo et al. 1969).

Repeated exposure

Numerous studies investigated the influence of arsenic exposure on intrauterine development, in particular the frequency of spontaneous abortions, congenital anomalies and stillbirths.

To test the hypothesis that the developmental toxicity of arsenic is due to its great affinity to thiol groups and glutathione and the resulting increase in lipid peroxidation in maternal and foetal compartments, the glutathione levels in blood and the lipid peroxidases in the placenta or blood were investigated in 49 mother–child pairs in hospitals near a copper mine and in a control group. The exposure was characterized by high levels of arsenic in the air (0.119 mg/m^3) and low levels in the water (about 12 µg/l). The incidence of complications during pregnancy, e.g. maternal toxaemia and foetal mortality as a result of congenital malformations, was higher in the region of the copper mines than in the national average. Taking into account maternal parity, smoking habits and the sex of the children, the average body weights and body lengths at birth were reduced. The average arsenic levels in the placentas of the exposed mothers were three times higher than those in the mothers not exposed (0.023 compared to 0.007 mg/kg, $p < 0.01$). In the group with the highest exposure the reduced glutathione was significantly decreased and the lipid peroxidases increased (Tabacova and Hunter 1995).

In a region in southeast Hungary with increased levels of arsenic in the water (> 100 µg/l, 25648 persons), a 1.4-fold increase in spontaneous abortions and a 2.7-fold increase in stillbirths ($p < 0.05$) were determined between 1980 and 1987 compared to in a region with low arsenic levels in the water (no data for the levels of arsenic, 20836 persons). There was no significant difference between the groups for the frequency of premature births and stillbirths (Borzsonyi *et al.* 1992).

To determine the relationship between the uptake of chemicals with the drinking water of the mother and congenital heart diseases of the offspring, all live births (270 cases, 665 controls) between 01.04.1980 and 31.03.1983 were investigated. The arsenic concentrations in the drinking water were between 0.8 and 22 µg/l. No relationship was found between the arsenic level in drinking water and all congenital heart diseases, but a three-fold increase in the frequency of coarctation of the aorta (OR = 3.4; 95 % CI = 1.3–8.9) was determined. There was no relationship with the frequency of patent *ductus arteriosus*, cono-truncal defects and ventricular septal defects. Non-differential errors in the analyses and the classification of the exposure, and the low exposure could be responsible in the opinion of the authors for the lack of positive findings (Zierler *et al.* 1988).

In a hospital-based study of the relationship between the quality of communal drinking water and the risk for spotaneous abortions, an increase in the frequency of spontaneous abortions (taking confounders into consideration) which was not significant was found for arsenic concentrations in drinking water of between 0.8 and 1.9 µg/l. As a result of the small amount of data, the authors state that further investigations are necessary to confirm these findings (Aschengrau *et al.* 1989).

In a hospital-based case–control study with an atmospheric distribution model to estimate the arsenic concentration in the air of a region in which agricultural products on an arsenic basis were produced for over 60 years, an increased risk was found for stillbirths with exposure to increased levels of arsenic (> 100 ng/m^3) after adjustment for demographic factors (POR = 4.0; 95 % CI = 1.2–13.7). In a separate analysis of the interaction between race/ethnic provenance and exposure, a significant relationship was found for Spanish inhabitants, who were exposed to the highest concentration (> 10 ng/m^3) (POR = 8.4; CI = 1.4–50.1). The risk for stillbirths was higher among African Americans in the adjusted model, but not significantly increased in the interaction model. Further studies are necessary to clarify the role of inhalation exposure via the environment in addition to the role of nutrition and other factors (Ihrig et al. 1998).

The subject of numerous studies was the possible influence of the smelting works in Ronnskar, Sweden, on the health of the employees and the local population. In addition to copper and lead, the smelting works produced arsenic, arsenic trioxide and a few other metals and their compounds. The arsenic emissions were given as 50 tons per year (Nordström et al. 1978, 1979a, 1979b). Significantly reduced birth weights and increased incidences of spontaneous abortions were determined in the employees and the local population. If women worked in the immediate area of the smelting processes, there was a significant increase in the frequency of abortions, which increased further if their husbands were employed in the smelting works (Nordström et al. 1979a). An increase in the incidence of spontaneous abortions and stillbirths together was found in women whose husbands worked in a smelter, relative to in women whose husbands were not exposed (Beckman and Nordstrom 1982). In children whose mothers worked near the smelting processes during pregnancy, unlike in children of mothers who lived in the vicinity of the smelting works, the frequency of single malformations (e.g. cleft palates) or multiple malformations was significantly increased (Nordström et al. 1979b). These studies are, however, unsuitable for evaluating developmental toxicity, as the exposure to arsenic was not sufficiently characterized.

In a retrospective study infant mortality was investigated in two regions of Chile between 1950 and 1996. In Antofagasta contamination of the drinking water with arsenic was documented, while in Valparaiso the levels were comparatively low. Investigation of the temporal development of late foetal mortality, mortality of newborn babies and mortality in early childhood revealed a relationship with the arsenic level in drinking water (Hopenhayn-Rich et al. 2000).

In another study, in Bangladesh two groups of 96 women aged between 15 and 49 were compared. One group had consumed ≥ 0.1 mg arsenic per litre drinking water (43.8 % of the women for 5 to 10 years), and the other group had not. The two groups were matched for age, social status, education and age at marriage. Only in the group of exposed women were the frequencies of spontaneous abortions, stillbirths and premature births significantly higher than in the control group (Ahmad et al. 2001).

4.6 Genotoxicity

Studies of the genotoxicity of arsenic compounds were carried out above all in persons who had taken up arsenic long-term with the drinking water. It was not differentiated between the different arsenic species in drinking water. As inorganic arsenic compounds can be transformed by micro-organisms into organic arsenic species by methylation, organic arsenic compounds are found in drinking water in addition to the inorganic arsenic compounds. Also in the human organism methylation of the absorbed inorganic arsenic compounds takes places to form MMA and DMA. The internal exposure in man after absorption of arsenic via the drinking water is therefore a mixture of inorganic and organic arsenic compounds. Studies of these exposed persons revealed clastogenic effects of arsenic (Table 1).

Table 1. Genotoxicity of arsenic compounds in man

Study population	Number of persons (age)	Exposure	Effects	References
single exposures				
test persons	1 volunteer per dose	ingestion (1×) 0.15 g $KAsO_2$ or 1, 10 or 20 g As_2O_3	in L: SCE significantly increased at 20 g As_2O_3	Hantson et al. 1996
long-term ingestion				
general population, Mexico	11 (21–62)	drinking water exposure; exposed persons: 390 µg/l, controls: 26 µg/l; As in urine: 1565 ± 916 µg/l (controls: 121 ± 90 µg/l)	in L: CA increased, no SCE, mutations at the HPRT locus increased in those with the highest exposure	Ostrosky-Wegman et al. 1991
general population, Mexico	35 (39.0)	drinking water exposure; exposed persons: 408 µg/l, controls: 30 µg/l; As in urine: 740 ± 361 µg/l (controls: 34 ± 35 µg/l)	in L: CA increased, in CC und UC: MN increased	Gonsebatt et al. 1997
general population, Nevada, USA	98 (not stated)	drinking water exposure; exposed persons: 109 µg/l, controls: 12 µg/l; ≥ 5 years	in L: no CA, no SCE	Vig et al. 1984
general population, Nevada, USA	18 (37.5)	drinking water exposure; exposed persons: 1313 µg/l, controls: 16 µg/l; ≥ 1 year, As in urine: 750 ± 160 µg/l (controls: 68 ± 23 µg/l)	in BC: MN increased	Warner et al. 1994

Table 1. continued

Study population	Number of persons (age)	Exposure	Effects	References
general population, Chile	70 (42.0)	drinking water exposure; exposed persons: 600 µg/l, controls: 15 µg/l; As in urine: 616, 84–1893 µg/l (controls: 66, 4–267 µg/l)	in BC: MN increased	Moore et al. 1997a
general population, Finland	32 (15–83)	drinking water exposure; exposed persons: 410 µg/l, controls: < 1 µg/l; As in urine: 180, 7–500 µg/l (controls: 7, 4–44 µg/l)	in L: CA significant correlations with level of As in urine	Maki-Paakkanen et al. 1998
general population, Argentina	22 (8–66)	lifelong drinking water exposure; 205 µg/l, As in urine: 260, 120–440 µg/l (controls: 8.4, 5.2–27 µg/l)	in L: MN significantly increased, no CA, no SCE	Dulout et al. 1996
general population, Argentina	282 (27–82)	drinking water exposure (> 20 years) (exposed persons: 130 µg/l, controls: 20 µg/l), As in urine: 160, 60–410 µg/l (controls: 70, 10–600 µg/l)	in L: SCE significantly increased	Lerda 1994
general population, Inner Mongolia	19 (38 ± 15)	drinking water exposure (exposed persons: 527.5 µg/l, controls: 4.4 µg/l)	in SC, CC and UC: MN significantly increased	Tian et al. 2001
patients with arsenic-induced skin tumours	(not stated)	(no other details)	8-OH-dG adducts in skin tumours significantly increased	Matsui et al. 1999
long-term inhalation				
smelters	9	exposure to a mixture of arsenic, lead and selenium	in L: CA significantly increased	Beckman et al. 1977
smelters	39 (21–62)	arsenic exposure 4–22 years, As in urine: 167–290 µg/l	in L: CA significantly increased	Nordenson et al. 1978

Table 1. continued

Study population	Number of persons (age)	Exposure	Effects	References
smelters	33 (20–65)	exposure probably to a mixture of substances; As in urine: 200 µg/l (1976), 80 µg/l (1979)	in L: CA significantly increased	Nordenson and Beckman 1982
smelters, Chile	15 (24–66)	copper factory; As in urine of employees with highest exposure > 260 µg/l	in L: HPRT mutations increased	Harrington-Brock et al. 1999
long-term dermal absorption				
tumour patients	6 (41–84)	patients treated with Fowler's solution (1 % $KAsO_2$ in H_2O) (no other details)	in L: SCE significantly increased, no chromosomal breaks	Burgdorf et al. 1977

BC = bladder epithelial cells, CA = chromosomal aberration, CC = epithelial cells of the cheek, HPRT= hypoxanthine phosphoribosyl transferase, L = lymphocytes, MN = micronuclei, SC = cells from sputum, SCE = sister chromatid exchange, UC = urothelial cells

In buccal and bladder epithelial cells, urothelial cells and lymphocytes an increase in the incidence of micronuclei was found and additionally in lymphocytes an increase in chromosomal aberrations and sister chromatid exchange (Dulout et al. 1996, Gonsebatt et al. 1997, Hantson et al. 1996, Lerda 1994, Maki-Paakkanen et al. 1998, Moore et al. 1997a, Ostrosky-Wegman et al. 1991, Tian et al. 2001, Vig et al. 1984, Warner et al. 1994). In another study, however, no genotoxic effects could be found in persons exposed to arsenic (Vig et al. 1984).

A significant increase in chromosomal aberrations in peripheral lymphocytes was detected in workers of the copper smelting plant in Ronnskar, Sweden, who were exposed to unspecified concentrations of arsenic trioxide, (Beckman et al. 1977, Nordenson and Beckman 1982, Nordenson et al. 1978). It is questionable whether there was additional exposure to other genotoxic substances.

A significant increase in sister chromatid exchange, but no chromosomal breaks were found in the peripheral lymphocytes of tumour patients who had applied a solution containing arsenic to the skin long-term (Burgdorf et al. 1977). A slight increase in HPRT mutations was described in peripheral lymphocytes of workers in a copper mine in Chile, in whom arsenic levels in urine of more than 260 µg/l were determined (Harrington-Brock et al. 1999).

4.7 Carcinogenicity

On the basis of epidemiological studies, national and international organisations have classified arsenic and arsenic compounds (with the exception of arsine) as carcinogenic in man (ACGIH 1999, EPA 2001, IARC 1980, 1987).

4.7.1 Inhalation

In several large cohort studies, an increased risk for lung cancer was observed in gold and tin miners and copper smelters in various countries and under different exposure conditions. The possibility of significant distortion of the increased lung cancer risk as a result of co-exposure to sulfur dioxide, various dusts and above all cigarette smoke could be ruled out in several studies (e.g. Enterline 1987b, Järup and Pershagen 1991). An extensive description of the epidemiological studies can be found in IARC (1980) and WHO (2001).

In view of the verified epidemiological evidence for the carcinogenicity of arsenic in man, the question arises whether a threshold concentration for the carcinogenic effects of arsenic can be deduced from the epidemiological studies. Only the relevant epidemiological studies are described below.

The dose–response relationship between exposure to arsenic and lung cancer risk was investigated in six published studies with quantitative data for the arsenic exposure. The studies investigated smelters in Tacoma, Washington (Enterline 1987a), Anaconda, Montana (Lee-Feldstein 1986), Ronnskar, Sweden (Järup and Pershagen 1991, Järup *et al.* 1989) and six smaller smelting works in the USA (Enterline 1987b), a cohort of workers from insecticide production in Midland, Michigan (Ott *et al.* 1974) and a cohort of miners and a few smelters in China (Taylor *et al.* 1989). In only two of the six studies (Lee-Feldstein 1986, Järup *et al.* 1989) was the relationship between arsenic exposure and lung cancer compatible with a linear dose–response relationship, while the relationship in the other studies was supralinear. The reasons suggested for this possible supralinear dose–response relationship were dose-dependent synergisms with smoking habits, competing dose-dependent causes of death or dose-dependent false classification of the arsenic exposure, but not distortion as a result of age or occupational co-exposures (Hertz-Picciotto and Smith 1993).

In the analysis of an extended follow-up to the Tacoma study (Enterline 1987a) the previously found supralinear dose-response relationship could be confirmed (Enterline *et al.* 1995).

An extended follow-up study of the cohort of 8014 smelters in Anaconda, Montana (see Lee-Feldstein 1986) investigated the relationship between arsenic exposure and lung cancer risk (Lubin *et al.* 2000). It was already seen in earlier analyses that sulfur dioxide and smoking habits probably did not represent relevant confounders for the relationship between arsenic exposure and lung cancer risk for this cohort (Lubin *et al.* 1981, Welch *et al.* 1982). Using internal controls and different weighting factors for the duration of work in areas with high levels of exposure to arsenic, the authors found a linear dose–response relationship between inhalation exposure to arsenic and lung cancer risk. They suggested that the supralinear dose–response relationships in earlier analyses of this cohort and in the analyses of the extended follow-up study of the Tacoma cohort (Enterline *et al.* 1987a) may be the result of a differential false classification of high exposure to arsenic (Lubin *et al.* 2000). Further data for the above-mentioned studies can be found in Table 2.

All in all, the data of the studies of the relationship between occupational inhalation exposure to arsenic and the lung cancer risk indicate there is most probably a linear, possibly also a supralinear, but not a sublinear dose–response relationship with a deducible threshold dose.

4.7.2 Ingestion

The evaluation of the EPA suggests that the re-analysis of a Taiwanese study of the relationship between oral exposure to arsenic via contaminated drinking water and skin cancer risk may indicate a non-linear dose–response relationship in the sense of a sublinear curve (EPA 1997). Even if the EPA do not attach much importance to this statement themselves, occupational mortality studies are not suitable for identifying increased skin cancer risks.

The study of Lubin *et al.* (2000) indicates a clear dose–response relationship between inhalation exposure to arsenic and lung cancer risk; an increase in other types of cancer was not observed. The cumulative inhalation exposure to arsenic in the cohort of Lubin *et al.* (2000) was much higher than in the Tacoma cohort and corresponded more or less to the long-term oral uptake of arsenic in endemic regions, e.g. Taiwan. Studies in these regions consistently associated the increased uptake of arsenic with an increased risk for cancer of the bladder, kidneys, liver and lungs. The incidences for lung cancer observed in these studies were increased to a much smaller extent than those for bladder, kidney and skin cancer (Bates *et al.* 1992).

4.7.3 Dermal absorption

Many early studies reported that the long-term dermal application of solutions containing arsenic, e.g. Fowler's solution (1 % $KAsO_2$ in H_2O), caused skin carcinomas (IARC 1980, WHO 2001).

Table 2. Epidemiological studies of exposure to arsenic

Study population (male)	Number of persons	Type of study	Exposure	Effects	References
workers in the packing of insecticides, USA	173 exposed persons, 1809 persons not exposed	proportional mortality study: causes of death of exposed persons compared with the cause of death of persons not exposed	IAEM < 1 to > 96	cancer of the respiratory passages increased two-fold (from 1–2 IAEM) to seven-fold (over 96 IAEM); dose–response relationship supralinear	Ott et al. 1974
smelters, Utah, USA	2288	cohort study: workers compared with the white male population of the USA, the state or the region	mean exposure 959.7 µg/m^3 × year; mean duration of employment 21.1 years	SMR for lung cancer 119.7 (not significant), 226.9 ($p < 0.01$) and 210.9 ($p < 0.01$); great differences between USA and regional values possibly as a result of the different smoking habits of the workers compared with the rest of the population in Utah	Enterline et al. 1987b
smelters, different companies, USA	unclear, several thousand (?)	cohort study: only male persons exposed to arsenic, comparison of different concentrations	< 100 to > 1000 µg/m^3 × year	lung cancer RR < 100 µg/m^3: 0.58; 100–249 µg/m^3: 0.85; 250–999 µg/m^3: 1.21; > 1000 µg/m^3: 1.60; $p = 0.059$ for trend; dose–response relationship supralinear	Enterline et al. 1987b
mine workers and smelters, China	107 exposed persons, 107 persons not exposed	embedded case–control study: workers with lung cancer compared with workers without lung cancer	IAEM median for cases: 45.9 (0–255.6); IAEM median for controls: 7.9 (0–132.9)	lung cancer in smokers with low cigarette consumption: OR 5.0 (IAEM 58.6–255.6); reference category: IAEM 0–10.4; no interaction with smoking	Taylor et al. 1989

Table 2. continued

Study population (male)	Number of persons	Type of study	Exposure	Effects	References
smelters, Sweden	3916	cohort study: workers compared with the population of the region	< 250 to > 100000 µg/m³ × year	SMR of 271 (lowest cumulative exposure) up to 1137 (highest cumulative exposure)	Järup et al. 1989
smelters, Sweden	107 exposed persons, 214 persons not exposed	embedded case–control study: workers with lung cancer compared with dead workers without lung cancer	< 250 to > 100000 µg/m³ × year	OR 1.0 for up to 5000 µg/m³ per year, then increased up to 8.7 for > 100000 µg/m³ × year (age and smoking checked)	Järup and Pershagen 1991
smelters, Tacoma, Washington, USA	2802	cohort study: workers compared with the white male population of the state of Washington	< 750 to > 45000 µg/m³ × year	SMR for cancer of the respiratory passages (bronchi, trachea, lungs): 188.1 after < 20 years exposure; SMR after ≥ 20 years 217.1; SMR also significantly increased for intestinal and bone cancer; relationship between the arsenic dose and cancer of the respiratory passages supralinear	Enterline et al. 1995
smelters, Anaconda, Montana, USA	8014	cohort study: workers compared with the population of the USA	time-weighted exposure 0.29, 0.58 and 11.3 mg/m³	SMR for cancer of the respiratory passages 183; RR for low dose group, exposure ≥ 35 years, 1.98; RR for highest dose group, exposure ≥ 10 years, 3.68	Lubin et al. 2000

RR: relative risk, OR: odds ratio, IAEM: time-weighted arsenic exposure multiplied by the duration of exposure (months)

5 Animal experiments and *in vitro* Studies

5.1 Acute toxicity

5.1.1 Inhalation

After mice inhaled arsenic trioxide (0.27, 0.50 or 0.94 mg/m^3) for 3 hours, pulmonary defence against bacteria was reduced in a dose-dependent manner, probably as a result of damage to the alveolar macrophages, and the susceptibility to infection was correspondingly increased (Aranyi *et al.* 1985). After mice were given intratracheal doses of sodium arsenite (5.7 mg arsenic per kg body weight), the humoral immune response to antigens was reduced, but not the resistance to bacteria or tumour cells (Sikorski *et al.* 1989).

In rats given intratracheal arsenic trioxide doses of 17 mg/kg body weight, increased lung weights, an increase in the ratio of lung weights to body weights, and an increase in the levels of pulmonary protein, 4-hydroxyproline and DNA were found. This indicates fibrogenic effects in the lungs. Intratracheal doses of gallium arsenide (100 mg/kg body weight) caused an increase in the lipid, protein and DNA levels in the lungs of rats, and the lung weights were increased. Smaller gallium arsenide particles caused greater effects, probably as a result of their better solubility. The effects of 65 mg/kg body weight were minor despite longer presence in the organism. All substances caused inflammatory changes in the lungs. The strength of the observed effects decreased in the order GaAs > As$_2$O$_3$ >> Ga$_2$O$_3$ (WHO 2001).

5.1.2 Ingestion

Sodium arsenite was found to be of higher acute toxicity than arsenic trioxide and the arsenic(V) compounds sodium arsenate, calcium arsenate and lead arsenate (see Table 3).

The symptoms of intoxication with arsenic were cramps, nausea and haemorrhage in the intestinal tract. In the case of arsenic trioxide the extent of toxicity depended on the purity of the substance. The pure substance caused less severe side effects, but the lethal effects were more pronounced (WHO 2001).

5.1.3 Dermal absorption

The dermal LD$_{50}$ for calcium arsenate and lead arsenate in rats was > 2400 mg/kg body weight (Gaines 1960).

5.1.4 Other routes of absorption

LD_{50} values of 21 and 8 mg arsenic per kg body weight were determined in mice for sodium arsenate and sodium arsenite after intramuscular injection of the substances (Bencko et al. 1978).

Dose-dependent nephrotoxic effects of sodium arsenate were determined in dogs given intravenous doses of the substance. A dose of 0.73 mg/kg body weight (0.26 mg arsenic per kg body weight) caused slight degeneration and the formation of vacuoles of the renal tubular epithelium, but no functional changes. 7.33 mg/kg body weight (2.64 mg arsenic per kg body weight) led to damage to the glomeruli and renal tubules. Vacuolar changes and tubular necrosis were also observed. After 14.77 mg/kg body weight (5.28 mg arsenic per kg body weight) one of three dogs died. In the kidneys severe damage to the glomerulus, proximal and distal tubules and collecting duct was found. Some of the damage in the surviving dogs was, however, reversible (WHO 2001).

80 % of starved rats did not survive a subcutaneous dose of arsenic trioxide of 10 mg/kg body weight, while the rats which were fed normally suffered no damage (Szinicz and Forth 1988). *In vitro* studies in which arsenic trioxide caused the dose-dependent inhibition of gluconeogenesis in isolated hepatocytes and kidney tubules, and arsenic pentoxide produced the same effects at higher concentrations, confirmed that acute intoxication with arsenic can cause disturbances in carbohydrate metabolism (Szinicz and Forth 1988).

Subcutaneous injection of arsenic trioxide doses of 10 mg/kg body weight caused a reduction in the level of fructose-1,6-diphosphate and glycerolaldehyde-3-phosphate and an increase in the concentration of phosphoenolpyruvate and pyruvate in the liver of guinea pigs (Reichl et al. 1988). The survival of mice given glucose or glucose and insulin together with 12 mg arsenic trioxide per kg body weight was significantly higher than that of mice given arsenic trioxide alone. All the mice in this last group died. The hepatic glucose and glycogen values were reduced in this group (Reichl et al. 1990).

Table 3. Acute toxicity of the arsenic compounds

Arsenic compounds	Species	LD$_{50}$ [mg/kg body weight]	LD$_{50}$ [mg arsenic/kg body weight]	References
ingestion				
arsenic trioxide	rat	20–385	15–293	Dieke and Richter 1946, Done and Peart 1971, Harrison et al. 1958
	mouse	34–53	26–39	Harrison et al. 1958, Kaise et al. 1985
sodium arsenite	rat	11–42	6–24	Done and Peart 1971, Schröder and Balassa 1966
	mouse	11	6	Schröder and Balassa 1966
sodium arsenate	rat	112	40	Schröder and Balassa 1966
	mouse	112	40	Schröder and Balassa 1966
calcium arsenate	rat	298	53	Gaines 1960
lead arsenate	rat	1050	231	Gaines 1960
dermal absorption				
calcium arsenate	rat	> 2400	> 400	Gaines 1960
lead arsenate	rat	> 2400	> 500	Gaines 1960
intramuscular absorption				
sodium arsenate	mouse	87	21	Bencko et al. 1978
sodium arsenite	mouse	14	8	Bencko et al. 1978

5.2 Subacute, subchronic and chronic toxicity

The results of studies of the toxicity of arsenic compounds after repeated administration are summarized in Table 4.

5.2.1 Inhalation

Repeated inhalation of arsenic trioxide caused effects on the immune system in mice after arsenic concentrations of 0.3 mg/m^3 and above (Aranyi et al. 1985); in rats the first lung damage and reductions in body weights occurred after arsenic concentrations of 7.6 mg/m^3 and above. After rats were given arsenic concentrations of 19.7 mg/m^3 their general condition was greatly impaired, with a dry, bloody exudate around the eyes, nose and urogenital region, yellow discoloration of the urogenital region and rales, dyspnoea and mortality (Holson et al. 1999). In hamsters, 80 % of the animals died after intratracheal administration of arsenic trioxide (0.3 mg arsenic) or arsenic trisulfide (0.5 mg arsenic), and 30 % of the animals died after administration of calcium arsenate (0.5 mg arsenic) (Pershagen et al. 1982).

5.2.2 Ingestion

Arsenic trioxide (11 mg arsenic per kg body weight) administered to rats for up to 14 days caused diarrhoea, bloody faeces and decreased mobility (Bekemeier and Hirschelmann 1989).

Dogs reacted most sensitively to repeated doses of sodium arsenite. After arsenic doses of about 1.2 mg/kg body weight and above, dose-dependent reductions in body weights were observed. The activity of alanine aminotransferase was increased after arsenic doses of about 1.2 mg/kg body weight and above and that of aspartate aminotransferase after about 4.6 mg/kg body weight. Histological examination, however, did not reveal liver damage (Neiger and Osweiler 1989). In another study sodium arsenite (2.4 mg arsenic per kg body weight) induced, in addition to a marked reduction in body weights, intestinal bleeding, and all 6 treated dogs died (Byron et al. 1967). In mice sodium arsenite (about 1.2 mg arsenic per kg body weight) caused increased mortality (Schroeder and Balassa 1967). In rats, after arsenic doses of about 7.2 mg/kg body weight, the bile ducts were found to be enlarged with thickened walls as a result of fibrosis, sometimes with infiltration of inflammatory cells and sometimes with weak to moderate hyperplasia (Byron et al. 1967).

Administration of sodium arsenate (up to 4 mg arsenic per kg body weight) for 4 to 6 weeks caused structural and functional changes in the liver of rats and mice (Fowler et al. 1977, Hughes and Thompson 1996). Repeated absorption of sodium arsenate by rats caused slight histological changes in the glomerulus and tubulus of the kidney and swollen hepatocytes in the area of the centrilobular vein after arsenic doses of about 2.5 mg/kg body weight (Carmignani et al. 1983). After arsenic doses of about 4.5 mg/kg body weight enlarged bile ducts were found and after 21 mg/kg body weight, reduced body weights and mortality (Byron et al. 1967, Kroes et al. 1974). This damage was induced by lead arsenate only after arsenic doses of about 93 mg/kg body weight and above (Kroes et al. 1974).

Of 3 rhesus monkeys given an arsenic(V) complex (3.7 mg arsenic per kg body weight) suspended in milk daily from the age of 3 days, one animal died from bronchopneumonia with severe bleeding, oedema and necrosis in the lungs, and with various aggregations of acute inflammatory cells in the brain and spinal cord after uptake of the substance for 7 days. The other animals survived the exposure to arsenic for one year with no damage visible on gross pathological examination. In 4 other rhesus monkeys, which were given this arsenic(V) complex (3.7 mg arsenic per kg body weight) from week 8 of life, no damage was found on gross pathological examination. One animal died after 273 days (Heywood and Sortwell 1979).

Of 20 cynomolgus monkeys given daily oral doses of sodium arsenate of 0.1 mg arsenic per kg body weight on 5 days a week for 17 years, 11 animals survived. 3 monkeys were found to have endometriosis and 3 others hyalinized islets of Langerhans of the pancreas. In one of these animals there was evidence of diabetes (Thorgeirsson et al. 1994).

Table 4. Effects of arsenic compounds after repeated administration to laboratory animals

Arsenic compound	Species	Duration	Dose/Concentration	Findings	References
inhalation					
arsenic trioxide	rat	6 h/day, 33 days (whole body inhalation chamber)	7.6 mg As/m^3 19.7 mg As/m^3	7.6 mg/m^3: rattling sounds in the lungs, body weights decreased 19.7 mg/m^3: dry red exudate around the eyes, nose, urogenital region; yellow discoloration of the urogenital region; sporadic wheezing, dyspnoea; mortality: 4/9; in one moribund animal: severe hyperaemia and discharge of plasma into the intestinal lumen	Holson et al. 1999
	mouse	3 h/day, 5 days/week, 1 or 4 weeks	0.3 mg As/m^3 0.5 mg As/m^3	pulmonary defence against bacteria decreased in a concentration-dependent manner, susceptibility to infection increased	Aranyi et al. 1985
	hamster	1 ×/week, 4 weeks	0.3 mg As, intratracheal	80 % mortality after 3 to 4 weeks	Pershagen et al. 1982
arsenic trisulfide	hamster	1 ×/week, 4 weeks	0.5 mg As, intratracheal	80 % mortality after 3 to 4 weeks	Pershagen et al. 1982
calcium arsenate	hamster	1 ×/week, 4 weeks	0.5 mg As, intratracheal	30 % mortality after 4 weeks	Pershagen et al. 1982
ingestion					
arsenic trioxide	rat	5 days/week, 4–14 days	11 mg As/kg body weight	diarrhoea, bloody faeces, mobility decreased	Bekemeier and Hirschelmann 1989

Table 4. continued

Arsenic compound	Species	Duration	Dose/Concentration	Findings	References
sodium arsenite	rat	2 years	up to 250 mg As/kg diet up to about 7.2 mg As/kg body weight	at about 7.2 mg As/kg body weight: bile ducts enlarged, with thickened walls as a result of fibrosis, sometimes with infiltration of inflammatory cells, sometimes with weak to moderate hyperplasia	Byron et al. 1967
	mouse	540 days	5 mg As/l drinking water about 1.2 mg As/kg body weight	survival decreased, mortality increased	Schroeder and Balassa 1967
	dog	1 ×/day, 58 days then 1 ×/day, 124 days	about 0.6 or 1.2 mg As/kg body weight then about 1.2 or 2.3 mg As/kg body weight	≥ 0.6/1.2 mg As/kg body weight: dose-dependent decrease in food consumption, body weight decreased, ALT activity increased	Neiger and Osweiler 1989
	dog	1 ×/day, 58 days then 1 ×/day, 61 days	about 2.3 then about 4.6 mg As/kg body weight	activity of AST and ALT increased, no lesions found in the liver after histological examination	Neiger and Osweiler 1989
	dog	2 years	up to 125 mg As/kg diet up to about 2.4 mg As/kg body weight	body weights decreased by 44 % to 61 %, intestinal bleeding, mortality: 6/6	Byron et al. 1967
sodium arsenate	rat	6 weeks	up to 85 mg As/l drinking water up to about 4 mg As/kg body weight	≥ about 1 mg As/kg body weight: mitochondrial membrane: activities of monooxidase and cytochromoxidase increased ≥ about 2 mg As/kg body weight: mitochondria: swelling at about 4 mg As/kg body weight: body weights decreased, liver: deposits in connective tissue, large lipid vacuoles in hepatocytes	Fowler et al. 1977

Table 4. continued

Arsenic compound	Species	Duration	Dose/Concentration	Findings	References
	rat	320 days	50 mg As/l drinking water about 2.5 mg As/kg body weight	liver and kidneys: slight histological changes	Carmignani et al. 1983
	rat	2 years	145 mg As/kg diet up to about 7.2 mg As/kg body weight	≥ 4.5 mg As/kg body weight: bile ducts enlarged, with thickened walls as a result of fibrosis, sometimes with infiltration of inflammatory cells, sometimes with weak to moderate hyperplasia	Byron et al. 1967
	rat	29 months	about 21 mg As/kg body weight	body weights decreased, mortality increased	Kroes et al. 1974
	mouse	28 days	about 0.3 mg As/kg body weight about 3.0 mg As/kg body weight	dose-dependent effects: liver: vacuoles in hepatocytes increased, non-protein-sulfhydryl levels decreased; plasma: glucose decreased, creatinine increased, triglyceride values increased	Hughes and Thompson 1996
	dog	2 years	up to about 2.4 mg As/kg body weight	at 2.4 mg As/kg body weight: body weights decreased; mortality: 1/6	Byron et al. 1967
	cynomolgus monkey	5 days/week, 17 years	0.1 mg As/kg body weight	endometriosis, hyalinized islets of Langerhans of the pancreas, mortality 11/20	Thorgeirsson et al. 1994
lead arsenate	rat	29 months	about 93 mg As/kg body weight	body weights decreased, mortality increased, bile ducts: enlarged, sometimes widened with inflammation and abscesses	Kroes et al. 1974

Table 4. continued

Arsenic compound	Species	Duration	Dose/Concentration	Findings	References
arsenic(V) complex	rhesus monkey	1 ×/day, 13 days	6 mg As/kg body weight	vomiting, changed faeces, weakness, increased salivation, uncontrolled shaking of the head	Heywood and Sortwell 1979
	rhesus monkey	1 year	3.7 mg As/kg body weight	after ingesting the arsenic complex suspended in milk, a 10-day-old rhesus monkey died after 7 days from bronchopneumonia with marked bleeding, oedema and necrosis of the lungs, with various aggregations of acute inflammatory cells in the brain and spinal cord; mortality: 2/7	Heywood and Sortwell 1979
parenteral uptake					
arsenic trioxide	guinea pig	2 ×/day, 5 days, subcutaneous	10 mg As/kg body weight	liver: glucose and glycogen levels decreased; mortality: 10/10	Reichl *et al.* 1988

ALT = alanine aminotransferase, AST = aspartate aminotransferase

5.2.3 Other routes of absorption

In guinea pigs given subcutaneous injections of arsenic trioxide of 2.5 mg/kg body weight (1.9 mg arsenic per kg body weight) twice a day for 5 days, the carbohydrate level in the liver was significantly decreased one hour and even 16 hours after the last dose. The level of glycogen was decreased most (Reichl *et al.* 1988). The carbohydrate depletion caused by arsenic was attributed to the inhibition of gluconeogenesis (Szinicz and Forth 1988).

Groups of 25 male and 25 female Swiss mice were given intravenous doses of sodium arsenate of 0.5 mg/kg body weight (0.2 mg arsenic per kg body weight) once a week for 20 weeks. After a follow-up period of a further 76 weeks there were no differences in the survival and body weights of the animals compared with the control animals, but a significant increase in hyperkeratosis was found in the male and female animals and in nephropathy only in the females (Waalkes *et al.* 2000). The preneoplastic and neoplastic effects that occurred are described in Section 5.7.2.3.

There are no data available for dermal absorption.

5.3 Local effects on skin and mucous membranes

5.3.1 Skin

There are no data available for the effects of inorganic arsenic compounds on the skin. After it was applied to the skin of rabbits, the arsenic metabolite MMA was described as slightly irritative (WHO 2001). Erythematous lesions developed on the skin of rats after exposure for two hours to the metabolite DMA (3746 mg/m^3) (ATSDR 2000).

5.3.2 Eyes

After rats were exposed to DMA concentrations of 2172 mg/m^3 for two hours, scab formation was observed round the eyes (ATSDR 2000).

5.4 Allergenic effects

In a maximization test, the sensitizing effects of disodium hydrogen arsenate (Na$_2$HAsO$_4$) and sodium arsenite were tested in Dunkin-Hartley guinea pigs. For intradermal induction 4 % disodium hydrogen arsenate and 0.2 % sodium arsenite, and for topical induction 20 % disodium hydrogen arsenate (after previous non-occlusive treatment for 24 hours with 10 % sodium dodecyl sulfate in petrolatum) and 0.5 % sodium arsenite were used, each in water. After provocation on day 21, 2 of 19 treated animals and 1 of 20 control animals reacted only after 24 hours to 1 % disodium hydrogen arsenate in physiological saline. At the 48-hour reading and on testing with 0.5 % and 0.1 % of the test substance none of the animals produced a reaction. After 24 hours 4 of 20 treated animals and 4 of 20 controls reacted to 0.1 % sodium arsenite in physiological saline, and after 48 hours only 1 of 20 control animals. None of the animals reacted to 0.05 % and 0.01 % sodium arsenite (Wahlberg and Boman 1986).

5.5 Reproductive and developmental toxicity

5.5.1 Fertility

Male hamsters and rats were given intratracheal doses of gallium arsenide of 7.7 mg/kg body weight, indium arsenide of 7.7 mg/kg body weight or arsenic trioxide of 1.3 mg/kg body weight twice a week for up to 8 weeks. Gallium arsenide reduced the sperm count in both species, indium arsenide only in rats. Gallium arsenide also caused sperm anomalies. No effects were observed with arsenic trioxide (WHO 2001).

No effects were seen on the frequency of mating and fertility in female rats after daily oral doses of arsenic trioxide of 8 mg arsenic per kg body weight (14 days before mating to day 19 of gestation) (ATSDR 2000).

Oral doses of MMA (sodium salt) of 0, 11.9 and 119 mg/kg body weight were given to male and female Swiss mice (1 male mouse with 5 female mice per cage; the number of treated animals is not given) every second day for 10 weeks. The animals were separated after 19 days. Body weights, the number of litters and some blood parameters, such as the erythrocyte and leukocyte counts, haematocrit value and total serum protein were determined. In the high dose group the body weight gains were significantly reduced. None of the female animals produced a litter. In the low dose group the number of litters was reduced by 50 %. The female animals in the low dose group which did not produce a litter were all mated with the same male animal. As the authors suspected effects on the male germ cells after the first experiment, another experiment was carried out in which only the male animals were treated. During the treatment with 119 mg/kg body weight every second day three times a week for 19 days, 10 male Swiss mice were mated with untreated female mice (1:1). The fertility, measured as the number of litters, was reduced from 90 % in the control group to 50 % in the group of treated male animals. There were no other significant differences in the litter size or the weights of the 3-week-old offspring (Prukop and Savage 1986).

5.5.2 Developmental toxicity

Inorganic arsenic compounds were found to be embryotoxic and foetotoxic in mice and hamsters after inhalation, ingestion and parenteral absorption. Arsenite is of greater toxicity than arsenate by a factor of 3 to 10. Hamsters were more sensitive than mice. Observed were reduced foetal weights and lengths, decreased embryonal protein levels, a reduced number of somites, delayed growth and lethal effects. The lethal effects depended on the dose and the day of gestation on which the dose was given. The most frequent malformations caused by inorganic arsenic compounds were dysrhaphic disturbances with exencephaly, encephalocele and incomplete closure of the neural tube. Less frequent were e.g. fused ribs, renal or gonadal agenesis, micromelia, fascial malformations, contortion of the hind limbs, microphthalmia or anophthalmia. Mice reacted most sensitively between days 7 and 9 of gestation. Exposure on these days led to damage to the neural tube and radioactive labelling revealed the highest arsenic concentrations to be in the neuroepithelium (WHO 2001). Very detailed descriptions of these studies can be found in several reviews (ATSDR 2000, WHO 2001). Even the most recent studies of the developmental toxicity of arsenic compounds, which are discussed below, did not yield any important new insights.

Inhalation

The toxic effects on reproduction were investigated in female rats after inhalation exposure to arsenic trioxide for six hours with arsenic concentrations of 0.1 to 20 mg/m^3 (14 days before mating to day 19 of gestation). At arsenic concentrations of 8 mg/m^3 toxic effects (breathing noises, dry, red exudate around the nose, reduced body weight gains) were observed in the dams. At concentrations of 20 mg/m^3 impairments in foetal development (early resorption of the foetuses) were found in addition to marked maternal toxicity (Holson et al. 1999).

Ingestion

The most sensitive species was found to be the rabbit. An increase in the incidence of resorption and fewer viable foetuses were found after daily oral doses of arsenic(V) acid during pregnancy even with arsenic doses as low as 1.5 mg/kg body weight (ATSDR 2000).

In Crl:CD(SD)BR rats given daily oral arsenic doses of about 1 to 8 mg/kg body weight as arsenic trioxide from day 14 before mating to the end of pregnancy, reduced foetal weights and various skeletal changes were observed after arsenic doses of 8 mg/kg body weight (Holson et al. 2000).

After Crl:CD(SD)BR rats were given oral arsenic doses of 4, 8, 15 or 23 mg/kg body weight as arsenic trioxide on day 9 of gestation, an increase in resorptions and a reduction in the number of foetuses per litter were found at the 23 mg/kg body weight dose (Stump et al. 1999).

An increase in resorptions, reduced foetal weights and fewer live offspring were found in CD1 mice given oral arsenic doses of 24 mg/kg body weight as arsenic(V) acid from days 6 to 15 of gestation (Nemec et al. 1998).

In a 3-generation study with mice given sodium arsenite with the drinking water in daily doses of 1 mg arsenic per kg body weight, a reduction in the frequency of litters was observed in all generations (ATSDR 2000).

In another publication, the evaluation of studies of reproductive toxicity in animals showed that no damage to the offspring is to be expected after inhalation or oral exposure of the dams to arsenic concentrations as can occur at the workplace (DeSesso 2001).

5.6 Genotoxicity

5.6.1 *In vitro*

There are extensive data available for the *in vitro* genotoxicity of inorganic arsenic compounds (WHO 2001). A selection of more recent studies can be found in Table 5. Inorganic arsenic(III) compounds caused sister chromatid exchange (Gebel *et al.* 1997, Hsu *et al.* 1997, Rasmussen and Menzel 1997), DNA–protein crosslinks (Dong and Luo 1993, Ramirez *et al.* 2000), inhibition of poly(ADP)-ribosylation (Yager and Wiencke 1997) and induction of DNA repair synthesis (Wiencke *et al.* 1997) and the inhibition of DNA repair processes (Bau *et al.* 2001, Hartwig *et al.* 1997). In addition DNA strand breaks (Dong and Luo 1993, Lynn *et al.* 1997, 2000, Mouron *et al.* 2001, Schaumlöffel and Gebel 1998, Sordo *et al.* 2001, Wang *et al.* 2001) and chromosome breaks (Rupa *et al.* 1997) were observed. Inorganic arsenic(III) compounds induced mutations in different genes (Hei *et al.* 1998, Moore *et al.* 1997b, Wiencke *et al.* 1997), structural and numeric chromosomal aberrations (Lee *et al.* 1985, Yih *et al.* 1997) and micronuclei (Liu and Huang 1997, Schaumlöffel and Gebel 1998) and hyperploidy (Ramirez *et al.* 1997, Rupa *et al.* 1997). There are, however, also some studies which do not confirm these effects of arsenic(III) compounds (Costa *et al.* 1997, Gibson *et al.* 1997, Lee *et al.* 1985). Arsenic(V) compounds were usually not genotoxic in the test systems investigated (Gebel *et al.* 1997, Lee *et al.* 1985, Rasmussen and Menzel 1997) or genotoxic only at cytotoxic concentrations (Lee *et al.* 1985, Moore *et al.* 1997b).

Treatment with 0.25 µM sodium arsenite induced in HeLa and CHO cells a significant increase in DNA breaks when a formamidopyridine-DNA-glycosylase was included (cleaves oxidized bases, such as formamidopyridine or 8-oxyguanine, from DNA), or a protein kinase K (can cut DNA out of DNA–protein crosslinks), or both together. It was shown in another study that catalase and inhibitors of calcium, nitrogen oxide synthase, superoxide dismutase or myeloperoxidase can influence the DNA-damaging effects of 2 µM sodium arsenite. The authors concluded that DNA damage caused by arsenic(III) compounds is the result of the calcium-mediated formation of peroxynitrite, hypochlorous acid and hydroxyl radicals (Wang *et al.* 2001). In another study with HeLa cells (short-term incubation 0.5 to 3 hours) oxidative DNA damage occurred with the addition of formamidopyridine-DNA-glycosylase even after as little as 0.01 µM arsenite (no other details) or 10 µM MMA or DMA (Schwerdtle *et al.* 2002).

5.6.2 *In vivo*

Soma cells

Studies of the genotoxic effects of inorganic arsenic compounds in animals are shown in Table 6.

Increased induction of DNA strand breaks (Banu *et al.* 2001) was observed after single exposures and chromosomal aberrations (Das *et al.* 1993, Jha *et al.* 1992,

Nagymajtenyi et al. 1985, Poma et al. 1981, RoyChoudhury et al. 1996) and micronuclei (Deknudt et al. 1986, Tice et al. 1997, Tinwell et al. 1991) after single and repeated doses of arsenic(III) compounds. In the bone marrow of offspring of the mice, which were given sodium arsenite doses of 0.1 mg/kg body weight once a week for 4 weeks, the number of chromosomal aberrations was still significantly increased (RoyChoudhury et al. 1996). A linear dose–response relationship was found for the micronuclei in polychromatic erythrocytes induced by sodium arsenite (Deknudt et al. 1986, Tinwell et al. 1991).

In mice the arsenic metabolite DMA caused aneuploid bone marrow cells only at very high doses (Kashiwada et al. 1998) and DNA strand breaks in lung cells, but not in those of the liver or kidneys (Yamanaka et al. 1989, Yamanaka and Okada 1994).

Germ cells

In the spermatogonia of Swiss albino mice, no increase in chromatid or chromosomal aberrations was found 12, 24, 36 and 48 hours after single intraperitoneal doses of arsenic trioxide (0, 4, 8 or 12 mg arsenic per kg body weight) (Poma et al. 1981). Nor was an increase in sister chromatid exchange or chromosomal aberrations in the spermatogonia of the Swiss albino mice found after the administration of 250 mg arsenic trioxide per litre drinking water (about 47 mg arsenic per kg body weight) for 2 to 8 weeks (Poma et al. 1987).

In BALB/c mice, tests for dominant lethal mutations after single intraperitoneal doses of sodium arsenite of 5 mg/kg body weight (2.9 mg arsenic per kg body weight) yielded negative results (Deknudt et al. 1986). No increase in sperm head anomalies was observed in BALB/c mice after sodium arsenite doses of 2.5, 5 or 7.5 mg/kg body weight (1.4, 2.9 or 4.3 mg arsenic per kg body weight) (Deknudt et al. 1986).

The dominant lethal test described in Section 5.5.1, in which mice were given oral MMA doses up to 119 mg/kg body weight (Prukop and Savage 1986), is invalid as a result of inadequate documentation and methodological shortcomings, such as the short duration of treatment and the small number of animals, and cannot, therefore, be used in the evaluation of the germ cell mutagenicity of arsenic and arsenic compounds.

Table 5. The genotoxicity of inorganic arsenic compounds *in vitro*

End point	Test system	Oxidation step	Results	Effective concentration [µg/ml]	Cytotoxicity [µg/ml]	References
SOS chromotest	*Escherichia coli* PQ37	As(III)	–	–	403 µM	Lantzsch and Gebel 1997
SCE	human lymphocytes	As(III)	+	10 µM	> 10 µM	Rasmussen and Menzel 1997
	human lymphocytes	As(III)	+	0.5 µM	5 µM	Gebel et al. 1997
	human lymphocytes	As(III)	+	65	not stated	Hsu et al. 1997
	human lymphocytes	As(III)	+	1 µM	not stated	Jha et al. 1992
	lymphoid cells	As(V)	–	–	not stated	Rasmussen and Menzel 1997
	human lymphocytes	As(V)	–	–	10 µM	Gebel et al. 1997
DNA–protein crosslinks	human lymphoma cells	As(III)	–	–	5 µM	Costa et al. 1997
	human foetal lung fibroblasts	As(III)	+	1 µM	not stated	Dong and Luo 1993
	human liver cell line	As(III)	+	0.001 µM	not stated	Ramirez et al. 2000
inhibition of poly(ADP)-ribosylation	Molt-3 cell line	As(III)	+	> 5 µM	> 7.5 µM	Yager and Wiencke 1997
inhibition of DNA repair synthesis	human fibroblasts	As(III)	+	2.5 µM	> 10 µM	Hartwig et al. 1997
	CHO cells	As(III)	+	4 µM	not stated	Bau et al. 2001
DNA repair synthesis	human fibroblasts	As(III)	+	1 µM	2.5 µM	Wiencke et al. 1997
DNA strand breaks	CHO cells	As(III)	+	10 µM	20 µM	Lynn et al. 1997
	human smooth muscle cells	As(III)	+	> 1 µM	not stated	Lynn et al. 2000
	human fibroblasts	As(III)	+	2.5 µM	not stated	Mouron et al. 2001

Table 5. continued

End point	Test system	Oxidation step	Results	Effective concentration [μg/ml]	Cytotoxicity [μg/ml]	References
	human lymphocytes	As(III)	+	>5 μM	not stated	Sordo et al. 2001
	human foetal lung fibroblasts	As(III)	+	1 μM	not investigated	Dong and Luo 1993
	human lymphocytes	As(III)	+	0.01 μM	5 μM	Schaumlöffel and Gebel 1998
	NB4 cells, HeLa cells	As(III)	+	> 0.25 μM	2 μM	Wang et al. 2001
	CHO-K1	As(III)	−	−	> 2 μM	Wang et al. 2001
chromosome breaks	human lymphocytes	As(III)	+	9 μM	not stated	Rupa et al. 1997
mutations: TK$^{+/-}$	L5178Y mouse lymphoma cells	As(III)	(+)	1–2	about 0.8	Moore et al. 1997b
		As(V)	(+)	10–14	about 7.5	Moore et al. 1997b
Na$^+$/K$^+$-ATPase	SHE cells	As(III)	−	−	> 3 μM	Lee et al. 1985
		As(V)	−	−	> 10 μM	Lee et al. 1985
HPRT	CHO cells with a copy of human chromosome 11	As(III)	(+)	1 μM	> 0.5 μM	Hei et al. 1998
	SHE cells	As(III)	−	−	> 3 μM	Lee et al. 1985
		As(V)	−	−	> 10 μM	Lee et al. 1985
supF reporter gene	human fibroblasts with shuttle vector pz189	As(III)	(+)	5 μM	2.5 μM	Wiencke et al. 1997
CA	human lymphocytes	As(III)	+	1 μM	not stated	Jha et al. 1992
	human skin fibroblasts	As(III)	+	5 μM	> 1.25 μM	Yih et al. 1997
	SHE cells	As(III)	(+)	6.2 μM	> 3 μM	Lee et al. 1985
		As(V)	(+)	64 μM	> 10 μM	Lee et al. 1985

Table 5. continued

End point	Test system	Oxidation step	Results	Effective concentration [µg/ml]	Cytotoxicity [µg/ml]	References
micronuclei	CHO cells	As(III)	+	80 µM	not stated	Liu and Huang 1997
	SHE cells	As(III)	+	–	7.5	Gibson et al. 1997
	human lymphocytes	As(III)	+	0.5 µM	5 µM	Schaumlöffel and Gebel 1998
hyperploidy	human lymphocytes	As(III)	+	0.01 µM	not stated	Ramirez et al. 1997
	human lymphocytes	As(III)	+	9 µM	not stated	Rupa et al. 1997

(+) positive at cytotoxic concentrations

5.7 Carcinogenicity

5.7.1 Short-term studies

Sodium arsenate and arsenite induced a dose-dependent transformation in Syrian hamster embryo cells (Lee *et al.* 1985) and BALB/3T3 cells (Bertolero *et al.* 1987). Arsenite was found to be four to ten times stronger than arsenate in the induction of transformations, which is probably the result of the greater cellular uptake of arsenite compared to arsenate (Bertolero *et al.* 1987). BALB/3T3 cells transformed by arsenite were tumour-promoting in mice after subcutaneous injection of the substance (Saffiotti and Bertolero 1989). The rapidly growing fibrosarcomas, however, did not metastasize. The transforming arsenic compounds are probably the arsenites (Bertolero *et al.* 1987, Saffiotti and Bertolero 1989).

Transgenic, v-Ha-ras oncogene-carrying mice (TG.AC) received arsenic doses of 48 mg/kg body weight with the drinking water for four weeks in the form of sodium arsenite. Afterwards, the tumour promotor 12-*O*-tetradecanoylphorbol-13-acetate (TPA) was applied to the shaved dorsal skin twice a day for 2 weeks. Skin papillomas were not found in the control animals nor in the animals treated only with arsenic. Only with the simultaneous application of sodium arsenite and TPA did the incidence of skin papillomas increase markedly. The skin of the animals treated with arsenic developed hyperkeratosis. In the epidermis an increase was found in the mRNA copies for the growth factor TGF-α and the granulocyte macrophage colony stimulating factor GM-CSF (Germolec *et al.* 1997, 1998).

In some studies with mice, after the administration of inorganic arsenic compounds a decrease was described in urethane-induced lung tumours (Blakley 1987), spontaneous breast tumours (Schrauzer and Ishmael 1974, Schrauzer *et al.* 1976) and tumours caused by the injection of mouse sarcoma cells (Kerkvliet *et al.* 1980). The arsenic compounds, however, decreased the lifetime of the tumour-carrying animals (Kerkvliet *et al.* 1980, Schrauzer and Ishmael 1974). If the animals formed tumours or received tumour transplants during the arsenic treatment, the growth of the tumours was significantly increased (Schrauzer and Ishmael 1974, Schrauzer *et al.* 1976). A tumour-promoting effect of sodium arsenite was demonstrated in Wistar rats after partial hepatectomy and pretreatment with diethylnitrosamine (30 mg/kg body weight) (Shirachi *et al.* 1983).

In a multiorgan carcinogenicity bioassay, the carcinogenic effects of the arsenic metabolite DMA was investigated in rats. Male F344/DuCrj rats were given single doses of diethylnitrosamine (DEN) (intraperitoneal, 100 mg/kg body weight) at the beginning of treatment, then four doses of *N*-methyl-*N*-nitrosourea (MNU) (intraperitoneal, 20 mg/kg body weight on days 5, 8, 11 and 14) and later four doses of 1,2-dimethyl-hydrazine (DMH) (subcutaneous, 40 mg/kg body weight on days 18, 22, 26 and 30). The animals were given 0.05 % *N*-butyl-*N*-(4-hyroxybutyl)-nitrosamine (BBN) with the drinking water for the first two weeks and 0.1 % *N*-bis(2-hyroxypropyl)-nitrosamine (DHPN) for the following two weeks. After a two-week pause, the rats were given 50, 100, 200 or 400 mg DMA per litre drinking water. At the end of week 30 the animals were killed. DMA significantly increased the formation of tumours in the bladder, kidneys, liver and thyroid gland in the pretreated animals. Also preneoplastic damage was markedly increased. In the animals exposed only to DMA no preneoplastic changes or tumours were found (Yamamoto *et al.* 1995).

In another study NBR rats were given 0.05 % BBN with the drinking water for 4 weeks and later 100 mg DMA per litre drinking water for 32 weeks. After week 36 a significant increase in simple, papillary or nodular hyperplasia and an increase in carcinomas were observed in the bladder (Li *et al.* 1998).

Male F344/DuCrj rats were given 0.05 % BBN with the drinking water for 1 to 4 weeks. Afterwards they received up to 100 mg DMA per litre drinking water for another 32 weeks. The administration of DMA after pretreatment with BBN caused a dose-dependent increase in the incidence of bladder papillomas and carcinomas, without pretreatment DMA did not increase the formation of tumours (Wanibuchi *et al.* 1996).

In a two-step carcinogenicity test male F344/DuCrj rats were given single intraperitoneal doses of DEN of 100 mg/kg body weight and 2 weeks later 0, 25, 50 or 100 mg DMA per litre drinking water for 6 weeks. In addition, 3 weeks after the DEN dose a partial hepatectomy was carried out. 8 weeks after the beginning of the study an increase in liver cancer was observed after DMA concentrations of 50 mg/l drinking water and above (Yamamoto *et al.* 1997).

Table 6. Genotoxicity of arsenic compounds in the mouse

End point	Species	Organ or cell type investigated	Arsenic compound	Administration, dose, duration	Results	References
single exposures						
DNA strand breaks	Swiss albino mouse	leukocytes	arsenic trioxide	oral, 0.1–4.9 mg As/kg body weight	+	Banu et al. 2001
	CD1 mouse	aneuploid bone marrow cells	DMA	intraperitoneal, 300 mg/kg body weight	+	Kashiwada et al. 1998
	CD1 mouse	lung, liver, kidneys	DMA	oral, 1500 mg/kg body weight	+	Yamanaka et al. 1989, Yamanaka and Okada 1994
CA	Swiss albino mouse	bone marrow cells	arsenic trioxide	intraperitoneal, 4–12 mg/kg body weight	–	Poma et al. 1981
	Swiss albino mouse	bone marrow cells	sodium arsenite	oral, 25 mg/kg body weight	+	Das et al. 1993
	Swiss albino mouse	bone marrow cells	sodium arsenite	subcutaneous, 0.1 mg/kg body weight, 1 ×/week, 4 weeks	+	RoyChoudhury et al. 1996
MN	C57BL mouse	polychromatic erythrocytes	sodium arsenite	intraperitoneal, 2.5–10 mg/kg body weight	+	Deknudt et al. 1986, Tinwell et al. 1991
	CBA mouse	polychromatic erythrocytes	sodium arsenite	intraperitoneal, 2.5–10 mg/kg body weight	+	Deknudt et al. 1986, Tinwell et al. 1991
	Balb/c mouse	polychromatic erythrocytes	sodium arsenite	intraperitoneal, 2.5–10 mg/kg body weight	+	Deknudt et al. 1986, Tinwell et al. 1991

Table 6. continued

End point	Species	Organ or cell type investigated	Arsenic compound	Administration, dose, duration	Results	References
repeated exposure						
CA	Swiss albino mouse	bone marrow cells	arsenic trioxide	oral, 250 mg/l drinking water, 2–8 weeks	+	Poma et al. 1987
	CFLP mouse	foetal liver cells	arsenic trioxide	inhalation, 0.26–28.5 mg/m^3, days 9 to 12 *post coitum*, caesarean section day 18 *post coitum*	(+)	Nagymajtenyi et al. 1985
MN	B6C3F$_1$ mouse	polychromatic erythrocytes	sodium arsenite	oral, 2.5–10 mg/kg body weight, 4 days	+	Tice et al. 1997

(+) positive only at foetotoxic concentrations, CA = chromosomal aberrations, MN = micronuclei

5.7.2 Long-term studies

Table 7 gives an overview of the carcinogenicity studies with arsenic compounds.

5.7.2.1 Inhalation

In hamsters, intratracheal administration of inorganic arsenic compounds for 15 weeks caused an increase in the frequency of lung tumours during a lifelong observation period. These were lung adenomas (Ishinishi *et al.* 1983, Pershagen and Björklund 1985, Yamamoto *et al.* 1987) and papillomas and carcinomas of the respiratory tract (Pershagen *et al.* 1984). Also many early stages of the lung tumours, such as hyperplasia (Yamamoto *et al.* 1987) and adenomatous changes (Pershagen and Björklund 1985, Pershagen *et al.* 1984, Yamamoto *et al.* 1987) were found. Benzo[a]pyrene amplified the carcinogenic effects on the lung induced by arsenic trioxide (Pershagen *et al.* 1984).

Grave disadvantages of the studies described above are the short duration of treatment and the fact that the investigations concentrated almost exclusively on the lungs of the animals. Valid long-term studies with inhalation exposure to inorganic arsenic compounds are not available.

5.7.2.2 Ingestion

A study in which mice were given the arsenic metabolite DMA for 50 weeks in concentrations of 0.05, 0.2 or 0.4 mg/l drinking water (about 0.01, 0.05 or 0.1 mg arsenic per kg body weight) showed that the tumour size and the total number of adenomas and carcinomas in the lungs increased in a dose-dependent manner. Administration for only 25 weeks did not yet induce lung tumours (Hayashi *et al.* 1998).

A dose-dependent increase in the incidence of bladder carcinomas was observed in F344 rats after administration of DMA with the drinking water (about 2.5 or 10 mg DMA per kg body weight) for 104 weeks. Other organs were not investigated (Wei *et al.* 1999).

Female C57BL/6J mice and female transgenic metallothionine knock-out mice were given sodium arsenate for 26 months with the drinking water (0.5 mg arsenic per litre). The daily arsenic dose was 2 to 2.5 µg arsenic per day, corresponding to 0.07 to 0.08 mg arsenic per kg body weight. In 41 % of the C57BL/6J mice and 26 % of the knock-out mice one or more tumours were found per animal. The tumours were situated in the gastrointestinal tract, lungs, liver, spleen, skin and gonads (Ng *et al.* 1999).

No malignant tumours were found in 11 of 20 cynomolgus monkeys which survived treatment with sodium arsenate in daily oral doses of 0.1 mg arsenic per kg body weight on 5 days a week for 17 years. 2 animals were found to have adenomas of the renal cortex (see Section 5.2.2) (Thorgeirsson *et al.* 1994).

Valid long-term studies with ingestion of inorganic arsenic compounds are not available.

Table 7. Carcinogenicity studies with arsenic compounds

Species	Type of administration, duration	Exposure*	Tumour incidence	Type of tumour	References
arsenic trioxide					
Syrian hamster ♀	intratracheal, 1 ×/week, 15 weeks, lifelong observation	controls	1/20 1/20	adenocarcinoma (kidneys) papilloma (forestomach)	Ishinishi et al. 1983
		0.25 mg (3 mg/kg body weight)	2/20[1] 1/20 1/20	adenomas (lungs) fibrous histiocytoma fibrosarcoma in subcutis with metastases in the ovary and lungs	
Syrian hamster ♀	intratracheal, 1 ×/week, 15 weeks, lifelong observation	controls	0/20		Ishinishi et al. 1983
		0.35 mg (4 mg/kg body weight)	3/10[2] 1/10	adenomas (lungs) leiomyoma in the uterus	
Syrian hamster ♂	intratracheal, 1 ×/week, 15 weeks, lifelong observation	controls	3/53	adenomas, adenomatous changes, and papillomas (respiratory tract)	Pershagen et al. 1984
		3 mg/kg body weight	21/47 3/47[3]	adenomas, adenomatous changes, papillomas (respiratory tract) carcinomas (respiratory tract)	
Syrian hamster ♀	intratracheal, 1 ×/week, 15 weeks, lifelong observation	controls	1/21 1/21 1/21	adenocarcinoma (lungs) adenoma (adrenal gland) adenocarinoma (adrenal gland)	Yamamoto et al. 1987
		0.25 mg (3 mg/kg body weight)	8/17 1/17 1/17 1/17	hyperplasia (lungs) adenocarcinoma (lungs) adenoma (adrenal gland) haemangiosarcoma (liver)	
arsenic trisulfide					
Syrian hamster ♂	intratracheal, 1 ×/week, 15 weeks, lifelong observation	controls	2/26	adenomatous changes (lungs)[#]	Pershagen and Björklund 1985
		3 mg/kg body weight	12/28 1/28	adenomatous changes (lungs) adenoma (lungs)	

Table 7. continued

Species	Type of administration, duration	Exposure*	Tumour incidence	Type of tumour	References
Syrian hamster ♀	intratracheal, 1 ×/week, 15 weeks, lifelong observation	controls	1/21 1/21 1/21	adenocarcinoma (lungs) adenoma (adrenal gland) adenocarcinoma (adrenal gland)	Yamamoto et al. 1987
		0.25 mg (3 mg/kg body weight)	3/22 1/22 1/22 1/22	hyperplasia adenoma (lungs) adenoma (adrenal gland) nephroblastoma (kidneys)	

sodium arsenate

Species	Type of administration, duration	Exposure*	Tumour incidence	Type of tumour	References
Swiss mice ♂	intravenous, 1 ×/week, 20 weeks, observation for a further 76 weeks	controls	8/25	interstitial cell hyperplasia (testes)	Waalkes et al. 2000
		0.2 mg/kg body weight	16/25	interstitial cell hyperplasia (testes)[7]	
Swiss mice ♀	intravenous, 1 ×/week, 20 weeks, observation for a further 76 weeks	controls	5/25 0/25 1/25 0/25	cystic hyperplasia (uterus) adenocarcinoma (uterus) preneoplastic foci (liver) adenoma (liver)	Waalkes et al. 2000
		0.2 mg/kg body weight	14/25 1/25 5/25 1/25	cystic hyperplasia (uterus)[7] adenocarcinoma (uterus) preneoplastic foci (liver)[7] adenoma (liver)	
C57BL/6J mice ♀	oral, 1 ×/day, up to 26 months	controls	0/90		Ng et al. 1999
		0.07–0.08 mg/kg	14.4 % 17.5 % 7.8 % 3.3 % 2.2 % 3.3 % 3.3 % 1.1 %	tumours in gastrointestinal tract lungs liver spleen bones skin reproductive system eyes (no other details)	
transgenic metallothionine knock-out mice ♀	oral, 1 ×/day, up to 26 months	controls	0/140		Ng et al. 1999
		0.07–0.08 mg/kg	12.9 % 7.1 % 5.0 % 0.7 % 1.4 % 5.0 %	tumours in gastrointestinal tract lungs liver spleen skin reproductive system (no other details)	

Table 7. continued

Species	Type of administration, duration	Exposure*	Tumour incidence	Type of tumour	References
sodium arsenate					
cynomolgus monkeys (*Macaca fascicularis*)	oral, 5 days/week, 17 years	controls	1/130 1/130	adenocarcinoma (kidneys) adenocarcinoma (intestine)	Thorgeirsson *et al.* 1994
		0.1 mg/kg body weight	2/11	adenoma (kidneys)	
calcium arsenate					
Syrian hamster ♂	intratracheal, 1 ×/week, 15 weeks, lifelong observation	controls	2/26	adenomatous changes in the lungs[#]	Pershagen and Björklund 1985
		3 mg/kg body weight	3/35 4/35 10/35	metaplasia (larynx, trachea) adenomatous changes in the lungs pulmonary adenomas[4]	
Syrian hamster ♀	intratracheal, 1 ×/week, 15 weeks, lifelong observation	controls	1/21 1/21 1/21	adenocarcinoma (lungs) adenoma (adrenal gland) adenocarcinoma (adrenal gland)	Yamamoto *et al.* 1987
		0.25 mg	7/25 6/25 1/25 2/25 1/25	hyperplasia (lungs) adenoma (lungs) adenocarcinoma (lungs) adenocarcinoma (adrenal gland) leukaemia	
DMA					
A/J mice ♂	oral, 50 weeks	controls	6/14 1/14	adenomas (lungs)[#] adenocarcinomas (lungs) lung tumours per animal: 0.50	Hayashi *et al.* 1998
		0.4 mg/l drinking water° (about 0.1 mg/kg body weight)	4/14 10/14 3/14 2/14	hyperplasia (lungs) adenomas (lungs) adenocarcinomas (lungs) tumours, not characterized (lungs) lung tumours per animal: 1.39[5]	

Table 7. continued

Species	Type of administration, duration	Exposure*	Tumour incidence	Type of tumour	References
DMA					
F344 rats ♂	oral 104 weeks	controls	0/28		Wei et al. 1999
		12.5 mg/l drinking water (about 0.6 mg DMA/kg body weight)	0/33		
		50 mg/l drinking water (about 2.5 mg DMA/kg body weight)	2/31 6/31[4]	papillomas (bladder)[#] carcinomas (bladder)	
		200 mg/l drinking water (about 10 mg DMA/kg body weight)	2/31 12/31[6]	papillomas (bladder) carcinomas (bladder)	

* exposure to arsenic if not otherwise stated, [#] other organs were not investigated, ° only data for the highest concentration group are listed in the table, [1] $p = 0.244$, [2] $p = 0.052$, [3] $p = 0.07$, [4] $p < 0.05$, [5] $p = 0.002$, [6] $p < 0.001$, [7] significant in the Fisher exact test

5.7.2.3 Other routes of absorption

Groups of 25 male and 25 female Swiss mice were given intravenous doses of sodium arsenate of 0.5 mg/kg body weight (0.2 mg arsenic per kg body weight) once a week for 20 weeks. After an observation period of a further 76 weeks, a significant increase in interstitial cell hyperplasia was observed in the testes of the male mice. In the female animals, the incidence and severity of cystic hyperplasia in the uterus were significantly increased. In one animal a rare adenocarcinoma was found in the uterus. Also the incidence of preneoplastic foci in the liver was significantly increased (see Table 7 and Section 5.2.4) (Waalkes et al. 2000).

6 Manifesto (MAK value/classification)

Arsenic and inorganic arsenic compounds, with the exception of arsine, are carcinogenic in man. After inhalation of the substance, carcinogenic effects are observed in the lungs and after ingestion, in the bladder, kidneys, skin and lungs. However, neither the studies with inhalation exposure to arsenic nor those with oral administration of the substance can be used to derive a NOAEL (no observed adverse effect level). As a NOAEL for carcinogenicity cannot be derived from the epidemiological studies, no MAK value can be established for arsenic and inorganic arsenic compounds. Arsenic and arsenic compounds therefore remain in Carcinogenicity category 1.

Inorganic arsenic compounds are clearly genotoxic in soma cells *in vitro* and in rodents and man.

No dominant lethal mutations were induced in mice after intraperitoneal administration of the substance in the only available study with germ cells. The negative results in this dominant lethal test cannot, however, dispel the suspicion that arsenic and inorganic arsenic compounds cause germ cell mutagenicity, as the test was carried out with only one single intraperitoneal treatment. As a result of the mutagenic effects in soma cells, the formation of genotoxic and systemically effective metabolites, the bioavailability of the substance in the gonads and the inadequate investigation in germ cells, arsenic and inorganic arsenic compounds are classified in Category 3A for germ cell mutagens.

Possible contact allergy after exposure to inorganic arsenic compounds cannot be deduced with certainty from the available data as no clear separation from unspecific or toxic effects is possible, despite evidence indicating such effects above all in the earlier literature. In addition, a study with animals yielded negative results. There are no data available for immunological effects on the respiratory passages. Despite the suspected contact allergy the substance is therefore not designated with an "S".

Arsenic acid penetrates the skin *in vivo* and *in vitro* in amounts of less than 10 % and sodium arsenate from aqueous solution or as the solid *in vitro* in amounts of about 30 % to 60 %. Several studies have shown that the internal exposure to arsenic correlates well with the external exposure. Dermal absorption under workplace conditions does not, therefore, seem to be relevant, and arsenic and inorganic arsenic compounds are not designated with an "H".

7 References

ACGIH (American Conference of Governmental Industrial Hygienists) (1999) Arsenic and inorganic compounds. in: *Documentation of TLVs and BEIs*, ACGIH, Cincinnati, OH, USA

Ahmad AS, Sayed MHSU, Barua S, Khan MH, Faruquee MH, Jalil A, Hadi S, Talukter K (2001) Arsenic in drinking water and pregnancy outcomes. *Environ Health Perspect 109*: 629–630

Aranyi C, Bradof JN, O'Shea WJ, Graham JA, Miller FJ (1985) Effects of arsenic trioxide inhalation exposure on pulmonary antibacterial defenses in mice. *J Toxicol Environ Health 15*: 163–172

Aschengrau A, Zierler S, Cohen A (1989) Quality of community drinking water and the occurrence of spontaneous abortion. *Arch Environ Health 44*: 283–290

ATSDR (Agency for Toxic Substances and Disease Registry) (2000) *Toxicological profile for arsenic*, US Department of Health and Human Services, Atlanta, Georgia

Banu BS, Danadevi K, Jamil K, Ahuja YR, Rao KV, Ishaq M (2001) In vivo genotoxic effect of arsenic trioxide in mice using comet assay. *Toxicol 162*: 171–177

Barbaud A, Mougeolle JM, Schmutz JL (1995) Contact hypersensitivity to arsenic in a crystal factory worker. *Contact Dermatitis 33*: 272–273

Bates MN, Smith AH, Hopenhayn-Rich C (1992) Arsenic ingestion and internal cancers: a review. *Am J Epidemiol 135*: 462–476

Bau DT, Gurr JR, Jan KY (2001) Nitric oxide is involved in arsenite inhibition of pyrimidine dimer excision. *Carcinogenesis 22*: 709–716

Bazarbachi A, El-Sabban ME, Nasr R, Quignon F, Awaraji C, Kersual J, Dianoux L, Zermati Y, Haidar JH, Hermine O, de Thé H (1999) Arsenic trioxide and interferon-α synergize to induce cell cycle arrest and apoptosis in human T-cell lymphotropic virus type I-transformed cells. *Blood 93*: 278–283

Beckett WS, Moore JL, Keogh JP, Bleecker ML (1986) Acute encephalopathy due to occupational exposure to arsenic. *Br J Ind Med 43*: 66–67

Beckman L, Nordstrom S (1982) Occupational and environmental risks in and around a smelter in northern Sweden. IX. Fetal mortality among wives of smelter workers. *Hereditas 97*: 1–7

Beckman G, Beckman L, Nordenson I (1977) Chromosome aberrations in workers exposed to arsenic. *Environ Health Perspect 19*: 145–146

Bekemeier H, Hirschelmann R (1989) Reactivity of resistance blood vessels *ex vivo* after administration of toxic chemicals to laboratory animals: arteriolotoxicity. *Toxicol Lett 49*: 49–54

Bencko V, Rossner P, Havrankova H, Puznova A, Tucek M (1978) Effects of the combined action of selenium and arsenic on mice versus suspension culture of mice fibroblasts. in: Fours JR, Gut I (Eds) *Industrial and environmental xenobiotics: in vitro versus in vivo biotransformation and toxicity*, Proceedings of an international conference held in Prague, Czechoslovakia, 13th–15th September 1977, 312–316

Bencko V, Wagner V, Wagnerova M, Batora J (1988) Immunological profiles in workers of a power plant burning coal rich in arsenic content. *J Hyg Epidemiol Microbiol Immunol 32*: 137–146

Bernstam L, Nriagu J (2000) Molecular aspects of arsenic stress. *J Toxicol Environ Health, B3*: 293–322

Bertolero F, Pozzi G, Sabbioni E, Saffiotti U (1987) Cellular uptake and metabolic reduction of pentavalent to trivalent arsenic as determinants of cytotoxicity and morphological transformation. *Carcinogenesis 8*: 803–808

Birmingham DJ, Key MM, Holaday DA, Perone VB (1965) An outbreak of arsenical dermatoses in a mining community. *Arch Dermatol 91*: 457–464

Blakley BR (1987) Alterations in urethan-induced adenoma formation in mice exposed to selenium and arsenic. *Drug Nutr Interact 5*: 97–102

Bolla-Wilson K, Bleecker ML (1987) Neuropsychological impairment following inorganic arsenic exposure. *J Occup Med 29*: 500–503

Borzsonyi M, Bereczky A, Rudnai P, Csanady M, Horvath A (1992) Epidemiological studies on human subjects exposed to arsenic in drinking water in southeast Hungary. *Arch Toxicol 66*: 77–78

Bourrain JL, Morin C, Beani JC, Amblard P (1998) Airborne contact dermatitis from cacodylic acid. *Contact Dermatitis 38*: 364–365

Brouwer OF, Onkenhout W, Edelbroek PM, de Kom JF, de Wolff FA, Peters AC (1992) Increased neurotoxicity of arsenic in methylenetetrahydrofolate reductase deficiency. *Clin Neurol Neurosurg 94*: 307–310

Burgdorf W, Kuvink K, Cerevenka J (1977) Elevated sister chromatid exchange rate in lymphocytes of subjects treated with arsenic. *Hum Genet 36*: 69–72

Byron WR, Bierbower GW, Brouwer JB, Hansen WH (1967) Pathologic changes in rats and dogs from two-year feeding of sodium arsenite or sodium arsenate. *Toxicol Appl Pharmacol 10*: 132–147

Carmignani M, Boscolo P, Iannaccone A (1983) Effects of chronic exposure to arsenate on the cardiovascular function of rats. *Br J Ind Med 40*: 280–284

Cavigelli M, Li WW, Lin A, Su B, Yoshioka K, Karin M (1996) The tumor promoter arsenite stimulates AP-1 activity by inhibiting a JNK phosphatase. *EMBO J 15*: 6269–6279

Chen CJ, Wu MM, Lee SS, Wang JD, Cheng SH, Wu HY (1988) Atherogenicity and carcinogenicity of high-arsenic artesian well water. Multiple risk factors and related malignant neoplasms of blackfoot disease. *Arteriosclerosis 8*: 452–460

Chen G-Q, Zhu J, Shi X-G, Ni J-H, Zhong H-J, Si G-Y, Jin X-L, Tang W, Li X-S, Xong S-M, Shen Z-X, Sun G-L, Ma J, Zhang T-D, Gazin C, Naoe T, Chen S-J, Wang A-Y, Chen Z (1996) In vitro studies on cellular and molecular mechanisms of arsenic trioxide (As_2O_3) in the treatment of acute promyelocytic leukemia: As_2O_3 induces NB_4 cell apoptosis with downregulation of Bcl-2 expression and modulation of PML-RARα/PML proteins. *Blood 88*: 1052–1061

Costa M, Zhitkovich A, Harris M, Paustenbach D, Gargas M (1997) DNA-protein cross-links produced by various chemicals in cultured human lymphoma cells. *J Toxicol Environ Health 50*: 433–449

Dai J, Weinberg RS, Waxman S, Jing Y (1999) Malignant cells can be sensitized to undergo growth inhibition and apoptosis by arsenic trioxide through modulation of the glutathione redox system. *Blood 93*: 268–277

Das T, Roychoudhury A, Sharma A, Talukder G (1993) Modification of clastogenicity of three known clastogens by garlic extract in mice *in vivo*. *Environ Mol Mutagen 21*: 383–388

Deknudt G, Leonard A, Arany J, Jenar-Du Buisson G, Delavignette E (1986) *In vivo* studies in male mice on the mutagenic effects of inorganic arsenic. *Mutagenesis 1*: 33–34

DeSesso JM (2001) Teratogen update: inorganic arsenic. *Teratology 63*: 170–173

Dieke SH, Richter CP (1946) Comparative assays of rodenticides on wild Norway rats. *Public Health Rep 61*: 672–679

Done AK, Peart AJ (1971) Acute toxicities of arsenical herbicides. *Clin Toxicol 4*: 343–355

Dong JT, Luo XM (1993) Arsenic-induced DNA-strand breaks associated with DNA-protein crosslinks in human fetal lung fibroblasts. *Mutat Res 302*: 97–102

Dulout FN, Grillo CA, Seoane AI, Maderna CR, Nilsson R, Vahter M, Darroudi F, Natarajan AT (1996) Chromosomal aberrations in peripheral blood lymphocytes from native Andean women and children from northwestern Argentina exposed to arsenic in drinking water. *Mutat Res 370*: 151–158

Eberhartinger C, Ebner H, Klotz L (1969) Zur Kenntnis und Interpretation follikulärer, papulopustulöser Reaktionen im Epikutantest (Knowledge and interpretation of follicular papulopustular reactions in epicutaneous tests) (German). *Berufs-Dermatosen 17*: 241–252

EC (European Commission DG Environment) (2000) *Ambient air pollution by As, Cd, and Ni compounds*. Working Group on Arsenic, Cadmium and Nickel Compounds, Brussels, Belgium

Enterline PE, Henderson VL, Marsh GM (1987a) Exposure to arsenic and respiratory cancer: a reanalysis. *Am J Epidemiol 125*: 929–938

Enterline PE, Marsh GM, Esmen NA, Henderson VL, Callahan CM, Paik M (1987b) Some effects of cigarette smoking, arsenic, and SO_2 on mortality among US copper smelter workers. *J Occup Med 29*: 831–838

Enterline PE, Day R, Marsh GM (1995) Cancers related to exposure to arsenic at a copper smelter. *Occup Environ Med 52*: 28–32

EPA (Environmental Protection Agency) (1997) *Report on the expert panel on arsenic carcinogenicity: review and workshop*. National Center for Environmental Assessment, Washington DC, 1–26

EPA (Environmental Protection Agency) (2001) *Arsenic in drinking water*. National Academic Press, Washington DC

Feldman RG, Niles CA, Kelly-Hayes M, Sax DS, Dixon WJ, Thompson DJ, Landau E (1979) Peripheral neuropathy in arsenic smelter workers. *Neurology 29*: 939–944

Fowler BA, Woods JS, Schiller CM (1977) Ultrastructural and biochemical effects of prolonged oral arsenic exposure on liver mitochondria of rats. *Environ Health Perspect 19*: 197–204

Gaines TB (1960) The acute toxicity of pesticides to rats. *Toxicol Appl Pharm 2*: 88–99

Gebel T, Christensen S, Dunkelberg H (1997) Comparative and environmental genotoxicity of antimony and arsenic. *Anticancer Res 17*: 2603–2607

Germolec DR, Spalding J, Boorman GA, Wilmer JL, Yoshida T, Simeonova PP, Bruccoleri A, Kayama F, Gaido K, Tennant R, Burleson F, Dong W, Lang RW, Luster MI (1997) Arsenic can mediate skin neoplasia by chronic stimulation of keratinocyte-derived growth factors. *Mutat Res 386*: 209–218

Germolec DR, Spalding J, Yu H-S, Chen GS, Simeonova PP, Humble MC, Bruccoleri A, Boorman GA, Foley JF, Yoshida T, Luster MI (1998) Arsenic enhancement of skin neoplasia by chronic stimulation of growth factors. *Am J Pathol 153*: 1775–1785

Gibson DP, Brauninger R, Shaffi HS, Kerchaert GA, LeBoeuf RA, Isfort RJ, Aarderma MJ (1997) Induction of micronuclei in Syrian hamster embryo cells: comparison of results in the SHE cell transformation assay for national toxicology program test chemicals. *Mutat Res 392*: 61–70

Goebel HH, Schmidt PF, Bohl J, Tettenborn B, Kramer G, Gutmann L (1990) Polyneuropathy due to acute arsenic intoxication: biopsy studies. *J Neuropathol Exp Neurol 49*: 137–149

Goering PL, Aposhian HV, Mass MJ, Cebrián M, Beck BD, Waalkes MP (1999) The enigma of arsenic carcinogenesis: role of metabolism. *Toxicol Sci 49*: 5–14

Goncalo S, Silva MS, Goncalo M, Baptista AP (1989) Occupational contact dermatitis to arsenic trioxide. in: Forsch PJ, Dooms-Goossens A (Eds) *Current topics in Contact Dermatitis* 333–336

Gonsebatt ME, Vega L, Salazar AM, Montero R, Guzman P, Bias J, Del Razo LM, Garcia-Vargas G, Albores A, Cebrian ME, Kelsh M, Ostrosky-Wegman P (1997) Cytogenetic effects in human exposure to arsenic. *Mutat Res 386*: 219–228

Greim H, Lehnert G (Eds) (1994). Arsentrioxid. in: *Biologische Arbeitsstoff-Toleranz-Werte (BAT-Werte) und Expositionsäquivalente für krebserzeugende Arbeitsstoffe (EKA)*, 7th issue, Wiley-VCH, Weinheim

Hantson P, Verellen-Dumoulin C, Libouton JM, Leonard A, Leonard ED, Mahieu P (1996) Sister chromatid exchanges in human peripheral blood lymphocytes after ingestion of high doses of arsenicals. *Int Arch Occup Environ Health 68*: 342–344

Harrington-Brock K, Cabrera M, Collard DD, Doerr CL, McConnell R, Moore MM, Sandoval H, Fuscoe JC (1999) Effects of arsenic exposure on the frequency of HPRT-mutant lymphocytes in a population of copper roasters in Antofagasta, Chile: a pilot study. *Mutat Res 431*: 247–257

Harrison JWE, Packman EW, Abbott DD (1958) Acute oral toxicity and chemical and physical properties of arsenic trioxides. *AMA Arch Ind Health 17*: 118–123

Hartwig A, Groeblinghoff UD, Beyersmann D, Natarajan AT, Filon R, Mullenders LH (1997) Interaction of arsenic(III) with nucleotide excision repair in UV-irradiated human fibroblasts. *Carcinogenesis 18*: 399–405

Hayashi H, Kanisawa M, Yamanaka K, Ito T, Udaka N, Ohji H, Okudela K, Okada S, Kitamara H (1998) Dimethylarsinic acid, a main metabolite of inorganic arsenics, has tumorigenicity and progression effects in the pulmonary tumors of A/J mice. *Cancer Lett 125*: 83–88

Hei TK, Liu SX, Waldren C (1998) Mutagenicity of arsenic in mammalian cells: role of reactive oxygen species. *Proc Natl Acad Sci USA 95*: 8103–8107

Heinrich-Ramm R, Mindt-Prüfert S, Szadkowski D (2001) Arsenic species excretion in a group of persons in northern Germany – contribution to the evaluation of reference values. *Int J Hyg Environ Health 203*: 475–477

Hertz-Picciotto I, Smith AH (1993) Observations on the dose-response curve for arsenic exposure and lung cancer. *Scand J Work Environ Health 19*: 217–226

Heywood R, Sortwell RJ (1979) Arsenic intoxication in the Rhesus monkey. *Toxicol Lett 3*: 137–144

Holmqvist I (1951) Occupational arsenical dermatitis – A study among employees at a copper ore smelting work including investigations of skin reactions to contact with arsenic compounds. *Acta Dermatol Venereol 31 (Suppl 26)*: 1–214

Holson JF, Stump DG, Ulrich CE, Farr CH (1999) Absence of prenatal developmental toxicity from inhaled arsenic trioxide in rats. *Toxicol Sci 51*: 87–97

Holson JF, Stump DG, Clevidence KJ, Knapp JF, Farr CH (2000) Evaluation of the prenatal toxicity of orally administered arsenic trioxide in rats. *Food Chem Toxicol 38*: 459–466

Hopenhayn-Rich C, Browning SR, Hertz-Picciotto I, Ferreccio C, Peralta C, Gibb H (2000) Chronic arsenic exposure and risk of infant mortality in two areas of Chile. *Environ Health Perspect 108*: 667–673

Hsu C-H, Li S-J, Chiou H-Y, Yeh P-M, Liou J-C, Hsueh Y-M, Chang S-H, Chen C-J (1997) Spontaneous and induced sister chromatid exchanges and delayed cell proliferation in peripher lymphocytes of Bowen's disease patients and matched controls of arseniasis-hyperendemic villages in Taiwan. *Mutat Res 386*: 241–251

Hsu C-H, Yang S-A, Wang YY, Yu H-S, Lin S-R (1999) Mutational spectrum of p53 gene in arsenic-related skin cancer from the blackfoot disease endemic area of Taiwan. *Br J Cancer 80*: 1080–1086

Hu Y, Su L, Snow ET (1998) Arsenic toxicity is enzyme specific and its effects on ligation are not caused by the direct inhibition of DNA repair enzymes. *Mutat Res 408*: 203–218

Huang S-C, Lee T-C (1998) Arsenite inhibits mitotic division and perturbs spindle dynamics in HeLa S3 cells. *Carcinogenesis 19*: 889–896

Huang S-Y, Chang C-S, Tang J-L, Tien H-F, Kuo T-L, Huang S-F, Yao Y-T, Chou WC, Chung C-Y, Wang C-H, Shen M-C, Chen Y-C (1998) Acute and chronic arsenic poisoning associated with treatment of acute promyelocytic leukaemia. *Br J Haematol 103*: 1092–1095

Huff J, Chan P, Nyska A (2000) Is the human carcinogen arsenic carcinogenic to laboratory animals? *Toxicol Sci 55*: 17–23

Hughes MF, Thompson DJ (1996) Subchronic dispositional and toxicological effects of arsenate administered in drinking water to mice. *J Toxicol Environ Health 49*: 177–196

IARC (International Agency for Research on Cancer) (1980) *Arsenic and arsenic compounds*. IARC monographs on the evaluation of the carcinogenic risk of chemicals to humans, Vol. 23, IARC Lyon, 39–141

IARC (International Agency for Research on Cancer) (1987) *Arsenic and arsenic compounds (Group 1)*. IARC monographs, Suppl 7: 100–107

Ihrig MM, Shalat SL, Baynes C (1998) A hospital-based case-control study of stillbirths and environmental exposure to arsenic using an atmospheric dispersion model linked to a geographical information system. *Epidemiology 9*: 290–294

Ishinishi N, Yamamoto A, Hisanaga A, Inamasu T (1983) Tumorigenicity of arsenic trioxide to the lung in Syrian golden hamsters by intermittent instillations. *Cancer Lett 21*: 141–147

Ishitsuka K, Hanada S, Suzuki S, Utsunomiya A, Chyuman Y, Takeuchi S, Takeshita T, Shimotakahara S, Uozumi K, Makino T, Arima T (1998) Arsenic trioxide inhibits growth of human T-cell leukaemia virus type I infected T-cell lines more effectively than retinoic acids. *Br J Haematol 103*: 721–728

Järup L, Pershagen G (1991) Arsenic exposure, smoking and lung cancer in smelter workers – a case-control study. *Am J Epidemiol 134*: 545–551

Järup L, Pershagen G, Wall S (1989) Cumulative arsenic exposure and lung cancer in smelter workers – a dose-response study. *Am J Ind Med 15*: 31–41

Jha AN, Noditi M, Nilsson R, Natarajan AT (1992) Genotoxic effects of sodium arsenite on human cells. *Mutat Res 284*: 215–221

Kaise T, Watanabe S, Itoh K (1985) The acute toxicity of arsenobetaine. *Chemosphere 14*: 1327–1332

Kashiwada E, Kuroda K, Endo G (1998) Aneuploidy induced by dimethylarsinic acid in mouse bone marrow cells. *Mutat Res 413*: 33–38

Kenyon EM, Hughes MF (2001) A concise review of the toxicity and carcinogenicity of dimethylarsenic acid. *Toxicol 160*: 227–236

Kerkvliet NI, Steppan LB, Koller LD, Exon JH (1980) Immunotoxicology studies of sodium arsenate – effects of exposure on tumor growth and cell-mediated tumor immunity. *J Environ Pathol Toxicol 4*: 65–79

Kitchin KT (2001) Recent advances in arsenic carcinogenesis: modes of action, animal model systems, and methylated arsenic metabolites. *Toxicol Appl Pharmacol 172*: 249–261

Kroes R, van Logten MJ, Berkvens JM, deVries T, van Esch GJ (1974) Study on the carcinogenicity of lead arsenate and sodium arsenate and on the possible synergistic effect of diethylnitrosamine. *Food Cosmet Toxicol 12*: 671–679

Lagerkvist BE, Linderholm H, Nordberg GF (1988) Arsenic and Raynaud's phenomenon. Vasospastic tendency and excretion of arsenic in smelter workers before and after the summer vacation. *Int Arch Occup Environ Health 60*: 361–364

Lai MS, Hsueh YM, Chen CJ, Shyu MP, Chen SY, Kuo TL, Wu MM, Tai TY (1994) Ingested inorganic arsenic and prevalence of diabetes mellitus. *Am J Epidemiol 139*: 484–492

Lantzsch H, Gebel T (1997) Genotoxicity of selected metal compounds in the SOS chromotest. *Mutat Res 389*: 191–197

Lee T-C, Oshimura M, Barrett JC (1985) Comparison of arsenic-induced cell transformation, cytotoxicity, mutation and cytogenetic effects in Syrian hamster embryo cells in culture. *Carcinogenesis 6*: 1421–1426

Lee-Feldstein A (1986) Cumulative exposure to arsenic and its relationship to respiratory cancer among copper smelter employees. *J Occup Med 28*: 296–302

Lerda D (1994) Sister-chromatid exchange (SCE) among individuals chronically exposed to arsenic in drinking water. *Mutat Res 312*: 111–120

Li JH, Rossman TG (1989) Inhibition of DNA ligase activity by arsenite: a possible mechanism of its comutagenesis. *Mol Toxicol 2*: 1–9

Li W, Wanibuchi H, Salim EI, Yamamoto S, Yoshida K, Endo G, Fukushima S (1998) Promotion of NCI-Black-Reiter male rat bladder carcinogenesis by dimethylarsenic acid an organic arsenic compound. *Cancer Lett 134*: 29–36

Li YM, Broome JD (1999) Arsenic targets tubulins to induce apoptosis in myeloid leukemia cells. *Cancer Res 59*: 776–780

Lindgren A, Vahter M, Dencker L (1982) Autoradiographic studies on the distribution of arsenic in mice and hamsters administered ^{74}As-arsenite or -arsenate. *Acta Pharmacol Toxicol 51*: 253–265

Liu J, Kadiiska MB, Liu Y, Lu T, Qu W, Waalkes MP (2001a) Stress-related gene expression in mice treated with inorganic arsenicals. *Toxicol Sci 61*: 314–320

Liu SX, Athar M, Lippai I, Waldren C, Hei TK (2001b) Induction of oxyradicals by arsenic: implication for mechanism of genotoxicity. *Proc Natl Acad Sci USA 98*: 1643–1648

Liu YC, Huang H (1997) Involvement of calcium-dependent protein kinase c in arsenite-induced genotoxicity in Chinese hamster ovary cells. *J Cell Biochem 64*: 423–433

Lubin J, Pottern LM, Blot WJ, Tokudome S, Stone BJ, Fraumeni JF (1981) Respiratory cancer among copper smelter workers: recent mortality statistics. *J Occup Med 23*: 779–784

Lubin JH, Pottern LM, Stone BJ, Fraumeni JF (2000) Respiratory cancer in a cohort of copper smelter workers: results from more than 50 years of follow-up. *Am J Epidemiol 151*: 554–565

Lugo G, Cassady G, Palmisano P (1969) Acute and maternal arsenic intoxication with neonatal death. *Am J Dis Child 117*: 173–182

Lynn S, Gurr JR, Lai HAT, Jan KY (2000) NADH oxidase activation is involved in arsenite-induced oxidative DNA damage in human vascular smooth muscle cells. *Circ Res 86*: 514–519

Lynn S, Lai HT, Gurr JR, Jan KY (1997) Arsenite retards DNA break rejoining by inhibiting DNA ligation. *Mutagenesis 12*: 353–358

Maki-Paakkanen J, Kurttio P, Paldy A, Pekkanen J (1998) Association between the clastogenic effect in peripheral lymphocytes and human exposure to arsenic through drinking water. *Environ Mol Mutagen 32*: 301–313

Mass MJ, Tennat A, Roop BC, Cullen WR, Styblo M, Thomas DJ, Kligerman AD (2001) Methylated trivalent arsenic species are genotoxic. *Chem Res Toxicol 14*: 355–361

Mass MJ, Wang L (1997) Arsenic alters cytosine methylation patterns of the promoter of the tumor suppressor gene p53 in human lung cells: a model for a mechanism of carcinogenesis. *Mutat Res 386*: 263–277

Matsui M, Nishigori C, Toyokuni S, Takada J, Akaboshi M, Ishikawa M, Imamura S, Miyachi Y (1999) The role of oxidative DNA damage in human arsenic carcinogenesis: detection of 8-hydroxy-2'-deoxyguanosine in arsenic-related Bowen's disease. *J Invest Dermatol 113*: 26–31

Menendez D, Mora G, Salazar AM, Ostrosky-Wegman P (2001) ATM status confers sensitivity to arsenic cytotoxic effects. *Mutagenesis 16*: 443–448

Mohamed KB (1998) Occupational contact dermatitis from arsenic in a tin-smelting factory. *Contact Dermatitis 38*: 224–225

Moore LE, Smith AH, Hopenhayn-Rich C, Biggs ML, Kalman DA, Smith MT (1997a) Micronuclei in exfoliated bladder cells among individuals chronically exposed to arsenic in drinking water. *Cancer Epidemiol Biomarkers Prev 6*: 31–36

Moore MM, Harrington-Brock K, Doerr CL (1997b) Relative genotoxic potency of arsenic and its methylated metabolites. *Mutat Res 386*: 279–290

Morton WE, Caron GA (1989) Encephalopathy: an uncommon manifestation of workplace arsenic poisoning? *Am J Ind Med 15*: 1–5

Mouron SA, Golijow CD, Dulout FN (2001) DNA damage by cadmium and arsenic salts assessed by the single cell gel electrophoresis assay. *Mutat Res 498*: 47–55

Murphy MJ, Lyon LW, Taylor JW (1981) Subacute arsenic neuropathy: clinical and electrophysiological observations. *J Neurol Neurosurg Psychiatry 44*: 896–900

Nagymajtenyi L, Selypes A, Berencsi G (1985) Chromosomal aberrations and fetotoxic effects of atmospheric arsenic exposure in mice. *J Appl Toxicol 5*: 61–63

Neiger RD, Osweiler GD (1989) Effect of subacute level dietary sodium arsenite on dogs. *Fundam Appl Toxicol 13*: 439–451

Nemec MD, Holson JF, Farr CH, Hood RD (1998) Developmental toxicity assessment of arsenic acid in mice and rabbits. *Reprod Toxicol 12*: 647–658

Ng JC, Seawright AA, Qi L, Garnett CM, Chiswell B, Moore MR (1999) Tumours in mice induced by exposure to sodium arsenate in drinking water. in: Abernathy C, Caldron R, Chapell W (Eds) *Arsenic exposure and health effects*, Oxford Elsevier Science, 217–223

Nordenson I, Beckman G, Beckman L (1978) Occupational and environmental risks in and around a smelter in northern Sweden: II. Chromosomal aberrations in workers exposed to arsenic. *Hereditas 88*: 47–50

Nordenson I, Beckman L (1982) Occupational and environmental risks in and around a smelter in northern Sweden. VII. Reanalysis and follow-up of chromosomal aberrations in workers exposed to arsenic. *Hereditas 96*: 175–181

Nordenson I, Beckman L (1991) Is the genotoxic effect of arsenic mediated by oxygen free radicals? *Hum Hered 41*: 71–73

Nordström S, Beckman L, Nordenson I (1978) Occupational and environmental risks in and around a smelter in northern Sweden. I. Variations in birth weight. *Hereditas 88*: 43–46

Nordström S, Beckman L, Nordenson I (1979a) Occupational and environmental risks in and around a smelter in northern Sweden. V. Spontaneous abortion among female employees and decreased birth weight in their offspring. *Hereditas 90*: 291–296

Nordström S, Beckman L, Nordenson I (1979b) Occupational and environmental risks in and around a smelter in northern Sweden. VI. Congenital malformations. *Hereditas 90*: 297–302

Oh SJ (1991) Electrophysiological profile in arsenic neuropathy. *J Neurol Neurosurg Psychiatry 54*: 1103–1105

Ostrosky-Wegman P, Gonsebatt ME, Montero R, Vega L, Barba H, Espinosa J, Palao A, Cortinas C, Garcia-Vargas G, del Razo LM (1991) Lymphocyte proliferation kinetics and genotoxic findings in a pilot study on individuals chronically exposed to arsenic in Mexico. *Mutat Res 250*: 477–482

Ott MG, Holder BB, Gordon HL (1974) Respiratory cancer and occupational exposure to arsenicals. *Arch Environ Health 29*: 250–255

Paschoud JM (1964) Notes cliniques au sujet des eczémas de contact professionnels par l'arsenic et l'antimoine (Clinical notes on occupational contact eczemas due to arsenic and antimony) (French). *Dermatologica 129*: 410–415

Pershagen G, Björklund NE (1985) On the pulmonary tumorigenicity of arsenic trisulfide and calcium arsenate in hamsters. *Cancer Lett 27*: 99–104

Pershagen G, Lind B, Björklund NE (1982) Lung retention and toxicity of some inorganic arsenic compounds. *Environ Res 29*: 425–434

Pershagen G, Nordberg G, Björklund NE (1984) Carcinomas of the respiratory tract in hamsters given arsenic trioxide and/or benzo[a]pyrene by the pulmonary route. *Environ Res 34*: 227–241

Poma K, Degraeve N, Kirsch-Volders M, Susanne C (1981) Cytogenetic analysis of bone marrow cells and spermatogonia of male mice after *in vivo* treatment with arsenic. *Experientia 37*: 129–130

Poma K, Degraeve N, Susanne C (1987) Cytogenetic effects in mice after chronic exposure to arsenic followed by a single dose of ethylmethane sulfonate. *Cytologia 52*: 445–449

Prukop JA, Savage NL (1986) Some effects of multiple sublethal doses of monosodium methanearsonate (MSMA) herbicide on hematology, growth and reproduction of laboratory mice. *Bull Environ Contam Toxicol 36*: 337–341

Rahman M, Axelson O (1995) Diabetes mellitus and arsenic exposure: a second look at case-control data from a Swedish copper smelter. *Occup Environ Med 52*: 773–774

Rahman M, Tondel M, Ahmad SA, Axelson O (1998) Diabetes mellitus associated with arsenic exposure in Bangladesh. *Am J Epidemiol 148*: 198–203

Ramirez P, Eastmond DA, Laclette JP, Ostrosky-Wegman P (1997) Disruption of microtubule assembly and spindle formation as a mechanism for the induction of aneuploid cells by sodium arsenite and vanadium pentoxide. *Mutat Res 38*: 291–298

Ramirez P, Del Razo LM, Gutierrez-Ruiz MC, Gonseblatt ME (2000) Arsenite induces DNA-protein crosslinks and cytokeratin expression in the WRL-68 human hepatic cell line. *Carcinogenesis 21*: 701–706

Rasmussen RE, Menzel DB (1997) Variation in arsenic-induced sister chromatid exchange in human lymphocytes and lymphoblastoid cell lines. *Mutat Res 386*: 299–306

Reichl FX, Szinicz L, Kreppel H, Fichtl B, Forth W (1990) Effect of glucose in mice after acute experimental poisoning with arsenic trioxide (As_2O_3). *Arch Toxicol 64*: 336–338

Reichl FX, Szinicz L, Kreppel H, Forth W (1988) Effect of arsenic on carbohydrate metabolism after single or repeated injection in guinea pigs. *Arch Toxicol 62*: 473–475

Rossberg J (1967) Berufsbedingte Arsen-Schädigungen der Haut (Skin lesions caused by occupation related to arsenic) (German). *Dermatol Wochenschr 153*: 977–987

RoyChoudhury A, Das T, Sharma A, Takukder G (1996) Dietary garlic extracts in modifying clastogenic effects of inorganic arsenic in mice: two-generation studies. *Mutat Res 359*: 165–170

Rupa DS, Schuler M, Eastmond DA (1997) Detection of hyperploidy and breakage affecting the 1cen-1q12 region of cultured interphase human lymphocytes treated with various genotoxic agents. *Environ Molec Mutagen 2*: 161–167

Saffiotti U, Bertolero F (1989) Neoplastic transformation of BALB/3T3 cells by metals and the quest for induction of a metastatic phenotype. *Biol Trace Elem Res 21*: 475–482

Salnikow K, Cohen MD (2002) Backing into cancer: effects of arsenic on cell differentiation. *Toxicol Sci 65*: 161–163

Schaumlöffel N, Gebel T (1998) Heterogeneity of the DNA damage provoked by antimony and arsenic. *Mutagenesis 13*: 281–286

Schrauzer GN, Ishmael D (1974) Effects of selenium and arsenic on the genesis of spontaneous mammary tumors in inbred C3H mice. *Ann Clin Lab Sci 4*: 441–447

Schrauzer GN, White DA, Schneider CJ (1976) Inhibition of the genesis of sponaneous mammary tumours in C3H mice: Effects of selenium-antagonistic elements and their possible role in human breast cancer. *Bioinorg Chem 6*: 265–270

Schroeder HA, Balassa JJ (1966) Abnormal trace metals in man: arsenic. *J Chron Dis 19*: 85–106

Schroeder HA, Balassa JJ (1967) Arsenic, germanium, tin and vanadium in mice: Effects on growth, survival and tissue levels. *J Nutrition 92*: 245–252

Schulz KH (1962) *Chemische Struktur und allergene Wirkung–unter besonderer Berücksichtigung von Kontaktallergenen* (Chemical structure and allergenic effect—particularly in consideration of contact allergens) (German). Editio Cantor, Aulendorf, 43–45

Schwerdtle T, Mackiw I, Bayerl A, Hartwig A (2002) Arsenite and its methylated metabolites MMAA and DMAA induce oxidative DNA damage in HeLa S3 cells. *Arch Pharmacol 365, Suppl 1*: 526

Seiwert M, Becker K, Friedrich C, Helm D, Hoffmann K, Krause C, Nöllke P, Schulz C, Seifert B (1999) *Umwelt-Survey 1990/92 Arsen-Zusammenhangsanalyse* (Environmental survey 1990/92, Arsenic multivariate statistical analysis) (German). WaBoLu 3/99, Umweltbundesamt, Berlin

Shen Z-X, Chen G-Q, Ni J-H, Li X-S, Xion S-M, Qui Q-Y, Zhu J, Tang W, Sun G-L, Yang K-Q, Chen Y, Zhou L, Fan Z-W, Wang Y-T, Ma J, Zhang P, Zhang T-D, Chen S-J, Chen Z, Wang Z-Y (1997) Use of arsenic trioxide (As_2O_3) in the treatment of acute promyelocytic leukemia (APL): II. Clinical efficacy and pharmacokinetics in relapsed patients. *Blood 89*: 3354–3360

Shirachi DY, Johansen MG, McGowan JP, Tu SH (1983) Tumorigenic effect of sodium arsenite in rat kidney. *Proc West Pharmacol Soc 26*: 413–415

Sikorski EE, McCay JA, White KL Jr, Bradley SG, Munson AE (1989) Immunotoxicity of the semiconductor gallium arsenide in female $B6C3F_1$ mice. *Fundam Appl Toxicol 13*: 843–858

Soignet SL, Maslak P, Wang Z-G, Jhanwar S, Calleja E, Dardashti LJ, Corso D, DeBlasio A, Gabrilove J, Scheinberg DA, Pandolfi PP, Warell RP (1998) Complete remission after treatment of acute promyelocytic leukemia with arsenic trioxide. *N Engl J Med 339*: 1341–1348

Sordo M, Herrera LA, Ostrosky-Wegman P, Rojas E (2001) Cytotoxic and genotoxic effects of As, MMA, and DMA on leukocytes and stimulated human lymphocytes. *Teratogen Carcinogen Mutagen 21*: 249–260

Stump DG, Holson JF, Fleeman TL, Nemec MD, Farr CH (1999) Comparative effects of single intraperitoneal or oral doses of sodium arsenate or arsenic trioxide during *in utero* development. *Teratology 60*: 283–291

Szinicz L, Forth W (1988) Effects of As_2O_3 on gluconeogenesis. *Arch Toxicol 61*: 444–449

Tabacova S, Hunter ES (1995) Pathogenic role of peroxidation in prenatal toxicity of arsenic. in: Zelikoff J, Bertin J, Burbacher T, Hunter S, Miller R, Silbergeld E, Tabacova S, Roger J: Health risks associated with prenatal exposure. *Experimental and Applied Toxicology 25*: 161–170

Taylor PR, Qiao YL, Schatzkin A, Yao SX, Lubin J, Mao BL, Rao JY, McAdams M, Xuan XZ, Li JY (1989) Relation of arsenic exposure to lung cancer among tin miners in Yunnan Province, China. *Br J Ind Med 46*: 881–886

Thomas DJ, Styblo M, Lin S (2001) The cellular metabolism and systemic toxicity of arsenic. *Toxicol Appl Pharmacol 176*: 127–144

Thorgeirsson UP, Dalgard DW, Reeves J, Adamson RH (1994) Tumor incidence in a chemical carcinogenesis study of nonhuman primates. *Regul Toxicol Pharm 19*: 130–151

Tian D, Ma H, Feng Z, Xia Y, Le XC, Ni Z, Allen J, Collins B, Schreinemachers D, Mumford JL (2001) Analyses of micronuclei in exfoliated epithelial cells from individuals chronically exposed to arsenic via drinking water in Inner Mongolia, China. *J Toxicol Environ Health, A 64*: 472–484

Tice RR, Yager JW, Andrews P, Crecelius E (1997) Effect of hepatic methyl donor status on urinary excretion and DNA damage in B6C3F$_1$ mice treated with sodium arsenite. *Mutat Res 386*: 315–334

Tinwell H, Stephens SC, Ashby J (1991) Arsenite as the probable active species in the human carcinogenicity of arsenic: mouse micronucleus assays on Na and K arsenite, orpiment, and Fowler's solution. *Environ Health Perspect 95*: 205–210

Trouba KJ, Wauson EM, Vorce RL (2000) Sodium arsenite-induced dysregulation of proteins involved in proliferative signaling. *Toxicol Appl Pharmacol 164*: 161–170

Vahter M (2000) Genetic polymorphism in the biotransformation of inorganic arsenic and its role in toxicity. *Toxicol Lett 112–113*: 209–217

Vig BK, Figueroa ML, Cornforth MN, Jenkins SH (1984) Chromosome studies in human subjects chronically exposed to arsenic in drinking water. *Am J Ind Med 6*: 325–338

Vogt BL, Rossman TG (2001) Effects of p53, p21 and cyclin D expression in normal human fibroblasts—a possible mechanism for arsenite's comutagenicity. *Mut Res 478*: 159–168

Waalkes MP, Keefer LK, Diwan BA (2000) Induction of proliferative lesions of the uterus, testes, and liver in Swiss mice given repeated injections of sodium arsenate: possible estrogenic mode of action. *Toxicol Appl Pharmacol 166*: 24–35

Wahlberg JE, Boman A (1986) Contact sensitivity to arsenical compounds—clinical and experimental studies. *Dermatosen Beruf Umwelt 34*: 10–12

Wang TS, Huang H (1994) Active oxygen species are involved in the induction of micronuclei by arsenite in XRS-5 cells. *Mutagenesis 9*: 253–257

Wang T-S, Hsu T-Y, Chung C-H, Wang ASS, Bau T-D, Jan K-Y (2001) Arsenite induces oxidative DNA adducts and DNA-protein cross-links in mammalian cells. *Free Radic Biol Med 31*: 321–330

Wanibuchi H, Yamamoto S, Chen H, Yoshida K, Endo G, Hori T, Fukushima S (1996) Promoting effects of dimethylarsinic acid on *N*-butyl-*N*-(4-hydroxybutyl)nitrosamine-induced urinary bladder carcinogenesis in rats. *Carcinogenesis 17*: 2435–2439

Warner ML, Moore LE, Smith MT, Kalman DA, Fanning E, Smith AH (1994) Increased micronuclei in exfoliated bladder cells of individuals who chronically ingest arsenic-contaminated water in Nevada. *Cancer Epidemiol Biomarkers Prev 3*: 583–590

Wauson EM, Langan AS, Vorce RL (2002) Sodium arsenite inhibits and reverses expression of adipogenic and fat cell-specific genes during *in vitro* adipogenesis. *Toxicol Sci 65*: 211–219

Wei M, Wanibuchi H, Yamamoto S, Li W, Fukushima S (1999) Urinary bladder carcinogenicity of dimethylarsinic acid in male F344 rats. *Carcinogenesis 20*: 1873–1876

Welch K, Higgins I, Oh M, Burchfiel C (1982) Arsenic exposure, smoking and respiratory cancer in copper smelter workers. *Arch Environ Health 37*: 325–335

WHO (World Health Organization) (2001) *Arsenic and arsenic compounds*. IPCS (International Programme on Chemical Safety) Environmental Health Criteria 224, WHO, Geneva

Wiencke JK, Yager JW, Varkonyi A, Hultner M, Lutze LH (1997) Study of arsenic mutagenesis using the plasmid shuttle vector pz189 propagated in DNA repair proficient human cells. *Mutat Res 386*: 335–344

Yager JW, Wiencke JK (1997) Inhibition of poly(ADP-ribose) polymerase by arsenite. *Mutat Res 386*: 345–351

Yamamoto A, Hisanaga A, Ishinishi N (1987) Tumorigenicity of inorganic arsenic compounds following intratracheal instillations to the lungs of hamsters. *Int J Cancer 40*: 220–223

Yamamoto S, Konishi Y, Matsuda T, Murai T, Shibata MA, Matsui-Yuasa I, Otani S, Kuroda K, Endo G, Fukushima S (1995) Cancer induction by an organic arsenic compound, dimethylarsinic acid (cacodylic acid), in GF344/DuCrj rats after pretreatment with five carcinogens. *Cancer Res 55*: 1271–1276

Yamamoto S, Wanibuchi H, Hori T, Yano Y, Matsui-Yuasa I, Otani S, Chen H, Yoshida K, Kuroda K, Endo G, Fukushima S (1997) Possible carcinogenic potential of dimethylarsinic acid as assessed in rat *in vivo* models: a review. *Mutat Res 386*: 353–361

Yamanaka K, Okada S (1994) Induction of lung-specific DNA damage by metabolically methylated arsenics via the production of free radicals. *Environ Health Perspect 102*: 37–40

Yamanaka K, Hasegawa A, Sawamura R, Okada S (1989) Dimethylated arsenics induce DNA strand breaks in lung via the production of active oxygen in mice. *Biochem Biophys Res Commun 165*: 43–50

Yih LH, Ho IC, Lee TC (1997) Sodium arsenite disturbs mitosis and induces chromosome loss in human fibroblasts. *Cancer Res 57*: 5051–5059

Zhang W, Ohnishi K, Shigeno K, Fujisawa S, Naito K, Nakamura S, Takeshita K, Takeshita A, Ohno R (1998) The induction of apoptosis and cell cycle arrest by arsenic trioxide in lymphoid neoplasms. *Leukemia 12*: 1383–1391

Zhao CQ, Young MR, Diwan B A, Coogan TP, Waalkes MP (1997) Association of arsenic-induced malignant transformation with DNA hypomethylation and aberrant gene expression. *Proc Natl Acad Sci USA 94:* 10907–10912

Zierler S, Theodore M, Cohen A, Rothman K (1988) Chemical quality of maternal drinking water and congenital heart disease. *Int J Epidemiol 17*: 589–594

completed 10.04.2002

Beryllium and its inorganic compounds

MAK value	–
Peak limitation	–
Absorption through the skin	–
Sensitization (2002)	Sah
Carcinogenicity (2003)	Category 1
Prenatal toxicity	–
Germ cell mutagenicity	–
BAT value	–

Substance	CAS number	Molecular formula	Molecular weight	Solubility in water
Beryllium	7440-41-7	Be	9.01	insoluble
Beryllium oxide	1304-56-9	BeO	25.01	barely soluble*
Beryllium nitrate	13510-48-0	$Be(NO_3)_2 \cdot 4H_2O$	205.08	$1.66 \cdot 10^6$ mg/l**
Beryllium sulfate	7787-56-6	$BeSO_4 \cdot 4H_2O$	177.13	$3.91 \cdot 10^6$ mg/l**
Beryllium fluoride	7787-49-7	BeF_2	47.01	readily soluble
Beryllium chloride	7787-47-5	$BeCl_2$	79.92	readily soluble
Beryllium hydroxide	13327-32-7	$Be(OH)_2$	43.03	insoluble
Beryllium citrate	no data	no data	no data	no data
Beryllium carbonate	66104-24-3	$Be_2CO_3(OH)_2$	112.05	insoluble
Beryllium acetate	543-81-7	$Be(C_2H_3O_2)_2$	127.10	insoluble
Beryllium stearate	no data	no data	no data	no data
Beryllium silicate	13598-00-0	Be_2SiO_4	110.07	no data
Bertrandite	12161-82-9	$4BeO \cdot 2SiO_2 \cdot H_2O$	238.24	no data
Beryl	1302-52-9	$3BeO \cdot Al_2O_3 \cdot 6SiO_2$	537.50	no data

* The solubility depends on the production temperature: beryllium oxide heated to 500°C is more soluble than if heated to 1000°C (no other details) (US EPA 1998), ** readily soluble at 20°C

This supplement to the documentation from 1977 (in Volume 3 of the present series) is essentially based on reviews, such as ATSDR 2002, US EPA 1998. WHO 1990 and WHO 2001. More recent literature and original studies are cited only if necessary for the evaluation.

Beryllium is a silvery white, shiny, hard and brittle metal, which oxidizes easily. It is manufactured via electrolysis or by reduction with magnesium from beryllium fluoride or beryllium chloride. The principal sources of these beryllium compounds are naturally occurring silicates or aluminium silicates (alumosilicates) containing beryllium, especially beryl and bertrandite. The precious beryls, aquamarine, emerald, and chrysoberyl also belong to the naturally occurring mineral forms of beryllium, coloured due to deposits of iron or chromium oxides.

Pure beryllium is used in the manufacture of X-ray windows. As a component in a number of alloys, beryllium increases their hardness, stability and resistance to corrosion, so that they can be used in making watch coils, surgical instruments, valve springs, aircraft brakes and high-temperature materials as well as for special applications in the air and space industries. Beryllium alloys have also found uses in automobile racing in the form of parts in motor, gear and brake systems. In addition, beryllium has been employed in such varied fields as special sports equipment, for example golf clubs and bicycle frames, and in dental alloys. As a result of their good conductive properties and great hardness, beryllium/copper alloys are also used in the electronics industry. Furthermore, beryllium can be used as a moderator in atomic reactors instead of graphite or heavy water.

Among other functions, beryllium oxide is used in high-temperature equipment such as special crucibles, in manufacturing aircraft ignition plugs, as an insulating material, in ceramic dyes and in production of luminescent materials for fluorescent lamps. In chemical synthesis, beryllium chloride is used as a catalyst, and beryllium nitrate is used to harden incandescent mantles in gas and acetylene lamps.

1 Toxic Effects and Mode of Action

Beryllium and its inorganic compounds are mainly absorbed by inhalation. The particle size and water solubility are decisive for the absorption. Non-ionized soluble beryllium compounds are rapidly removed from the lungs, while the ionized forms remain in the lungs. Beryllium is transferred from the lungs into the blood in two phases. The rapid elimination of the substance via mucociliary clearance during the first one to two weeks is followed by a slower elimination in which beryllium is absorbed by alveolar macrophages and transported to the tracheobronchiolar lymph nodes. In man beryllium often remains in the lungs for several years and also serves as evidence of previous exposure in cases of chronic beryllium disease. Beryllium and its inorganic compounds are usually not metabolized and are mainly eliminated with urine and also with faeces.

Acute beryllium disease occurs after inhalation exposure to high beryllium concentrations ($> 100 \, \mu g/m^3$) and is characterized by symptoms of acute pneumonitis. Above all, it is the result of the direct toxicity of water-soluble beryllium compounds.

Unlike acute beryllium disease, chronic beryllium disease (berylliosis) also may be triggered by insoluble beryllium compounds, as immunological mechanisms in the sense of an allergic reaction of delayed type play a role here. It is characterized by the formation of non-caseating granulomas and is accompanied by functional restrictions of the lungs, such as reduction of the vital and total capacity and a reduced diffusion capacity. Repeated inhalation exposure to beryllium may also lead to cardiovascular, renal, hepatic and haematological effects, and to weight loss.

As a result of direct irritation, water-soluble beryllium compounds may cause poorly healing dermatitis. If undissolved particles enter the skin as a result of wounds or impaired skin barriers, ulcers or necrosis may develop. Beryllium and its inorganic compounds may be the cause of allergic contact eczema or granulomatous skin reactions of immunological origin.

Soluble beryllium compounds induce sister chromatid exchange, chromosomal aberrations and gene mutations in mammalian cells. Genotoxic effects are probably the result of the induction of DNA protein complexes and the influence on DNA polymerases. Also an indirect mechanism, the hypermethylation of promotor sequences, which leads to the inactivation of the corresponding genes, has been suggested.

Beryllium and its inorganic compounds are carcinogenic in man. Increased mortality from lung cancer was determined in different retrospective studies of workers in beryllium production plants. Death from lung cancer was disproportionately frequent in persons who had previously suffered from acute beryllium disease.

Beryllium deposits in the lungs were detected in animal experiments after acute inhalation of beryllium and its inorganic compounds. Pneumonitis with granulomatous damage, necrosis, fibrosis, hyperplasia, thickening of the alveolar walls and progressive lymphocyte and macrophage infiltration were found in the lungs. In addition, granulomas, abscesses, necrosis, congestion, emphysema and proliferative changes appeared after repeated exposure.

Increased incidences of adenocarcinomas or squamous cell carcinomas, epidermoid carcinomas and pleural mesotheliomas were observed in rats and monkeys after inhalation of beryllium metal, sulfate, chloride, fluoride or oxide. Although the studies do not meet present-day standards, they document that, like in man, lung tumours develop in animals after inhalation of beryllium and its inorganic compounds. There is evidence that beryllium and its inorganic compounds also provoke lung tumours in rats and mice after intratracheal or intraperitoneal administration. Osteosarcomas were observed in rabbits after intravenous and intramedullary (into the bone marrow) injection. No carcinogenic effects were induced in rats, mice and dogs after long-term administration of beryllium sulfate—probably because of the poor gastrointestinal absorption.

2 Mechanism of Action

Lung toxicity: The lung damage caused by beryllium and its inorganic compounds is mainly provoked by an inflammatory process. After inhalation exposure to beryllium, alveolar macrophages excrete lysosomal enzymes and tissue swelling occurs. The acidic properties of the water-soluble beryllium compounds increase their toxic effects (Zorn and Fischer 1998). Increased lipid and protein values, and increased activities of acid and alkaline phosphatases, lysozyme, and lactate dehydrogenases in the lung lavage fluid from rats exposed to beryllium indicate cell damage (ATSDR 2002). Long-term effects probably depend more on the chemical and crystallographic properties and the surface properties of the particles. Soluble beryllium compounds are converted into colloidal orthophosphates and hydroxides by the buffer properties of the lung fluid (Vorwald 1966). In chronic beryllium disease, in which the insoluble compounds are as effective or even more effective than the soluble ones, there is an allergic reaction of the delayed type, in which beryllium appears to function as a hapten and stimulates the local proliferation and accumulation of beryllium-specific T-cells in the lungs (Nikula et al. 1997). Suggested mediators were, among others, pro-inflammatory cytokines such as TNF-α, interleukin-6, interleukin-2, and interferon-γ, and an elevated concentration of neopterin in the serum as a possible additional marker for the disease. As a rule, berylliosis is progressive, even if further exposure to beryllium is avoided, and may lead to complete pulmonary failure (see Section 4.4.2).

Cell cycle: Beryllium sulfate influences the cell cycle in human fibroblasts with participation of the tumour suppressor protein (TP53) and the inhibitor of cyclin-dependent kinases CDKN1a (p21$^{\text{Waf-1, cip-1}}$) by inhibiting the G_0-G_1/pre-S-phase transition. In this case beryllium induces an increase in the p53 protein that serves as a transcription factor for CDKN1a in the cell nucleus. The inhibition of cyclin-dependent kinases prevents the phosphorylation of the protein Rb, which in unphosphorylated form prevents the activation of the E2F transcription factor. The transcription factor is necessary for the expression of proteins for DNA synthesis in the S-phase. CDKN1a inhibits DNA replication by binding to PCNA (proliferating cell nuclear antigen), activation of the replicative DNA polymerase δ being prevented. A further inhibitor of cyclin-dependent kinases, the protein p27^{KIP}, was increasingly formed in cells treated with beryllium. The underlying basic mechanism of the increased beryllium-induced formation of these inhibitors is unclear. DNA damage was considered possible (Lehnert et al. 2001).

Genotoxicity: The genotoxic effects of beryllium are based on the induction of DNA protein complexes (WHO 1990). In addition, beryllium binds to DNA polymerases, and the accuracy of DNA synthesis is reduced by the inhibition of the correction function (Focher et al. 1990, Luke et al. 1975, Sirover and Loeb 1976). The resulting increased presence of base substitutions is proportional to the beryllium concentration (Sirover and Loeb 1976). In lung tumours induced by beryllium metal in F344/N rats, GGT-GTT transversions in the K-ras gene were detected in 2 of 12 tumours. Mutations in the p53-cDNA or the c-raf-1-cDNA were not detected (Nickell-Brady et al. 1994). An indirect genotoxic mechanism, the hypermethylation of promotor sequences, which leads to the inactivation of the corresponding genes, is suggested. Hypermethylation of the promotor sequences of the p16 gene and of the oestrogen receptor α gene were detected in lung tumours of rats which were produced by 4-methylnitrosamino-1-(3-pyridyl)-1 butanol, carbon black or diesel engine emissions. In rats, it was possible to document hypermethylation of the p16 gene in 80 %, and of the oestrogen receptor α gene in 50 % of the lung tumours caused by chronic beryllium metal exposure, and the inhibiting effects of hypermethylation on gene transcription were confirmed. p16 gene expression in tumour tissues was found to be 30 to 60 times lower compared with that in tumour-free tissues.

The switching-off of the tumour suppressor p16 and the oestrogen receptor α genes by hypermethylation of promotor sequences are steps in the multi-stage process of carcinogenesis (Belinsky et al. 2002). It is evident that DNA hypermethylation arises because the above-mentioned carcinogens activate the methylation apparatus. In alveolar type II cells it was shown that increased activity of cytosin DNA methyl transferase caused comprehensive DNA methylation (Belinsky 1998). The change in the DNA methylation pattern possibly plays a key role in beryllium-induced carcinogenesis (Goodman and Watson 2002). Hypermethylation of promotor sequences acts here as a mutation causing functional loss. The inactivation of tumour suppressor genes by hypermethylation also was discussed as a cause of the carcinogenic effects of arsenic (Goering et al. 1999).

3 Toxicokinetics and Metabolism

3.1 Absorption, distribution, elimination

Absorption

Inhalation is the main intake route for beryllium and its inorganic compounds. There are no data for absorption rates after inhalation of the substance. Less than 1 % of orally administered beryllium is absorbed via the gastrointestinal tract. Absorption of beryllium via the skin may be considered as insignificant (ATSDR 2002, US EPA 1998, WHO 1990).

Distribution

In addition to the dose and the particle size, the water solubility of beryllium and its inorganic compounds is also important for uptake of the substance into the body. The higher the solubility, the faster a relatively uniform distribution in the organism takes place, with moderate retention in the lungs (US EPA 1998). In the case of poorly soluble beryllium compounds, 70 % to 99 % remains in the lungs (Zorn and Fischer 1998). Ionization of the beryllium compound also plays a role. Non-ionized soluble beryllium compounds, for example citrate, are rapidly removed from the lungs, while the ionized forms remain in the lungs (ATSDR 2002). Animal studies show that the transfer of beryllium from the lungs into the blood takes place in two phases. The rapid elimination of beryllium via mucociliary clearance during the first one to two weeks is followed by a slower phase in which the beryllium is taken up by alveolar macrophages and transported to the tracheobronchiolar lymph nodes. The biphasic elimination of beryllium from the bronchi in rats yielded a half-life (time) of one to 60 days (rapid phase) and of 0.6 to 2.3

years (slow phase) (US EPA 1998). In man, beryllium often remains in the lungs for several years and is also evidence of earlier exposure in cases of chronic beryllium disease. The elimination rate of beryllium from the lungs is greater in hamsters than in rats, and higher in male animals than in females (ATSDR 2002).

After 10 to 20 hours exposure to beryllium dust in unknown concentrations, 25 persons had serum beryllium concentrations of 3.5 ± 0.47 ng/l on the day after exposure (1.0 ng/l in persons not exposed), 2.4 ± 0.3 ng/l after six days; after six to eight weeks values had returned to normal (ATDSR 2002). In eight workers exposed to beryllium concentrations of around 8 ng/m^3 in the form of beryllium chloride for ten days, four to six hours per day, the beryllium concentrations in blood and urine were four times higher in comparison with persons not exposed. 60 % to 70 % of the beryllium in blood was bound to pre-albumin or γ-globulin (ATSDR 2002). In different animal studies, after inhalation of beryllium sulfate, fluoride or chloride the greatest amounts of beryllium were deposited in the lungs and the hilar lymph nodes, followed by the bones, liver and kidneys, with a particularly long retention time in the bones (Zorn and Fischer 1998). In human organs the average beryllium concentrations were: 0.21 µg/g lung, 0.08 µg/g brain, 0.07 µg/g kidney and spleen, 0.04 µg/g liver, muscle and vertebra, 0.03 µg/g heart and 0.02 µg/g bone. There are no data for the type and duration of the exposure or the origin of the organs (ATSDR 2002). In the case of acute beryllium disease, 1 to > 2 µg beryllium/g was determined in lung tissue and 0.12 µg/g in liver tissue. For chronic beryllium disease the concentrations in the lungs were 0.001 to 3.1 µg/g and in the lymph nodes up to 8.5 µg/g dry weight. An above average beryllium exposure was assumed at beryllium concentrations in lung tissue of 0.02 µg/g dry weight and above (Zorn and Fischer 1998). There is evidence that beryllium passes through the placenta and also into breast milk (ATSDR 2002).

Elimination

20 % to 70 % of the absorbed beryllium is excreted with urine, the rest with faeces or sputum. Twelve hours after inhalation exposure to beryllium there is already a clear increase in the beryllium concentration in urine. After two weeks exposure to beryllium concentrations of 10 ng/m^3 the beryllium concentration in faeces was 20 % lower than in the urine (Zorn and Fischer 1998).

Jewel polishers with long-term exposure to beryllium concentrations of 20 µg/m^3 for more than four hours per week were found to have urinary beryllium levels of 0.13 ± 0.12 µg/l (median 0.09 µg/l) before the shift and 0.08 ± 0.07 µg/l (median 0.003 µg/l) after the shift (see Section 4.6). The lack of an increase in the beryllium concentration in the urine after the end of the shift was attributed to the retarded beryllium transfer into the urine. Thus, the beryllium detected originated from the previous shift (Wegner *et al.* 2000).

After intravenous administration of radioactive beryllium to rabbits, 28.8 % of the initial amount was excreted with urine within the first 24 hours. After this first rapid elimination phase, the amount excreted daily with the urine in the second phase was between 0.5 % and 1.85 % and that with the faeces 0.2 % to 0.5 % (ATSDR 2002).

Rats excreted orally administered, radioactively labelled beryllium chloride with the urine (0.11 %) and with faeces (104.7 %). Similar values were found for mice, dogs and monkeys (ATSDR 2002).

3.2 Metabolism

Usually, beryllium and its inorganic compounds are not metabolized. Only soluble beryllium salts are partially converted in the lungs into less soluble forms (ATSDR 2002).

4 Effects in Man

4.1 Single exposures

Single inhalation exposure to high beryllium concentrations (> 100 µg/m^3) may result in acute beryllium disease, characterized by symptoms of acute pneumonitis such as progressive coughing, shortness of breath, loss of appetite, weight loss, fatigue and cyanosis. Acute beryllium disease is defined as beryllium-induced lung disease with a duration of less than one year and a latency period of around three days and, above all, is attributed to the direct toxicity of the water-soluble beryllium compounds. As a rule, acute beryllium disease completely disappears after one to four weeks. Only in rare cases does it lead to death as a result of the formation of severe pulmonary oedema. Another exposure to beryllium may result in renewed pneumonitis. In isolated cases transition to the chronic form is also possible without further exposure and after several years (WHO 1990).

Evaluation of the US Beryllium Case Registry showed that the number of cases of acute beryllium disease has decreased drastically since the introduction of the threshold limit value for beryllium of 2 µg/m^3 in 1959 (WHO 1990). The Beryllium Case Registry is a register existing since 1952, in which all diagnosed cases of acute or chronic beryllium disease are recorded. Criteria for entry into this register are demonstrated exposure to beryllium or the detection of beryllium in the lung tissue, in each case in connection with the clinical symptoms of beryllium disease (ATSDR 2002).

Directly and ten months after exposure to beryllium chloride (concentration unknown) for 10 to 20 hours, no hepatic effects could be detected in 25 laboratory workers (ATSDR 2002).

4.2 Repeated exposure

Chronic beryllium disease, berylliosis, usually occurs as a result of long-term inhalation exposure to low beryllium concentrations. In addition, systemic effects also are observed. The LOAEL (lowest observed adverse effect level) for berylliosis in man is between 0.00052 and 0.0012 mg/m^3 (ATSDR 2002).

A comprehensive evaluation of the studies of berylliosis has been drawn up in the description of the allergenic effects of beryllium and its compounds (see Section 4.4).

Lungs: Berylliosis is defined as beryllium-induced lung disease with a duration of more than one year and a latency period of several weeks to 20 years after exposure (WHO 1990). Unlike acute beryllium disease, berylliosis may also be triggered by insoluble beryllium compounds, as immunological mechanisms in the sense of an allergic reaction of the delayed type play a role here. Chronic berylliosis, which is very difficult to differentiate from sarcoidosis, is characterized by the formation of non-caseating granulomas, in which, above all, epithelioid cells and CD^{4+} T-lymphocytes are to be found. Functional restrictions of the lungs are, among other things, a reduction of vital and total capacity as well as a reduced diffusion capacity. The disease is usually progressive, even if further exposure to beryllium is avoided, and may lead to complete pulmonary failure. The lymphocyte transformation test (LTT) is a valid method for detecting beryllium sensitization (see Section 4.4.2).

Heart: Autopsies of 17 persons exposed to beryllium revealed hypertrophy of the right atrium and the ventricle (ATSDR 2002). Increased mortality as a result of heart diseases (SMR 1.06, 95 % confidence interval (CI) 1.00–1.12), in particular ischemic heart diseases (SMR 1.08, 95 % CI 1.01–1.14), was found among 9225 workers of US beryllium plants (retrospective cohort study) (Ward *et al.* 1992).

Kidneys: Kidney stones were found in approximately 10 % of the cases of chronic berylliosis documented in the Beryllium Case Registry (ATSDR 2002). Increased mortality as a result of non-specific nephritis, kidney failure or nephrosclerosis was detected among 9225 workers of US beryllium plants (SMR 1.49, 95 % CI 1.00–2.12) (Ward *et al.* 1992).

Liver: Autopsies of 17 persons exposed to beryllium revealed liver necrosis only in one worker (ATSDR 2002).

Body weight: Weight losses occurred in 128 of 170 berylliosis patients (van Ordstrand *et al.* 1945) and in 17 workers exposed to beryllium (no other details) (ATSDR 2002).

Blood: There were no unusual findings in one machinist with chronic berylliosis and 170 workers exposed to beryllium (ATSDR 2002).

4.3 Local effects on skin and mucous membranes

4.3.1 Skin

As a result of their irritant effects, soluble beryllium compounds are capable of producing pronounced, poorly healing dermatitis. If, due to injuries or an impaired skin barrier, beryllium salt solutions or undissolved particles penetrate the skin, ulcers or necrosis may develop (DeNardi *et al.* 1953).

4.3.2 Eyes

Conjunctivitis, possibly in connection with contact dermatitis on the face, developed in 42 workers exposed to beryllium after direct eye contact with beryllium solutions (van Ordstrand *et al.* 1945). A pair of twins exposed to beryllium complained about the reduced formation of tear fluid (ATSDR 2002).

4.4 Allergenic effects

4.4.1 Sensitizing effects on skin

Up to the middle of the 20th century, beryllium-extracting or beryllium-processing plants frequently reported allergic contact dermatitis from beryllium salts and, more rarely, from beryllium oxide and beryllium (Curtis 1951, DeNardi *et al.* 1953). Patch tests using 2 % and 1 % beryllium fluoride in water (pH 5–5.5) yielded a positive result in all 13 patients with dermatitis from contact with beryllium fluoride or metallic beryllium. Twelve of 13 and 5 of 13 patients reacted to test preparations containing 0.1 % and 0.01 % beryllium fluoride. Reactions to equivalent preparations of beryllium sulfate, beryllium chloride and beryllium nitrate were observed less frequently (Curtis 1951). In a later investigation, 111 cases from an industrial plant refining beryllium (salts) were reported in whom patch testing was carried out with 1 % disodium tetrafluoroberyllate (Na_2BeF_4) in a hydrophilic ointment base (Nishimura 1966). More recent communications report cases of sensitization in dental technicians and individual patients from dental prosthesis material containing beryllium (Haberman *et al.* 1993, Schönherr and Pevny 1985, Vilaplana *et al.* 1992, 1994).

Up to the middle of the 20th century, occupational injuries particularly in the form of cuts from fluorescent lamps containing e.g. zinc-manganese-beryllium silicates, were held responsible for keloid-like, poorly healing or cicatrizing granulomas produced by beryllium salts or beryllium oxides which had entered the skin (DeNardi *et al.* 1953, Dutra 1949, Fisher 1953, Folesky 1967, Nichol and Dominguez 1949, Rietschel and Fowler 1995). In other cases, beryllium granulomas also occurred as sequelae of skin injuries due to processing beryllium or alloys containing beryllium (Dutra 1949, Sneddon 1955).

Reactions to patch tests can also develop into poorly healing granulomatous skin changes, at times making topical or intradermal administration of steroids necessary (Bobka et al. 1997). Patch tests using 0.5 mol aqueous solutions of beryllium sulfate, beryllium chloride and beryllium nitrate produced sensitization in 2 investigators who used themselves as controls (Zschunke and Folesky 1969). In another investigation, testing with 2 %, 1 % and 0.1 % beryllium fluoride in water produced, in 8 of 16 control persons, a "flare up" indicative of sensitization resulting from the patch test. Here also, corresponding reactions to beryllium sulfate, beryllium chloride and beryllium nitrate were observed less frequently (Curtis 1951). Instead of the patch test, the lymphocyte transformation test (LTT) can demonstrate *in vitro* cutaneous sensitization to beryllium; however, in the meantime this test is principally employed as an important diagnostic criterion for chronic (pulmonary) berylliosis.

In a maximization test, 22 volunteers were subjected five times to an occlusive, 48-hour induction treatment with 5 % beryllium sulfate. Two weeks later, after occlusive provocation with 1 % beryllium sulfate in petrolatum lasting for 48 hours, sensitization was demonstrated in 18 of these 22 volunteers (Kligman 1966).

4.4.2 Sensitization of the respiratory passages

Acute berylliosis

Inhalation of beryllium compounds—particularly beryllium fluoride or beryllium oxide—at high concentrations mostly above 100 µg/m^3 (DeNardi et al. 1953, Eisenbud 1998, Eisenbud et al. 1948) can result in acute inflammation of the respiratory tract in the form of tracheobronchitis and pneumonitis, producing pulmonary oedema in severe cases. Apart from particularly severe cases, symptoms including dyspnoea, coughing and pain in the chest are usually completely reversible. This form of disease was first observed in workers employed in the extraction and enrichment of beryllium and/or beryllium compounds from ores containing beryllium; but, in the meantime, it no longer occurs apart from cases of high accidental exposure. After a latency period of up to 5 years and more (up to 30 years in individual cases), chronic pulmonary disease may also occur in a small number of those affected, which was, in particular, first observed in persons employed in the manufacture of fluorescent lamp tubes (Eisenbud and Lisson 1983, Hardy and Tabershaw 1946). As a rule, however, this chronic berylliosis occurs in employees without preceding acute pulmonary disease caused by high beryllium contamination and is consequently, in most cases, the result of chronic exposure to low beryllium concentrations.

Chronic berylliosis

Even exposure to beryllium dating back over many years can be the cause of a chronic pulmonary condition which, under certain circumstances, may at first be falsely diagnosed as sarcoidosis, a condition which cannot be distinguished clinically from chronic berylliosis. In certain cases, it is not possible to determine an initially unrecognised low-level exposure to beryllium until after establishing a particularly detailed occupational case history including several decades prior to diagnosing suspected berylliosis. Demonstrating the presence of beryllium in tissue (Williams and Wallach 1989) or in urine (Apostoli and Schaller 2001) is complicated and, as a criterion, contested due to the fact that the causal exposure can be minimal and reach far back into the past and very low, though pathogenetically relevant, concentrations may not be detected (Schreiber et al. 1999).

Chronic berylliosis is characterized by the formation of non-caseating granulomas in which, histologically, the presence of epithelioid cells and $CD4^+$-T lymphocytes in particular is detectable. For diagnosis, X-ray examination is not sufficiently sensitive. Neither does computer tomographic investigation, as sole criterion, cover all clinically relevant cases, nor does it allow for a clear delineation from sarcoidosis. Functional restrictions, which are also manifestly similar to those found with other interstitial pulmonary conditions, include a reduction in vital capacity and total capacity of the lungs as well as a reduced diffusion capacity (Newman et al. 1996, Rossman 1996, Schreiber et al. 1999). Among other factors, proinflammatory cytokines such as tumour necrosis factor-α, interleukin-6, interleukin-2 and interferon-γ (Tinkle and Newman 1997, Tinkle et al. 1996, 1997, 1999) are being discussed as possible mediators of berylliosis and, as a possible additional marker for the disease, an increased concentration of serum neopterin (Harris et al. 1997). Chronic berylliosis is thus comparable to sarcoidosis from an immunopathological viewpoint as well. The disease is, as a rule, progresssive—even when further exposure to beryllium is avoided—and may result in complete pulmonary collapse (Williams 1996). It may be possible that beryllium stored in the tissue is responsible for the progressive immunological process.

Susceptibility to the disease is evidently also determined via a special genotype for the human leukocyte antigen HLA-DP: the coding of this antigen, which probably plays a major part in the presentation of beryllium to the sensitized T-lymphocytes (Amicosante et al. 2001), can take place via numerous alleles which, phenotypically, result mainly in a variation in the β-chain. In approximately 73 %–97 % of the patients with contracted berylliosis, alleles were found coding for a glutaminic acid residue at position 69 of the β-chain in the HLA-DP molecule (HLA-DPB1Glu69), but only in approximately 30 %–48 % of those who had been exposed but had not contracted the disease (Richeldi et al. 1993, 1997, Rossman 2001, Saltini et al. 1998, 2001, Stubbs et al. 1996, Wang et al. 2001).

Lymphocyte transformation test (LTT)

As, pathogenetically, berylliosis is manifest in the clinical picture of an allergic reaction of delayed type in the lungs, mediated by Th1 cells, detection of sensitization to beryllium can confirm the clinical diagnosis. Sensitization is detectable in approximately 1 %–5 % of those exposed to beryllium and, depending on the workplace involved, manifest in 3 %–16 % of these (Schreiber et al. 1999). Although, in earlier reports (Curtis 1959, DeNardi et al. 1953), the patch test was also used to provide evidence of (pulmonary) sensitization to beryllium, the lymphocyte transformation test with peripheral lymphocytes has in the meantime acquired validity as the most suitable method for demonstrating sensitization. A number of publications have reported on the high sensitivity and specificity of the test, and in 2 investigations a positive predictive value of 100% was cited, in contrast to more recent findings (Maier 2001, Newman 1996). According to a number of other investigations, false-negative findings, but also sporadic false-positive findings, were frequently obtained (Kreiss *et al.* 1997, Markham 1996, Schönherr and Pevny 1985, Schreiber *et al.* 1999), and divergent findings are frequently obtained in different laboratories (Deubner *et al.* 2001a, Stange *et al.* 1996a). According to one investigation involving a small number of cases, a false-negative result was obtained in up to 54 % of the cases (Markham 1996). In another investigation, a great variability in the frequency of consistent findings was obtained from 3 different laboratories. The highest positive predictive value (49 %) of the test for diagnosed berylliosis was found in patients whose LTT yielded a positive result in the first test in 2 laboratories (Deubner *et al.* 2001a). Although greater sensitivity can be obtained using lymphocytes from the bronchoalveolar lavage fluid (Paustenbach *et al.* 2001, Rossman 2001), false-negative findings are here also possible when there is a high proportion of macrophages, particularly in the case of smokers (Newman 1996, Schreiber *et al.* 1999). Epicutaneous and intracutaneous tests conceal the danger of iatrogenic sensitization (and thus of false-positive findings), and also possible aggravation of the pulmonary symptoms (Schönherr and Pevny 1985). Their diagnostic importance cannot be assessed conclusively (Müller-Quernheim *et al.* 1996, Rossman 1996, Schreiber *et al.* 1999). In 6 patients, in whom LTT investigations with lymphocytes from bronchoalveolar lavage fluid yielded positive results but not, however, LTT with peripheral lymphocytes, a reaction was nevertheless observed in a patch test using a 1 % beryllium sulfate solution in water. No subsequent effects on pulmonary symptoms occurred in the patients tested in this investigation (Bobka *et al.* 1997).

Case reports

Berylliosis was diagnosed clinically and histologically in 2 employees who processed beryllium/copper alloy with a beryllium content of 2 %. Pulmonary biopsies from both patients showed non-caseating granulomas and LTT investigations with both peripheral lymphocytes and lymphocytes from bronchoalveolar lavage yielded positive results. Both patients reported multiple skin injuries incurred occupationally, which healed only slowly. Analyses in the relevant work areas revealed beryllium concentrations of less than $2\,\mu g/m^3$ at all times. However, no information was given on possible exposure peaks, or the extent of exposure to beryllium from other sources at a time (or times) preceding the investigation (Balkissoon and Newman 1999).

A dental technician, who had been working regularly for 20 years with beryllium and a chromium/cobalt/molybdenum alloy as well as other substances, developed occupational dyspnoea with cough. His pulmonary function was reduced: the total capacity, forced vital capacity, forced expiratory volume in the first second and diffusion capacity were around 50 % to 60 % of the expected value. Histologically, non-caseating granulomas with incorporation of foreign bodies were found in the material from pulmonary biopsy. Electron microscopy revealed silicates, aluminium compounds and the alloy used. LTT investigation of lymphocytes obtained from bronchoalveolar lavage fluid yielded positive results, yet negative results in the case of peripheral lymphocytes (Brancaleone et al. 1998).

Sarcoidosis was diagnosed in a dental technician who was a smoker and had initially been free of symptoms. Non-productive cough, a progressive loss in vital capacity and diffusion capacity required medication with corticoids, which could be completely discontinued only after 14 years. As, in the meantime, exposure to beryllium for at least 16 years was assumed (no other details), histological investigations were carried out on transbronchial biopsy material, which revealed the presence of epithelioid cell granulomas. LTT investigations of lymphocytes from bronchoalveolar lavage fluid resulted in borderline findings, while the LTT with peripheral lymphocytes was, in contrast, positive. In intracutaneous tests with 0.1 ml 0.01 % and 0.05 % beryllium sulfate respectively, papulous-erythematous reactions with a diameter of 3 or 6 mm appeared after 4 days. After 4 weeks, in material from excisions carried out in the test areas, epithelioid cell granulomas were found as manifestations of beryllium sensitization (Müller-Quernheim et al. 1996).

Cohort studies

In a large-scale beryllium-processing plant in the USA (Rocky Flats, Colorado) LTT investigations were performed between 1991 and 1994 in 1885 of the currently employed 8772 workers. Positive results were obtained in 28 cases, 6 of these being diagnosed as berylliosis. The tests then carried out in 452 workers over the subsequent 3 years yielded positive results in 8 cases, of which one was found to have the clinical symptoms of berylliosis. Of the total 9865 formerly employed persons, 2512 were investigated using the LTT, 47 of these cases with positive results and 22 of these with clinical symptoms (Stange et al. 1996b).

In around 290–432 air samples taken every year from the main building of a processing plant, the following values were determined in the proximity of the machines concerned: an average beryllium concentration of 0.23–0.42 µg/m^3 between 1970 and 1974, of 0.10–0.16 µg/m^3 between 1975 and 1982, of 0.16 to 0.27 µg/m^3 from 1983 to 1986 and, after installation of a new air filter system, concentrations of 0.03 and 0.05 µg/m^3 in 1987 and 1988. In the years 1984, 1985, 1986 and 1987, personal air monitoring in the breathing zone produced average values of 1.09, 1.20, 0.46 and 0.19 µg/m^3 and, in 15 % of the determinations, exposure levels of 2 µg/m^3 and more (Barnard et al. 1996, Stange et al. 1996a, 1996b).

In another beryllium-processing plant, positive LTT results were obtained in 15 of 235 employees investigated between 1995 and 1997, chronic berylliosis being diagnosed in 8 of these 15 and, in one further case, a diagnosis of suspected berylliosis being given. Sensitization was found within the first 3 months of employment in 4 of the 15. Repetitions of the tests were carried out on a biennial basis up to 1997 in 187 and up to 1999 in 109 initially non-sensitized employees. These provided evidence of sensitization in 5 of the 187 and 2 of the 109 investigated persons, respectively, in 5 of whom the clinical picture of berylliosis and, in one, suspected berylliosis were diagnosed. Analyses of the air in the proximity of the machines yielded a median exposure of 0.3 µg/m^3 (Newman et al. 2001). Beryllium sensitization or chronic berylliosis was observed particularly in those employees machining materials containing beryllium (e.g. lathe operating, grinding, milling). Thus, in an investigation complementing the study cited, beryllium sensitization or berylliosis was determined in 11.5 % of those persons working at machines of these types, and in 2.9 % of those not occupied in such a manner. For the 8 sensitized employees and 12 employees with chronic berylliosis a median cumulative exposure of 2.6 µg/m^3 per year was determined and for 206 control persons a median cumulative exposure of 1.2 µg/m^3 per year. Personal air sampling in a total of 100 cases showed that those persons working with machines were exposed to markedly higher levels as well as an increased proportion of particles with diameters of 1 µm and below than the remaining employees. In both groups, the median lifetime-weighted exposure (LTWE) was 0.25 µg/m^3. No case of sensitization was found among the 22 employees with an LTWE below 0.02 µg/m^3, whereas 1 employee of the 33 with an LTWE below 0.035 µg/m^3 was sensitized. An LTWE of above 0.2 µg/m^3 was cited for 12 of 20 sensitized persons (Kelleher et al. 2001).

An increased prevalence of beryllium sensitization was also observed, in 1992, in workers employed at machines in a ceramics plant processing beryllium oxide; this correlated qualitatively with the median results of the exposure data: among the 136 persons investigated, 8 were found to be sensitized, 6 of these with clinical symptoms, of whom 7 were from the collective of 49 persons working at the machines. The daily weighted average estimates of exposure were below 1 µg/m^3, although 8 % of the activities involved exposure to levels of more than 2 µg/m^3 and up to 14 µg/m^3 (Kreiss et al. 1996). A further investigation carried out in 1998 on 151 of the total of 167 employees concerned yielded similar results, although measures to reduce exposure had been carried out in the meantime. Positive LTT results were obtained in 8 of the 77 persons employed before 1992 and in 7 of the 74 persons employed subsequent to 1992. Berylliosis was diagnosed in only 1 of the 74 subjected to short-term exposure, but in 7 of 77 employees subjected to long-term exposure. Granulomas and positive LTT results with cells from BAL fluid were reported for 4 of 7 persons with long-term exposure, granulomas were found in one employee, but only one borderline positive BAL-LTT result, and one positive BAL-LTT result in 2 employees without evidence of granulomas. In 12 of the 15 sensitized employees, a positive result in the LTT was determined for the first time in 1998. Also in this investigation, a high level of sensitization was determined using the LTT in employees working at the machines on a long-term basis (7 of 40, as compared with 2 of 36 in those employed since 1992). However, most of the 8 sensitized long-term employees also performed work in other areas associated with an increased prevalence of sensitization. A median exposure of 0.28 µg/m^3 (peak exposure: 6.1 µg/m^3) was reported for the 74 persons employed after 1992, and of 0.39 µg/m^3 (peak exposure: 14.9 mg/m^3) for the 77 persons employed prior to 1992. The average exposure was above 2 µg/m^3 in only 4 workers, also including the one worker not employed until after 1992 who had contracted the disease (Henneberger et al. 2001).

In an industrial plant processing beryllium, beryllium alloys and beryllium oxide, positive LTT results were found in 59 of the 627 persons investigated between 1984 and 1993, which, however, were not reproducible in all cases. Bronchoscopy was carried out in 47/59 resulting in a diagnosis of berylliosis in 24 cases, of which in 4 cases, however, no histological evidence was found for granulomas. The presence of sensitization was assumed in 19 further cases. Including 5 cases already diagnosed, the authors give 29/632 (4.6 %) as the prevalence for the disease. A higher prevalence was associated particularly with employment in the manufacture of ceramics (9.3 %; 14 of 150 employees) as well as in beryllium production (work in this field started after cessation of ceramics production: 4 of 50 employees compared with 1 of 140 employees from other fields) (Kreiss *et al.* 1997). A later analysis of 55 workplace and 53 personal air samples in the vicinity of 5 smelting furnaces provided evidence of a correlation between the frequency of sensitization and/or disease, and the amount of particles able to reach the alveoli (diameter less than 10 µm or 3.5 µm, respectively) in the samples taken on site. However, regarding the amount of fine dust particles in the personal air samples and the extent of total dust exposure, the authors were not able to demonstrate a correlation of this kind (Kent *et al.* 2001). It is possible that the total dust exposure is not so much responsible for development of the disease as is the relative number of respirable particles (McCawley *et al.* 2001).

In contrast to the findings listed above from production plants, in an investigated collective consisting of 75 workers in a plant for processing beryllium ores employing a total of 87 persons, sensitization was found in 2 employees and berylliosis in one case (which, however, can possibly be attributed to previous exposure in another plant). The 3 sensitized employees, but also 36 of the remaining 72 employees, reported high exposure as a result of accidents. Exposure levels determined between 1970 and 1999 in the relevant work areas revealed a median exposure of 0.1–0.6 µg/m^3 with peaks of 5.6–234.5 µg/m^3. Breathing zone and personal air samples produced median values of 0.3–2.5 µg/m^3 and 0.05–0.8 µg/m^3 with maximum values of 4.3–271.2 µg/m^3 and 0.05–165.7 µg/m^3, respectively (Deubner *et al.* 2001b).

Of a total of 57 gemstone cutters and gemstone grinders from 12 different workshops, 27 were employed for an average 25.5 ± 15.8 years every week for an average 21.1 ± 12.5 hours per week in the processing of beryl, an aluminium/beryllium silicate, the remaining 30 over approximately the same period for an average of 0.7 ± 1.2 hours per week. An exposure range from below 0.4 to a maximum of 20 µg/m^3 was reported for beryllium. In the workshop with the highest exposure (about 4 or 20 µg/m^3 in two of 4 analyses), beryllium was found in the urine of all 9 employees examined (an average of 0.18 ± 0.15 µg/l before starting work; an average of 0.12 ± 0.10 mg/l after the end of work). No indications of berylliosis were found in any of the employees after clinical, radiological and spirometric investigations. The LTT result was positive in one employee without pulmonary symptoms who had been employed for 47 years. On the other hand, in the employees in whose urine beryllium had been found (detection limit 0.06 µg/l), no significant differences were found in the LTT when they were compared with those employees in whom no beryllium was found in the urine (Wegner *et al.* 2000).

Small collectives of 6–25 employees from two companies were exposed to average beryllium concentrations of 0.01 to 1.85 µg/m^3 during the manufacture or processing of beryllium/copper alloys between 1992 and 1995. An increase in lymphocyte proliferation of more than 200 % was determined in each individual year in up to 17 % of those tested using the LTT with beryllium sulfate (a total of 12/83 employees). No increase could be found, however, in any of the 28 employees who had been exposed, between 1993 and

1994, to beryllium concentrations below 0.01 µg/m³ (Yoshida et al. 1997). However, as the findings obtained in the LTT over the individual years were reproducible in only a few cases, and as no details were given on previous exposure to beryllium (where relevant), no correlation can be derived from the exposure levels and percentage of sensitizations.

According to an evaluation of 217681 personal air samples from 194 persons employed between 1981 and 1997 at a beryllium processing plant in which nuclear weapons were manufactured, where a total of more than 400 persons were employed between 1961 and 1997, an average exposure of 0.11–0.72 µg/m³ (95 % percentile: 0.22–1.89 µg/m³) was determined in the individual years, with an average of up to 2.04 µg/m³ in the foundry area (95 % percentile up to 8.61 µg/m³). Analyses of 367757 workplace-related samples yielded values of 0.02–0.32 µg/m³ (95 % percentile: below 0.5 µg/m³) with a maximum value of 1128 µg/m³ in the foundry area. In most of the determinations, values below 2 µg/m³ were obtained, although, in up to 18.5 % of the personal air samples in the foundry area, values over 2 µg/m³ were also obtained. The highest workplace-related exposure (0.87 µg/m³ on average; 95 % percentile: up to 2.9 µg/m³) was determined in the foundry workers. Berylliosis was diagnosed using clinical parameters in only one case. Because LTT investigations were not carried out in the employees of this plant on a routine basis (Johnson et al. 2001), it is not possible to give a definitive sensitization status for this collective.

Furthermore, cases of sensitization and disease were also reported in persons living in close proximity to beryllium-processing plants but who had no occupational contact with beryllium. Exposure to beryllium at concentrations between 0.01 and 0.1 µg/m³ was already discussed as an induction threshold for the disease as early as 1949 (Eisenbud et al. 1949). The observed cases of disease can, however, at least in persons related to workers in beryllium processing plants, be attributed to exposure to beryllium (and beryllium compounds), resulting, for example, from particles on employees' clothing or in vehicles used by them (DeNardi et al. 1953, Eisenbud and Lisson 1983, Eisenbud et al. 1949, Sanderson et al. 1999). For this reason as well, but especially due to the different methods of determination and an inhomogenous composition of the dust involved, comparison of the prevalences observed from the different exposure data cited in these investigations is hardly possible.

4.5 Reproductive and developmental toxicity

There are no data available for toxic effects on fertility.

In a case control study of 2096 mothers and 3170 fathers of still-born babies, 363 mothers and 552 fathers of premature babies, and 218 mothers and 371 fathers of "small-for-gestational-age" babies, it was not possible to determine a connection between occupational exposure of the parents to beryllium and the risk of a still or premature birth or the birth of a child that was small for its gestational age (Savitz et al. 1989).

4.6 Genotoxicity

Investigations of the induction of sister chromatid exchange and micronuclei in jewel polishers exposed to beryllium for more than four hours per week did not reveal any differences compared with persons exposed for four hours or less per week. A control group not exposed to beryllium was not investigated. Personal air sampling yielded beryllium concentrations of up to 20 µg/m^3 in individual cases (Wegner et al. 2000).

4.7 Carcinogenicity

Until 1966 the US Beryllium Case Registry did not yield any evidence of carcinogenic effects of beryllium in man, although the difficulty of providing such evidence was brought to attention. Of the 60 persons with chronic berylliosis treated in two hospitals in Massachusetts between 1944 and 1966, 18 patients had died by 1969, without one tumour having been observed in them (ATSDR 2002).

4.7.1 Case-control studies

In a population-based case-control study of lung cancer in Hawaiian men, 261 newly diagnosed cases of primary lung carcinomas in a population-based tumour register were evaluated between September 1979 and July 1982. 444 controls corresponding in age and sex were chosen via randomly selected telephone numbers (RDD). Information about the occupational activity and smoking habits of the persons was collected in interviews. The level of occupational exposure to coal tar, petroleum, arsenic, chromium, asbestos, nickel and beryllium was evaluated by means of a job exposure matrix, and was classified in one of three exposure groups (no, low, or high exposure). Occupational groups with beryllium exposure were mechanics, electricians, toolmakers, metal workers, automobile mechanics, welders, heating installers, transportation workers, physicians and maintenance workers in the transportation business. The relationship between the individual substances and lung cancer was given as odds ratios (OR), which were determined on the basis of a multiple regression analysis taking account of age, ethnic group and smoking habits, but without adjustment for the other occupational carcinogens. For beryllium a relationship was found between lung cancer and low exposure (OR 1.62, 95 % CI 1.04–2.51) and also between lung cancer and high exposure (OR 1.57, 95 % CI 0.81–3.01) (Hinds et al. 1985).

4.7.2 Cohort studies

Table 1 contains an overview of the most important cohort studies.
In a first cohort study of the cancer-inducing effects of beryllium in man, 3685 male workers employed between 1937 and 1948 in beryllium plants in Lorain, Ohio and Reading, Pennsylvania, were investigated. The reports of the US Social Security Admini-

stration up to 1966, registered beryllium diseases and cases of death from cancer were evaluated (Mancuso and El-Attar 1969). In a second phase, additional statistical analyses were carried out, which considered the length of employment of the workers and extended the observation period to the year 1967 (Mancuso 1970). In a third phase, the data of a cohort with an employment period from 1942 to 1948 and an observation period up to 1975 were evaluated. In employees who had been employed for 15 years or more, significantly increased mortality from lung cancer was found (Lorain: SMR 2.0, 95 % CI 1.3–3.1; Reading: SMR 1.5, 95 % CI 1.0–2.1) (Mancuso 1979). A fourth study included an extended employment and observation period. If these cohorts are subdivided according to the duration of employment, mortality from lung cancer is significantly increased for workers employed for less than 12 months (SMR 1.38, $p < 0.05$) or longer than 49 months (SMR 2.22, $p < 0.01$), but not for workers employed for between 13 and 48 months (SMR 1.06) (Mancuso 1980).

Table 1. Cohort studies (including embedded case–control study) of the carcinogenicity of beryllium compounds (according to IARC 1993), concrete data for the exposure concentrations are not evident in the literature

Cohorts	Cohort size	Length of employment	End of observation	Comparison population	SMR	95 % CI	Cases of lung cancer [n]	References
Lorain	3685	1942–48	1974	white, male Americans	1.8[a]	1.2–2.7	25	Mancuso 1979
Reading		1942–48	1975		1.25[a]	0.9–1.7	40	
both together					1.42[a]	1.1–1.8	65	
Lorain Reading	3685	1937–48	1976	viscose-rayon workers	1.40	[1.1–1.7]	80	Mancuso 1980
Reading	3055	1942–67	1975	white, male Americans	1.25[a]	0.9–1.7	47	Wagoner et al. 1980
BCR: total	421	entries in the register 1952–75	1975	white, male Americans	[1.93]	[0.8–4.0]	7	Infante et al. 1980
acute BD					2.86[a]	1.0–6.2	6	
chronic BD					0.66[a]	0.1–3.7	1	
BCR: total	689	entries in the register 1952–80	1988	Americans; ♂,♀	2.00	1.33–2.89	28	Steenland and Ward 1991
acute BD					2.32	1.35–3.72	17	
chronic BD					1.57	0.75–2.89	10	
7 beryllium plants	9225	1940–69	1988	Americans; ♂	1.26	1.12–1.42	280	Ward et al. 1992
Reading	3569	1940–1996	1992	5 controls corresponding in age and sex per lung cancer case	1.22	1.03–1.43	142	Sanderson et al. 2001b

BD = beryllium disease, BCR = US Beryllium Case Registry, CI = confidence interval, SMR = standard mortality ratio, [ ] value calculated by the IARC working group, [a]Saracci adjustment

A National Institute for Occupational Safety and Health (NIOSH) group evaluated company reports and the causes of death connected with beryllium from the study of the Reading (Pennsylvania) factory. The 3055 white men were employed in the beryllium plant in the period between 1942 and 1967 and were observed until the end of 1975. The highest mortality from malignant neoplasms of the lungs, trachea, and bronchi was found in the workers employed for more than 5 years, and in those for whom more than 25 years had passed since the beginning of their employment (Wagoner et al. 1980).

The studies of Mancuso (Mancuso 1970, 1979, 1980, Mancuso and El-Attar 1969) and Wagoner et al. (1980) were limited by the lack of data for interfering factors such as smoking status and exactly defined areas of activity and were not able to clearly demonstrate a relationship between beryllium exposure and lung cancer (MacMahon 1994, US EPA 1998).

In the evaluation of the case reports of 421 white male workers that were registered in the Beryllium Case Registry between 1952 and 1975, the mortality from lung cancer was compared with the corresponding mortality in American white men. Death from lung cancer in registered persons with beryllium disease was not statistically more frequent. However, if only persons who had previously suffered from acute beryllium disease were considered, mortality from lung cancer was significantly increased (Infante et al. 1980).

A follow-up study over 13 more years also considered women. The cohorts included 689 patients (66 % male), 34 % of whom suffered from acute beryllium disease and 64 % from chronic beryllium disease (2 % uncertain diagnoses). The number of cases of death from lung cancer among the patients was increased, but no trend could be found either with respect to the duration of employment or the time span since the beginning of employment. On the basis of an investigation of the smoking habits of 32 % of the cohort members in 1965, smoking was not considered to be an interfering factor. Death from lung cancer was more frequent in workers who had acute beryllium disease (SMR 2.32, 95 % CI 1.35–3.72) than in those with chronic beryllium disease (SMR 1.57, 95 % CI 0.75–2.89) (Steenland and Ward 1991).

The employees of the factories in Lorain and Reading as well as those of five further plants in Pennsylvania and Ohio were included in a comprehensive retrospective cohort study. 9225 men (8905 Caucasian, 320 of other ethnic groups) who had worked in the plants for at least two days between 1940 and 1969 were considered. The vital status of the participants in the study was determined up to 1988 by means of reports from different authorities and the national death register and was known for up to 3.3 % of the participants in the study; death certificates were available for up to 1.4 % of all of the participants who had died. It was possible to supplement information concerning the smoking habits for 1466 participants in the study (15.9 %) on the basis of an investigation from 1968. These data were compared with the information for the smoking habits of the entire US population between 1965 and 1970, in order to examine the influence of smoking on the mortality. For malignant neoplasms of the trachea, the bronchi and the lungs, the SMR for the entire cohort was 1.26 (95 % CI 1.12–1.42). In the Lorain plant the highest SMR was determined to be 1.69 (95 % CI 1.28–2.19). The Reading plant also had a significantly increased SMR of 1.24 (95 % CI 1.03–1.48). The SMR for lung cancer was slightly increased in four of the five other plants, but not significantly so. The lung cancer SMR did not increase with the duration of employment, but with the time interval since beginning employment. The SMR was 1.32 (p < 0.01) for workers who were employed for less than one year. A time span of over 30 years since beginning employment resulted in the highest SMR of 1.46 (p < 0.01). Neither the use of the local cancer death frequencies nor consideration of the smoking habits decisively changed the SMR. The majority of deaths from lung cancer originated in the plants in Lorain and Reading. The reason for this was, in part, that 54 % and 58 %, respectively, of the persons employed there had latency periods of more than 15 years. The beryllium exposure in these two plants was markedly higher than in the five others. In the 1940s determination of the beryllium exposure in the Lorain plant yielded values of 411 µg/m^3 at a mixing workplace and up to 43300 µg/m^3 in the breathing zone at an alloying workplace. Probably for this reason the start of employment also had an influence on the lung cancer rate. When the cohorts were divided into three groups corresponding to the start of employment, the group employed before 1950 had the highest SMR with 1.42 (95 % CI 1.25–1.62), compared with those employed between 1950 and 1959 (SMR 1.24, 95 % CI 1.03–1.49), and between 1960 and 1969 (SMR 0.62, 95 % CI 0.4–0.92). It is of interest that of 98 workers at the Lorain plant who suffered from beryllium disease (91 acute, six chronic, one unclear), eleven died from lung cancer (SMR 3.33, 95 % CI 1.66–5.95). Among the other 1094 Lorain workers there were 46 deaths from lung cancer (SMR 1.51, 95 % CI 1.11–2.02) (Ward et al. 1992).

The lack of data for beryllium exposure from the 1940s and 50s was regarded as a limitation (Sanderson et al. 2001a) of the previously described retrospective cohort studies (Mancuso 1970, 1979, 1980, Wagoner et al. 1980, Ward et al. 1992). A job exposure matrix was developed based on the results of over 7300 determinations in air at the Reading beryllium plant, around 7150 of which came from the years after 1971. This was carried out without previous knowledge of the time at which the workers who later died of lung cancer were employed, the plant departments in which they worked or what activities they performed. It was possible to estimate the number of days of exposure, the cumulative, average and maximum exposure to beryllium, as well as to all compounds and types of exposure (fume, vapour, aerosol) concerned for each case of lung cancer and for the corresponding controls by means of this job exposure matrix. For beryllium the cumulative exposure appears to be a good index for chronic berylliosis, as this is caused by repeated or continuous relatively low exposure. However, short-term or peak exposures are also of importance, as acute beryllium disease is caused by high beryllium concentrations (> 100 µg/m^3) (Sanderson et al. 2001a).

In another study the relationship between the level of beryllium exposure and deaths from lung cancer among the workers of the beryllium plant in Reading employed between 1940 and 1969 was investigated in an embedded case–control study. The cases of death from lung cancer were monitored in the cohort of 3569 male workers until the year 1992. Up until the extended observation period the characteristics of this cohort were the same as those in the study of Ward et al. (1992). Each of the 142 cases of death from lung cancer was compared with respect to age and ethnic group with five correspondingly selected controls. The cumulative, average and maximum exposure of each worker was estimated by means of the previously created job exposure matrix (Sanderson et al. 2001a). As was already found by Ward et al. (1992), an increased incidence of death from lung cancer was found among the beryllium workers of the

Reading plant (SMR 1.22, 95 % CI 1.03–1.43). Considering the smoking habits did not change the result. In the cases of death from lung cancer the employment period was found to be shorter and the cumulative beryllium dose lower than in the controls. However, if the last 10 and 20 years are not considered in the exposure evaluation, for the cases of death from lung cancer the exposure duration was on average longer and the cumulative beryllium doses higher than in the controls (see Table 2).

Table 2. Comparison of the mean geometric exposure of lung cancer cases and controls using different exposure parameters (Sanderson *et al.* 2001b)

Exposure parameter	Lung cancer cases (n = 142)		Controls (n = 710)		
	Geometric mean	(GSD)	Geometric mean	(GSD)	p-value
duration of employment [days]	202.1	(9.4)	328.0	(9.4)	0.019
duration of employment [days], without considering the last 10 years	178.4	(19.7)	133.0	(19.7)	0.284
duration of employment [days], without considering the last 20 years	58.4	(40.6)	31.3	(40.6)	0.067
cumulative exposure [µg/m^3·days]	4606	(9.3)	6328	(9.3)	0.123
cumulative exposure [µg/m^3·days], without considering the last 10 years	4057	(38.9)	2036	(38.9)	0.041
cumulative exposure [µg/m^3·days], without considering the last 20 years	844	(134)	305	(134)	0.024
average exposure [µg/m^3]	22.8	(3.4)	19.3	(3.4)	0.142
average exposure [µg/m^3], without considering the last 10 years	22.6	(6.6)	12.3	(6.6)	0.0005
average exposure [µg/m^3], without considering the last 20 years	10.2	(11.9)	5.3	(11.9)	0.004
maximum exposure [µg/m^3]	32.4	(3.8)	27.1	(3.8)	0.150
maximum exposure [µg/m^3], without considering the last 10 years	30.8	(7.6)	16.1	(7.6)	0.0005
maximum exposure [µg/m^3], without considering the last 20 years	13.1	(13.9)	6.5	(13.9)	0.004

GSD = geometric standard deviation

The persons who died from lung cancer had, therefore, been exposed to high beryllium concentrations over a certain period of time more than 10 or 20 years before their death. This is supported by the fact that the majority of the persons who died from lung cancer were employed in the beryllium plants in the 1940s and 50s, a time when the beryllium concentration in the air at the Reading plant was extremely high, among other things, because there was no threshold limit value. The average and maximum levels of exposure to beryllium in the persons who died from lung cancer were higher than those in the controls. When the last 10 or 20 years were disregarded, this difference was significant ($p < 0.01$). Thus, the average beryllium exposure gives a better indication of the lung cancer risk than the cumulative beryllium exposure, as highly exposed persons often worked at the beryllium plant for only a short time. In these studies the authors were not able to show a direct relationship between lung cancer deaths and previous acute or chronic beryllium disease as a result of the limited information available concerning the number of cases of beryllium disease (Sanderson *et al.* 2001b).

The studies of Sanderson *et al.* (2001a, 2001b), in which the relevant exposure periods and levels were evaluated and taken into account, show that the lack of a relationship between the duration of employment and the incidence of deaths from lung cancer, criticised by MacMahon (1994), could be due to the fact that, in the case of beryllium, high short-term exposure seems to be a relevant risk factor for lung cancer.

A reanalysis of the retrospective cohort study of Ward *et al.* (1992) of 9225 employees from 7 beryllium plants in Pennsylvania and Ohio criticised that in this study the number of expected cases of death from lung cancer was too low, as persons from the entire county were selected as the control group, although the majority of the employees in the beryllium plants lived in the surrounding cities. Using correction factors evaluated from the estimated incidences only in towns, an SMR was calculated for the Lorain plant of 1.14 (95 % CI 0.86–1.48) instead of 1.69 (95 % CI 1.28–2.19) and for the Reading plant of 1.07 (95 % CI 0.89–1.28) instead of 1.24 (95 % CI 1.03–1.48) (Levy *et al.* 2002).

These SMRs from the reanalysis by Levy *et al.* (2002) must be regarded as too low, as 32 % of the employees of the Reading plant did not live immediately in the city, and Lorain City was described by the authors as highly industrialized, and thus other occupationally related cases of lung cancer could lead to a falsely high reference rate. The reanalysis also used other smoke correction factors for calculating a lung cancer risk adjusted for smoking. In the reanalysis, taking into account the above points led to lower SMRs for lung cancer and only in a few scenarios to increased SMRs for lung cancer which were statistically significant.

The authors of the reanalysis cited the work on exposure evaluation by Sanderson *et al.* (2001a), but not the embedded case-control study (Sanderson *et al.* 2001b) published in the same volume, which takes account of the above criticism. This analysis (Sanderson *et al.* 2001b), on which the evaluation of beryllium as a human carcinogen is essentially based, showed that in the case of employees of the Reading beryllium plant the persons who died of lung cancer had been exposed to high beryllium concentrations 10 or 20 years before their death.

5 Animal Experiments and *in vitro* Studies

5.1 Acute toxicity

5.1.1 Inhalation

Inhaled beryllium is highly toxic. In rats the four-hour LC_{50} for **beryllium sulfate** was 0.15 mg beryllium/m^3 and for **beryllium phosphate** 0.86 mg beryllium/m^3. In guinea pigs the four-hour LC_{50} for **beryllium phosphate** was 4.02 mg beryllium/m^3 (Venugopal and Luckey 1978). The LC_{50} for **beryllium acetate** mist was 3 mg beryllium/m^3 for mice after exposure for two hours (WHO 1990).

After inhalation of beryllium and inorganic beryllium compounds, beryllium deposits were detected in the lungs. In the lungs of rats, mice, hamsters and dogs, inflammatory and granulomatous damage, necrosis, fibrosis, hyperplasia, thickening of the alveolar walls, or progressive peribronchial and perivascular lymphocyte and macrophage infiltration were found. In the bronchoalveolar lavage increased lactate dehydrogenase, β-glucuronidase, acid and alkaline phosphatase, and lysozyme activities were found as well as an increased protein content. These changes persisted in some cases for weeks to months (ATSDR 2002, Haley *et al.* 1989, 1990, Robinson *et al.* 1968, Sanders *et al.* 1975).

The studies of the inhalation toxicity of beryllium and its inorganic compounds are presented in Table 3.

5.1.2 Ingestion

In rats the LD_{50} for **beryllium sulfate** was 7.0, for **beryllium phosphate** 6.5, for **beryllium fluoride** 18.8, for **beryllium oxyfluoride** 18.3 and for **beryllium chloride** 9.8 mg beryllium/kg body weight. Similar values were obtained for mice (Venugopal and Luckey 1978).

5.1.3 Intravenous, intraperitoneal, subcutaneous and intramuscular injection

Independently of whether administered intravenously, intraperitoneally, intramuscularly or subcutaneously, LD_{50} values between 0.04 and 1.3 mg beryllium/kg body weight were described for rats, mice and monkeys for **beryllium nitrate**, **sulfate, fluoride, chloride, hydroxide, carbonate** and **phosphate**. The LD_{50} for guinea pigs was between 1.2 and 6.8 mg beryllium/kg body weight (Venugopal and Luckey 1978, WHO 1990).

Table 3. Acute inhalation toxicity of beryllium and its inorganic compounds

Animal species, number/group	Exposure	Findings	References
Beryllium metal			
F344 rat, 74♂	0.8 mg Be/m^3 50 min (nose only)	initial dose to the lungs: 625 µg Be; mortality 20/74 (12–15 days after exposure), acute necrotizing haemorrhagic exudative interstitial pneumonia (peak 14 days after exposure), progressing to necrosis (peak around 59 days after exposure) and chronic inflammation (115–171 days after exposure); bronchoalveolar lavage: lactate dehydrogenase increased, β-glucuronidase increased, protein levels increased (3–14 days after exposure)	Haley et al. 1990
Beryllium oxide			
F344 rat, (no other details)	0.447 mg Be/m^3 (calcined at 560°C) 1 h	lungs: inflammation with an increased number of interstitial monocytes, thickening of the alveolar septa, damage to the type-II cells; bronchoalveolar lavage: protein levels increased, lipids increased, activities of acid and alkaline phosphatases, lysozyme and lactate dehydrogenase increased	ATSDR 2002
Wistar rat, 70♀, 70♂	about 1–100 mg Be/m^3 (calcined at 1000°C) 30–180 min (nose only)	initial alveolar deposition of 62–104 µg Be; lungs: moderate formation of granulomatous lesions, dust-laden foamy macrophages, peribronchiolar and subpleural lymphocyte infiltration	Sanders et al. 1975
Wistar rat, 35♀, 35♂	about 1–100 mg Be/m^3 (calcined at 1000°C) 30–180 min (nose only)	initial alveolar deposition of 138–156 µg Be; no influence on body weights, spleen, liver, kidney or lung weights, number of red and white blood corpuscles; lungs: no subpleural or peribronchial areas with granulomatous changes, functional and structural damage to the macrophages; bronchoalveolar lavage: no unusual findings	Sanders et al. 1975
Syrian golden hamster, 30♀, 30♂	about 1–100 mg Be/m^3 (calcined at 1000°C) 30–180 min (nose only)	initial alveolar deposition of 16–17 µg Be; lungs: small areas of granulomatous reaction and structurally changed macrophages	Sanders et al. 1975

Table 3. continued

Animal species, number/group	Exposure	Findings	References
Beagle dog, 2–4 (no other details)	10 mg Be/m^3 (calcined at 500°C or 1000°C) 5–40 min (nose only)	lungs: peribronchial and perivascular lymphocyte and macrophage infiltration progressing to microgranulomas and granulomatous pneumonia, interstitial fibrosis and epithelial hyperplasia; stronger effects and shorter retention in the lungs with BeO calcined at 500°C; bronchoalveolar lavage: number of lymphocytes increased (still 3 months after exposure)	Haley *et al.* 1989
Beryllium sulfate			
F344 rat, (no other details)	≥ 3.3 mg Be/m^3 1 h (nose only)	lungs: inflammation with hyperplasia and vacuolization of the type-II cells, thickened interstitium and macrophage and leukocyte infiltration; bronchoalveolar lavage: activities of lactate dehydrogenase and alkaline phosphatase increased	ATSDR 2002
BALB/c mouse, (no other details)	7.2 mg Be/m^3 1 h (nose only)	lungs: inflammation with hyperplasia and vacuolization of the type-II cells, thickened interstitium and macrophage and leukocyte infiltration; bronchoalveolar lavage: activities of lactate dehydrogenase and alkaline phosphatase increased	ATSDR 2002
Beryllium fluoride (40 %), beryllium oxide (50 %), beryllium chloride			
Beagle dog, 1♂, 1♀	115 mg Be/m^3 20 min	transient loss of appetite, weight loss in the first 7 days after the end of exposure; lungs: inflammation with granulomatous foci, ultrastructural changes of the interstitium	Robinson *et al.* 1968

5.2 Subacute, subchronic, and chronic toxicity

5.2.1 Inhalation

In rats, mice, hamsters, guinea pigs, rabbits, cats, dogs, pigs and monkeys repeatedly exposed to **beryllium oxide, sulfate, fluoride** or **beryllium hydrogen phosphate**, above all lung damage was observed, such as pneumonitis, in some cases with infiltration of foamy, dust-laden macrophages or lymphocytes, granulomas, abscesses, necrosis, lung congestion, thickening of the alveolar walls, emphysema, fibrosis or proliferative changes (Hall *et al.* 1950, Reeves *et al.* 1967, Schepers 1964, Stokinger *et al.* 1950, Vorwald and Reeves 1959, Wagner *et al.* 1969).

The inhalation studies after repeated administration are presented in detail in Table 4.

Beryllium concentrations as low as 0.006 mg/m^3 administered as **beryllium oxide** led to pneumonitis and later to fibrotic changes (Vorwald and Reeves 1959) in rats after long-term exposure. Long-term exposure to beryllium concentrations of ≥ 0.034 mg/m^3 administered as **beryllium sulfate** also led to lung damage (Reeves *et al.* 1967, Schepers *et al.* 1957). If there was a follow-up period after the end of the exposure, the lung damage was seen to progress and later reached its maximum (Schepers *et al.* 1957). The administration of **beryllium sulfate, fluoride** or **hydrogen phosphate** caused *cor pulmonale* in some animals (Schepers 1964), which possibly represents compensatory growth of the heart muscle as a result of the fibrotic and oedematous lung damage.

Delayed body weight gains or decreased body weights occurred as a symptom of systemic toxicity following high concentrations. Long-term administration of beryllium concentrations of around 0.2 mg/m^3 as **beryllium oxide, fluoride, sulfate** or **hydrogen phosphate** led to increased mortality in rats, mice, hamsters, and squirrel and rhesus monkeys (Hall *et al.* 1950, Schepers 1964, Stokinger *et al.* 1950, Wagner *et al.* 1969).

After short-term exposure of dogs to beryllium concentrations of 30 mg/m^3, or medium-term administration of 3.6 mg/m^3 as **beryllium oxide** (Hall *et al.* 1950) or 0.04 mg/m^3 as **beryllium sulfate** for 100 days (Stokinger *et al.* 1950), a decrease in the arterial oxygen partial pressure was diagnosed. The drop in the arterial oxygen partial pressure reflects the reduced capability of the lungs to oxygenate blood.

Clear effects on the liver, including hepatocyte degeneration, sinusoid inflation and Kupffer cell migration, were found only at concentrations that also resulted in high mortality (Schepers 1964). Changes in the serum proteins were described in some studies (Hall *et al.* 1950). The decrease in the serum protein concentration and the albumin/globulin ratio in blood observed in dogs exposed to beryllium concentrations of 3.6 mg/m^3 as **beryllium oxide** was interpreted as an indication of liver damage (ATSDR 2002). However, increased serum albumin and serum globulin levels were also found in rats (2.0 mg beryllium/m^3 as **beryllium sulfate**) and dogs (0.04 mg beryllium/m^3 as **beryllium sulfate**) (ATSDR 2002).

In some studies macrocytic anaemia, with a decrease in the number of erythrocytes, an increase in the average erythrocyte volume (MCV) and transient leukocytosis, was observed (Hall *et al.* 1950).

Apart from after lethal concentrations, which caused damage to glomeruli and tubules, only slight effects on the kidneys of rhesus monkeys were found (Schepers 1964). After the exposure of dogs to beryllium concentrations of 0.43 mg/m^3 as **beryllium sulfate** and of rats to 2.0 mg/m^3 as **beryllium sulfate** for 51 to 100 days, an increase in the protein levels in urine was found (Stokinger *et al.* 1950).

Lethal concentrations of **beryllium fluoride** and **beryllium hydrogen phosphate** caused hypertrophy or hypoplasia of the adrenals in rhesus monkeys (Schepers 1964).

In the case of **beryllium oxide** the toxicity depends on the calcining temperature. **Beryllium oxide** calcined at low temperatures (400°C) is more toxic than that calcined at high temperatures (≥ 1000°C) because of its better solubility *in vivo*. Clearance from the lungs of **beryllium oxide** calcined at high temperatures is lower.

Table 4. Effects of beryllium and its inorganic compounds after repeated inhalation exposure

Species number/group	Exposure	Findings	References
Beryllium oxide			
Wistar rat, 20♀	32 mg Be/m^3 (calcined at 1350°C) 6 h/day, 5 days/week, 10 days	no significant effects on lungs, liver, kidney, spleen and blood count	Hall et al. 1950
Wistar rat, 10♂	31 mg Be/m^3 (calcined at 1150°C) 6 h/day, 5 days/week, 60 days	no histological changes in lungs, liver, kidney and spleen; in a 2nd investigation with BeO (fluorescence grade): moderate leukocytosis after 10 days in 3/5 animals	Hall et al. 1950
Wistar rat, 23♂ and ♀	30 mg Be/m^3 (calcined at 400°C) 6 h/day, 5 days/week, 15 days	mortality: ♂: 6/13, ♀: 9/10; respiratory insufficiency, leukocytosis, body weights decreased	Hall et al. 1950
Charles River rat, 60♂	0.210 mg Be/m^3 (bertrandite ore); 0.620 mg Be/m^3 (beryl ore) 6 h/week, 5 days/week, up to 17 months	0.210 mg Be/m^3: mortality increased; lungs: granulomas from large, tightly packed, dust-laden macrophages and a few lymphocytes; no influence on haematological parameters, liver, kidney, thymus, spleen; with exposure up to 17 months: bronchial lymphocyte infiltration, abscesses, thickened lung lobes; 0.620 mg Be/m^3: foamy macrophages, lymphocyte infiltration, abscesses, thickened lung lobes; no influence on haematology, liver, kidneys, thymus, spleen; body weights decreased if exposure > 12 to 17 months	Wagner et al. 1969
Sherman rat (no other details)	0.006 mg Be/m^3 6 h/day, 5 days/week, for life	lung inflammation, later fibrotic changes	Vorwald and Reeves 1959
Guinea pig, 10♀	31 mg Be/m^3 (calcined at > 1000°C) 6 h/day, 5 days/week, 10 days	no significant effects on the lungs, liver, kidney, spleen and blood count	Hall et al. 1950
Syrian golden hamster, 48♂	0.210 mg Be/m^3 (bertrandite ore); 0.620 mg Be/m^3 (beryl ore) 6 h/day, 5 days/week, up to 17 months	0.210 mg Be/m^3: mortality increased; lungs: granulomas from tightly packed, dust-laden macrophages and a few lymphocytes; no influence on liver, kidneys, thymus, spleen, body weight; 0.620 mg Be/m^3: no lesions comparable to those with bertrandite; no influence on liver, kidneys, thymus, spleen, body weight	Wagner et al. 1969

Table 4. continued

Species number/group	Exposure	Findings	References
New Zealand rabbit, 3♂	31 mg Be/m^3 (calcined at 1350°C) 6 h/day, 5 days/week, 10 days	number of erythrocytes decreased; no significant effects on lungs, liver, kidneys, spleen, blood parameters	Hall et al. 1950
New Zealand rabbit, 3♂	30 mg Be/m^3 (calcined at 1150°C) 6 h/day, 5 days/week, 60 days	macrocytic anaemia; number of erythrocytes decreased; mean corpuscular volume increased; no histological damage to the lungs or other organs; mortality not increased	Hall et al. 1950
mixed-breed cat, 2♂	30 mg Be/m^3 (calcined at 1150°C) 6 h/day, 5 days/week, 15 days	loss of appetite, body weights decreased, slight lung damage; no histological changes in other organs	Hall et al. 1950
mixed-breed dog, 1♀, 1♂	30 mg Be/m^3 (calcined at 400°C) 6 h/day, 5 days/week, 15 days	loss of appetite, body weights decreased, moderate progressive leukocytosis; lungs: moderate damage; loss of the epithelium in terminal bronchi, intraalveolar oedema; no histological changes in other organs	Hall et al. 1950
mixed-breed dog, 2♀, 2♂	31 mg Be/m^3 (calcined at 1150°C) 6 h/day, 5 days/week, 17.5 days	lungs: moderate inflammation, areas with emphysema and atelectasis; no histological changes in other organs	Hall et al. 1950
mixed-breed dog, 4♀	3.6 mg Be/m^3 (calcined at 400°C) 6 h/day, 5 days/week, 40 days	loss of appetite, ≥ 25 % of body weights decreased; increased effort of breathing; lungs: marked damage with proliferation of the bronchial epithelium and foci with infiltrating cells, collapsed or emphysematic alveoli, interstitial inflammation; serum protein decreased, arterial oxygen partial pressure decreased; progressive macrocytic anaemia; number of erythrocytes decreased, mean corpuscular volume increased	Hall et al. 1950
rhesus monkey, 2♂	30 mg Be/m^3 (calcined at 1150°C) 6 h/day, 5 days/week, 15 days	despite severe dyspnoea in one monkey 2 months after the end of exposure only slight lung damage; body weights decreased, no changes in other organs	Hall et al. 1950
squirrel monkey, 12♂	0.210 mg Be/m^3 (bertrandite ore); 0.620 mg Be/m^3 (beryl ore) 6 h/day, 5 days/week, up to 23 months	0.210 and 0.620 mg Be/m^3: mortality increased; lungs: inflammation with aggregates of dust-laden macrophages and lymphocytes, plasma cells near respiratory bronchioles and small vessels; no influence on haematological parameters, liver, kidneys, thymus, spleen, body weights	Wagner et al. 1969

Table 4. continued

Species number/group	Exposure	Findings	References
Beryllium sulfate			
F344 rat, 20 (sex not stated)	2.59 mg Be/m^3 2 h/day, 14 days (nose only)	mortality: 20/20 animals died 3 days after the end of exposure	Sendelbach and Witschi 1987
mouse, 38 (no other details)	2.0 mg Be/m^3 51 days 4.3 mg Be/m^3 14 days 6 h/day, 5 days/week	2.0 mg Be/m^3: mortality: 4/38; 4.3 mg Be/m^3: mortality: 0/38; body weights decreased; lung findings only described in summary	Stokinger et al. 1950
rat, 10–47♂	0.04 mg Be/m^3 100 days 0.43 mg Be/m^3 95 days 2.0 mg Be/m^3 51 days 4.3 mg Be/m^3 14 days 6 h/day, 5 days/week	0.04 mg Be/m^3: mortality: 0/20 0.43 mg Be/m^3: mortality: 23/47; erythrocyte count decreased, macrocytic anaemia, transient leukocytosis; 2.0 mg Be/m^3: mortality: 13/15; erythrocyte count decreased, macrocytic anaemia, transient leukocytosis, mild thrombocytosis; proteinuria; 4.3 mg Be/m^3: mortality: 10/10; body weights decreased; transient leukocytosis; proteinuria; lung findings only described in summary	Stokinger et al. 1950
Sherman and Wistar rat, 136–139 ♀ and ♂	0.035 mg Be/m^3 4–8 h/day, 5–6 days/week, 180 days; follow-up 180 days	mortality: 46/136 during exposure, 12/136 after the end of exposure; lungs: progressive damage, accumulation of foamy macrophages, thickened alveolar walls from proliferation or macrophage infiltration, granulomas	Schepers et al. 1957
Sherman rat (no other details)	0.0547 mg Be/m^3 6 h/day, 5 days/week, for life	lung inflammation and lung fibrosis	Vorwald and Reeves 1959
Sprague Dawley rat, 150 (no other details)	0.034 mg Be/m^3 7 h/day, 5 days/week, up to 18 months	mortality: ♀ increased; body weight gain decreased; lungs: inflammation with accumulation of histiocytes, thickened alveolar walls, macrophage accumulation in alveolar spaces, proliferation	Reeves et al. 1967

Table 4. continued

Species number/group	Exposure	Findings	References
guinea pig, 10–34♀	0.04 mg Be/m³ 100 days 0.43 mg Be/m³ 95 days 2.0 mg Be/m³ 51 days 4.3 mg Be/m³ 14 days 6 h/day, 5 days/week	0.04 mg Be/m³: mortality: 0/23; 0.43 mg Be/m³: mortality: 2/34; 2.0 mg Be/m³: mortality: 7/12; 4.3 mg Be/m³: mortality: 3/10; lung findings described only in summary	Stokinger et al. 1950
hamster, 10–83 (no other details)	0.04 mg Be/m³ 100 days 0.43 mg Be/m³ 95 days 2.0 mg Be/m³ 51 days 4.3 mg Be/m³ 14 days 6 h/day, 5 days/week	0.04 mg Be/m³: mortality: 0/83; 2.0 mg Be/m³: mortality: 5/10; body weights decreased; 4.3 mg Be/m³: mortality: 2/10; lung findings only described in summary	Stokinger et al. 1950
rabbit, 3–24 (no other details)	0.04 mg Be/m³ 100 days 0.43 mg Be/m³ 95 days 2.0 mg Be/m³ 51 days 4.3 mg Be/m³ 14 days 6 h/day, 5 days/week	0.04 mg Be/m³: mortality: 0/23; 0.43 mg Be/m³: mortality: 2/24; erythrocyte count decreased, macrocytic anaemia, transient leukocytosis; 2.0 mg Be/m³: mortality: 1/10; erythrocyte count decreased, macrocytic anaemia, transient leukocytosis, slight thrombocytosis; 4.3 mg Be/m³: mortality: 0/3; transient leukocytosis; lung findings described only in summary	Stokinger et al. 1950
cat, 4–5 (no other details)	0.04 mg Be/m³ 100 days 0.43 mg Be/m³ 95 days 2.0 mg Be/m³ 51 days 6 h/day, 5 days/week	0.04 mg Be/m³: mortality: 0/4; body weights decreased by 20 %; 0.43 mg Be/m³: mortality: 1/5; 2.0 mg Be/m³: mortality: 4/5; body weights decreased by 43 %; lung findings described only in summary	Stokinger et al. 1950
pig, 2 (no other details)	2.0 mg Be/m³ 51 days 6 h/day, 5 days/week	2.0 mg Be/m³: body weights decreased by 28 %; lung findings described only in summary	Stokinger et al. 1950
dog, 5 (no other details)	0.04 mg Be/m³ 100 days 0.43 mg Be/m³ 95 days 2.0 mg Be/m³ 51 days 6 h/day, 5 days/week	0.04 mg Be/m³: mortality 0/5; transient body weight decreases of 10 %; erythrocyte count decreased; macrocytic anaemia; phospholipid and cholesterol levels in the blood increased, arterial oxygen partial pressure decreased in 3/5 animals; 0.43 mg Be/m³: mortality 0/5; body weights decreased by 11 %; transient leukocytosis, proteinuria in 1/5 animals; 2.0 mg Be/m³: mortality 4/5; body weights decreased by 4 %; transient leukocytosis, proteinuria; lung findings described only in summary	Stokinger et al. 1950

Table 4. continued

Species number/group	Exposure	Findings	References
rhesus monkey, 4♀	0.198 Be/m³ 6 h/day, 7 days	mortality 1/4; body weights decreased; anorexia, dyspnoea; lungs: emphysema, slight fibrosis, *cor pulmonale* in 1/4; kidneys: degeneration of the glomeruli in 1/4	Schepers 1964
monkey, 1–5 (no other details)	0.04 mg Be/m³ 100 days 0.43 mg Be/m³ 95 days 2.0 mg Be/m³ 51 days 6 h/day, 5 days/week	0.04 mg Be/m³: mortality: 0/2; 0.43 mg Be/m³: mortality: 0/5; body weights decreased by 31 %; 2.0 mg Be/m³: mortality: 1/1; body weights decreased by 25 %; lung findings described only in summary	Stokinger *et al.* 1950

Beryllium fluoride

Species number/group	Exposure	Findings	References
rhesus monkey, 4♀	0.184 Be/m³ 6 h/day, 7–18 days	mortality 2/4; body weights decreased; severe dyspnoea; lungs: pneumonitis, oedema, emphysema, alveolar septa with infiltration, epithelialization, *cor pulmonale* in 2/2 survivors; liver: hepatocellular degeneration; kidneys: degeneration of the tubules in 3/4, damaged glomeruli in 2/4; adrenals: hypotrophy	Schepers 1964

Beryllium hydrogen phosphate

Species number/group	Exposure	Findings	References
rhesus monkey, 4♀	0.198 mg Be/m³ 30 days 1.13 mg Be/m³ 10 days 8.3 mg Be/m³ 8 days 6 h/day	dose-dependent effects, with one exception pneumonitis was the cause of all deaths; 0.198 mg Be/m³: mortality 1/4; body weights decreased; anorexia, dyspnoea; lung oedema; no influence on the liver, kidneys, adrenals; 1.13 mg Be/m³: mortality 4/4; body weights decreased; anorexia, dyspnoea; lungs: oedema, infiltration, granulomas; alveolar walls: epithelialization, fibrosis; *cor pulmonale*; liver: hepatocellular degeneration, inflation of the sinusoids, Kupffer cell migration; adrenals: hypoplasia; 8.3 mg Be/m³: mortality 4/4; body weights decreased: extreme anorexia and dyspnoea; lungs: marked lung oedema, infiltration, granulomas; alveolar walls: epithelialization, fibrosis; moderate *cor pulmonale*; liver: hepatocellular degeneration, inflation of the sinusoids, Kupffer cell migration; kidneys: damage to glomeruli and tubules; adrenal damage; hypotrophy of the pancreatic acinus; thyroid oedema; spleen damage.	Schepers 1964

5.2.2 Oral administration

After three to four weeks of **beryllium carbonate** administration with the diet (35–840 mg beryllium/kg body weight and day) rats developed dose-dependent rachitis. The induction of the rachitis was not interpreted as a direct effect of beryllium on the bones. It was assumed that beryllium binds to soluble phosphates in the intestine, which are transformed into insoluble beryllium phosphates, and absorption of the phosphates from the intestine is thus reduced. This mechanism also was held responsible for the reduced serum phosphate level and the reduced activity of the alkaline phosphatases in rats that had received beryllium doses of at least 70 mg/kg body weight and day as **beryllium carbonate** with the diet (ATSDR 2002).

In rats given beryllium doses of 2 mg/kg body weight and day as **beryllium nitrate** with the diet every 3 days for 40 days, thickening of the alveolar epithelium with necrotic areas was observed. It was suggested that beryllium nitrate particles were inhaled during eating and thus the lung damage can be attributed to inhalation exposure (ATSDR 2002).

The administration of beryllium doses of 12 mg/kg body weight and day for 33 days, or up to 1.0 mg/kg body weight and day as **beryllium sulfate** with the diet for 172 weeks caused inflammatory and ulcerative damage to the stomach and small and large intestine, which in the case of 12 mg/kg body weight and day also led to body weight losses and anorexia in dogs (ATDSR 2000).

Beryllium sulfate (about 0.25–0.75 mg beryllium/kg body weight and day) administered with the diet or the drinking water for life resulted in no further substance-related effects in mice and rats apart from glucosuria (see also Section 5.7) (Schroeder and Mitchener 1975a, 1975b).

There are no data available for the dermal absorption of beryllium and its inorganic compounds, nor for the local effects on skin and mucous membranes.

5.3 Allergenic effects

5.3.1 Sensitizing effects on the skin

To investigate the provocation of an allergic reaction by beryllium alloys, 5 groups of 20 female albino Dunkin-Hartley guinea pigs were each subjected, in a maximization test, to intradermal induction treatment by injecting 2×100 µl Freund's complete adjuvant (FCA), 1 % **beryllium sulfate** in water and a 1:1 mixture made up of 2 % **beryllium sulfate** and FCA. This was followed after 7 days by a 48-hour occlusive topical induction treatment with 5 % **beryllium sulfate**. After 24-hour occlusive provocation carried out in 2 of the groups 14 days later using 3 % **beryllium sulfate**, positive results were obtained in 11/20 and 10/20 animals. Provocation carried out with beryllium/copper alloys (containing 0.4 % and 2 % beryllium) after a further 7 days produced a reaction in

13/20 and 8/20 animals, respectively. In a third group, 6/19 and 8/19 animals also reacted to these two alloys and, in a fourth group, 12/20 animals in each case to pure beryllium and an alloy consisting of beryllium and aluminium in equal parts. In another group, none of the 20 animals reacted to pure aluminium and pure copper. None of 10 control animals per group pretreated with FCA reacted to either the **beryllium sulfate** solution or one of the metals or alloys (Zissu *et al*. 1996).

In another maximization test with **beryllium chloride**, documented as $BeCl_5$, intradermal and epicutaneous induction were carried out with 0.001 % **beryllium chloride** in water and 10 % **beryllium chloride** in petrolatum, respectively. After provocation with 0.5 % and 1 % **beryllium chloride** in petrolatum, 7/20 and 11/20 animals reacted after 24 hours, and 11/20 and 13/20 animals after 48 hours. In the control group pretreated with FCA, reactions were observed in 0/20 and 5/20 animals after 24 hours, and in 3/20 and 7/20 after 48 hours (Boman *et al*. 1979, 1980).

Contradicting results were obtained in the local lymph node assay: whereas a test with groups of 3 female BALB/c mice with 0.25 %, 1 % and 4 % **beryllium sulfate tetrahydrate** in dimethyl sulfoxide produced a negative result in each case (Mandervelt *et al*. 1997), (non-concentration-dependent) positive results were obtained with 2.5 %, 5 % and 10 % **beryllium sulfate tetrahydrate** in dimethyl formamide in a study with groups of 4 CBA/Ca mice (Basketter *et al*. 1999).

After the application of 0.05 ml 20 % **beryllium fluoride** in Polysorbate 80 (Tween® 80) three times to one ear of each animal, only animals from 3 of the 5 strains used reacted after provocation carried out 14 days later by applying 0.05 ml 1 % **beryllium fluoride** solution in 1 % Triton® X-100. Positive results were obtained in 26/33 Hartley guinea pigs, in 5/10 Pirbright guinea pigs, and in 8/11 animals of an inbred strain not further specified. 10 Himalaya guinea pigs and 10 animals of another inbred strain did not react (Polak *et al*. 1973).

Due to the fact that only 2 guinea pigs from one of the two inbred strains, not further specified, were examined, as well as the contradictory documentation given, it is not possible to evaluate the positive findings obtained in an experimental investigation using FCA. Provocation was carried out via intradermal injection of 5.0 µg (documented elsewhere as 0.5 µg) **beryllium sulfate** 6 weeks after the beginning of the induction treatment. For induction, each of the animals received 2 intradermal injections of 1.0 ml 1 % **beryllium oxide** in physiological saline, FCA, and a mixture of **beryllium oxide** preparation and FCA. A second intradermal treatment with 10 mg **beryllium oxide** in 1 ml physiological saline was carried out after 2 weeks, which was repeated after a further 2 weeks (Barna *et al*. 1984a).

In another, earlier investigation, positive findings were also determined in 10 male guinea pigs (no other details). Provocation was carried out intracutaneously with 0.1 ml 0.025 % **disodium tetrafluoroberyllate** (Na_2BeF_4) and epicutaneously with 1.0 % Na_2BeF_4 in a hydrophilic ointment base following weekly induction treatment over 10 and 20 weeks, respectively, via subcutaneous administration of a suspension of 0.5 mg beryllium particles with an average diameter of 0.1 µm in physiological saline (Nishimura 1966).

Although plausible, these results cannot be evaluated definitively on account of the non-standardized treatment protocol.

5.3.2 Sensitization of the respiratory passages

In different animal models in rats, mice, dogs and monkeys under different exposure conditions, the effects of pulmonary exposure to beryllium, beryllium oxide or beryllium salts were investigated without finding, histopathologically, a close correlation with the respective pathological conditions in man (Nikula *et al.* 1997). The instillation of 50 or 150 µg **beryllium powder** with a median diameter of 1.4 µm in physiological saline into the bronchi of one lung lobe each in 3 macaques (*Macaca fascicularis*), but not the instillation of 1.0 µg **beryllium** into the bronchi of a third lung lobe, produced interstitial fibrosis, lymphocytic infiltrates and granulomas. Similar findings were obtained after instilling 50, 100 or 150 µg **beryllium** into one lung lobe each in 4 other animals. After only 14 days, an increased number of lymphocytes was detected in the bronchoalveolar lavage fluid, the proliferation of which could be stimulated *in vitro* by beryllium. After instillation of 0.5, 12.5 and 37.5 µg **beryllium oxide**, corresponding histological findings were obtained only at the highest administered dose in only 1 of 3 animals and to a far less marked extent. According to the authors, the changes were to a less marked extent in the 4 animals treated with 12.5, 25.0 or 37.5 µg **beryllium oxide**. There was only a slight increase in the number of lymphocytes in these animals after 60 days, and *in vitro* stimulation of proliferation was insignificant (Haley *et al.* 1994). It is not possible to clearly assign the findings described in this incompletely documented investigation to the quantities administered.

After only 6 and 8–10 weeks after intratracheal administration of 5 mg **beryllium oxide** in 1 ml physiological saline, moderately pronounced granulomatous changes in the lungs were found in the guinea pigs in one of the 2 inbred strains used, which were, however, not further specified. An increase in lymphocyte proliferation was determined by LTT investigations with **beryllium sulfate**. Skin tests with **beryllium sulfate** carried out after 6 weeks were likewise positive only in the animals of this strain; in addition, the proportion of T lymphocytes in the bronchoalveolar lavage fluid was increased after 2 weeks in the animals of this strain only (Barna *et al.* 1984a, Barna *et al.* 1984b).

After 90 minutes nose-only exposure to an aerosol containing beryllium, female A/J and C3H/HeJ mice absorbed on average around 50–60 µg of the metal. Histological investigation of the pulmonary tissue carried out after 6 months yielded similar findings in both strains: interstitial infiltrates with mononuclear cells, microgranulomas, multinuclear giant cells of the Langhans' cell type and, in most of the animals, slight to moderate interstitial fibrosis. Interstitial, compact accumulation of lymphocytes, especially of B lymphocytes in the central area, together with an increase in peripheral $CD4^+$ cells, in which a less extensive *in situ* lymphocyte proliferation was detected than in the microgranulomas, were found in the lungs of all animals. Nevertheless, specific *in vitro* lymphocyte proliferation due to **beryllium sulfate** was found neither in peripheral lymphocytes of the blood nor in lymphocytes from the spleen or tracheobronchial lymph nodes (Nikula *et al.* 1997).

5.4 Reproductive and developmental toxicity

5.4.1 Fertility

In male and female rats given beryllium doses of ≤ 31 mg/kg body weight in the form of **beryllium sulfate** with the diet for 2 years, no changes were observed in the testes, seminal vesicles, epididymis, prostate, ovaries, uterus or oviducts (ATSDR 2002).

5.4.2 Developmental toxicity

Single intratracheal doses of beryllium of 0.2 mg/kg body weight administered as **beryllium oxide** to male and female rats had no influence on the litter size or mortality of young animals within the follow-up period of 15 months (Clary et al. 1975). Intratracheal administration of doses of 50 mg/kg body weight as **beryllium chloride** on days 3, 5 or 20 of gestation caused increased mortality and reduced body weights in the foetuses, and internal malformations were observed in the offspring (no other details) (WHO 2001).

In male and female dogs kept together, which received long-term doses of beryllium with the diet of 0.1 mg/kg body weight and day as **beryllium sulfate** from the time of mating, no external or skeletal malformations and no unusual findings with regard to litter size, the number of living offspring, body weights and survival were observed (ATSDR 2002, WHO 2001).

5.5 Genotoxicity

5.5.1 *In vitro*

Soluble beryllium compounds, such as **beryllium nitrate**, **sulfate** and **chloride**, yield positive results in indicator tests (Kada et al. 1980, Kanematsu et al. 1980, Kuroda et al. 1991, US EPA 1998) and bacterial mutagenicity tests (Arlauskas et al. 1985, Kuroda et al. 1991, Ulitzur and Barak 1988, Zakour and Glickman 1984). They also may produce sister chromatid exchange (Kuroda et al. 1991, Larramendy et al. 1981), chromosomal aberrations (Larramendy et al. 1981) and gene mutations (Hsie et al. 1979a, 1979b, Miyaki et al. 1979, Zakour and Glickman 1984). However, negative results were also found in genotoxicity investigations with yeast (Kuroda et al. 1991, Nishioka 1975), bacteria (Arlauskas et al. 1985, Dunkel et al. 1984, Ogawa et al. 1987, Rosenkranz and Poirer 1979, Rossman and Molina 1986, Rossman et al. 1984, Tso and Fung 1981) and mammalian cells (Ashby et al. 1990, Brooks et al. 1989, Paton and Allison 1972). In primary rat hepatocytes **beryllium sulfate** did not induce DNA repair synthesis (Williams et al. 1989). An overview of the studies is given in Table 5.

The different results possibly depend on variations in the test conditions, the different physical and chemical properties of the inorganic beryllium compounds, and the test strains used. Also the binding of beryllium to phosphate and hydroxide ions or proteins of the culture media plays a role (ATSDR 2002). High concentrations of magnesium salts or citrate ions in the culture medium probably reduce the uptake of inorganic beryllium salts in bacteria (WHO 1990).

The potential mechanisms at the basis of the genotoxic effects were already discussed in Section 2.

Table 5. Genotoxicity of inorganic beryllium compounds *in vitro*

End point	Test system	Substance	Concentration range [μg/plate]*	Effective concentration*	Cyto-toxicity*	Result +S9	Result −S9	References
rec assay	*B. subtilis* M45(rec⁻), H17(rec⁺)	BeSO$_4$	n.s.	10 mM	n.s.	+		Kada *et al.* 1980, Kanematsu *et al.* 1980
		BeCl$_2$	50 mM		–	–		Nishioka 1975
		BeCl$_2$	375–1500	1500	n.s.	+		Kuroda *et al.* 1991
		BeO	0.1		n.s.	–		Kuroda *et al.* 1991
		Be(NO$_3$)$_2$	375–1500		n.s.	–		Kuroda *et al.* 1991
	E. coli	BeSO$_4$	n.s.	n.s.	n.s.	+		US EPA 1998
	E. coli pol A$^{+/−}$	BeSO$_4$	250		–	–	–	Rosenkranz and Poirer 1979
SOS chromotest	*E. coli* B/r WP2	BeCl$_2$	5 mM		–	–	–	Rossman *et al.* 1984
gene mutation	*S. typhimurium* TA1535, TA1538	BeSO$_4$	100–250		–	–	–	Rosenkranz and Poirer 1979
	E. coli WP2 uvr A	BeSO$_4$	0.5–333		–	–	–	Dunkel *et al.* 1984
	S. typhimurium TA98, TA100, TA1535, TA1536, TA1537, TA1538	BeSO$_4$	up to 250		–	–	–	Simmon 1979
	S. typhimurium TA98, TA100, TA1535, TA1537, TA1538	BeSO$_4$	0.5–333		–	–	–	Dunkel *et al.* 1984
	S. typhimurium TA100	BeSO$_4$	0.5–50 mM	≥ 0.5 mM	–	–	+^{F}	Arlauskas *et al.* 1985
	S. typhimurium TA98, TA100, TA1535, TA1537, TA1538	BeSO$_4$·4H$_2$O	2–50000		–	–	–	Ashby *et al.* 1990
	S. typhimurium TA1537, TA2637	BeCl$_2$	0.08–200		–	–	–	Ogawa *et al.* 1987
	S. typhimurium TA98	BeCl$_2$	313–5000	–	–	–	–	Kuroda *et al.* 1991
	S. typhimurium TA100			625		+	–	

Table 5. continued

End point	Test system	Substance	Concentration range [μg/plate]*	Effective concentration*	Cyto-toxicity*	Result +S9	Result −S9	References
	S. typhimurium TA98	BeO	0.11–0.44	–	> 0.43	–	–	Kuroda et al. 1991
	S. typhimurium TA100			0.11		+	–	
	S. typhimurium TA98	Be(NO$_3$)$_2$	1250–5000	–	–	–	–	Kuroda et al. 1991
	S. typhimurium TA100			1250	–	+	–	
bioluminescence photobacterium fischeri (UV induced) (lacI gene)		BeCl$_2$	1.25 mM	1.25 mM	n.s.		+	Ulitzur and Barak 1988
	E. coli WP2 uvr A	BeCl$_2$	up to 2 mM	–	n.s.		–	Rossman and Molina 1986
	E. coli KMBL 3835	BeCl$_2$	0.005–0.5 mM	0.01 mM	≥ 0.05 mM		+	Zakour and Glickman 1984
	S. typhimurium LT$_2$, TA100	Be(NO$_3$)$_2$·4 H$_2$O	0.1–100 mM	–	100 mM		–	Tso and Fung 1981
sister chromatid exchange	V79 cells	BeCl$_2$	c. 0.39–3.13 mM	c. 0.39–0.11 mM	c. 3.13 mM		+	Kuroda et al. 1991
	V79 cells	Be(NO$_3$)$_2$	c. 0.23–3.76	c. 0.23–0.46 mM	c. 3.76 mM		+	Kuroda et al. 1991
	V79 cells	BeO	c. 0.8–3.6 μM	–	c. > 3.6μM		–	Kuroda et al. 1991
	human lymphocytes	BeSO$_4$·4 H$_2$O	0.006–0.028 mM	0.028 mM	n.s.		+	Larramendy et al. 1981
	SHE cells	BeSO$_4$·4 H$_2$O	0.006–0.028 mM	0.028 mM	n.s.		+	
DNA repair synthesis (UDS)	rat hepatocytes	BeSO$_4$·4 H$_2$O	7.3 mM	–	–		–	Williams et al. 1989
chromosome aberrations	human lymphocytes	BeSO$_4$·4 H$_2$O	0.028 mM	0.028 mM	n.s.		+	Larramendy et al. 1981
	SHE cells	BeSO$_4$·4 H$_2$O	0.028 mM	0.028 mM	n.s.		+	
	CHL cells	BeSO$_4$·4 H$_2$O	78–625 mg/ml	–	–	–	–	Ashby et al. 1990
	CHO cells	BeSO$_4$·4 H$_2$O	0.2–1.0 mM	–	–		–	Brooks et al. 1989
	human fibroblasts	BeSO$_4$	< 10 mM	–	10 mM		–	Paton and Allison 1972
	human leukocytes		< 0.001 mM	–	0.001 mM		–	
gene mutation HPRT	CHO cells	BeSO$_4$·4 H$_2$O	n.s.	n.s.	n.s.		+	Hsie et al. 1979a, 1979b
	V79 cells	BeCl$_2$	2–3 mM	2 mM	3 mM		+	Miyaki et al. 1979

* unless otherwise stated the data refer to [μg/plate], F fluctuation test, n.s. not specified

5.5.2 In vivo

Oral administration of beryllium doses of 71.2 or 117 mg/kg body weight as **beryllium sulfate** (1.4 and 2.3 g/kg body weight) did not induce micronuclei in the bone marrow in CBA mice. Reduced erythropoiesis after 24 hours was viewed as a toxic effect (Ashby et al. 1990).

5.6 Carcinogenicity

5.6.1 Short-term studies

In BALB/c-3T3 cells, **beryllium sulfate** concentrations of 50 to 200 µg/ml caused an increase in the transformation frequency by 9 to 41 times depending on the concentration. After subcutaneous transfusion these transformed cells induced tumours in naked mice within 50 days, in which an amplification of the K-ras and c-jun genes was found. A change in the expression of the genes or the amount of protein was not observed (Kesheva et al. 2001). In another investigation, however, the expression of 18 genes was significantly altered in the tumours induced in naked mice exposed in the same way. The expression of oncogenes, such as c-fos, c-jun, c-myc and R-ras was increased, while the expression of those genes was reduced whose proteins play a role in DNA synthesis, repair or recombination (Joseph et al. 2001).

Beryllium sulfate concentrations of 2.5 or 5 µg/ml produced morphological transformations in embryonic cells of the Syrian hamster. The marker for toxicity, the cloning efficiency, decreased at these concentrations to 63 % and 46 %, respectively, of the control value (DiPaolo and Casto 1979).

Syrian golden hamsters were given intraperitoneal doses of **beryllium sulfate** of 25 or 50 mg/kg body weight on day 11 of gestation. Transformations were detected in the embryonic cells studied 48 to 60 hours after administration of the substance (DiPaolo and Casto 1979).

The increased transformation frequency of the embryonic cells of the Syrian hamster induced by adenovirus SA 7 was further increased by the administration of 0.56 mM **beryllium sulfate** solution (Casto et al. 1979)

5.6.2 Long-term studies

Many of the available animal studies of the carcinogenicity of beryllium and its inorganic compounds are to be viewed critically as a result of their conceptual weaknesses. Some of the studies are old and were often conducted only with one dose and without control groups, were documented inadequately, and frequently lack information on the strain and number of animals used, the follow-up period and the evaluation of the results.

Inhalation

After inhalation of **beryllium metal, sulfate, chloride, fluoride** or **oxide**, increased incidences of lung cancer were observed in rats and monkeys. The malignant tumours which developed in the lungs were adenocarcinomas or squamous cell carcinomas, epidermoid carcinomas or pleural mesotheliomas (Nickell-Brady et al. 1994, Reeves et al. 1967, Schepers 1961, Schepers et al. 1957, Vorwald 1968, Vorwald et al. 1955, Wagner et al. 1969, WHO 1990). No increase in the frequency of lung cancer was determined in hamsters and monkeys exposed to bertrandite or beryl (Wagner et al. 1969). The studies of the carcinogenic effects of beryllium and its inorganic compounds after inhalation are shown in Table 6.

Table 6. Carcinogenicity after inhalation of beryllium and its inorganic compounds

Animal species, number/group	Exposure	Concentration	Tumours	References
Beryllium metal				
F344/N rat, 30♂, 30♀	1×, 14 months follow-up	controls; 410 mg/m^3, 30 min; 500 mg/m^3, 8 min; 830 mg/m^3, 48 min; 980 mg/m^3, 39 min	not investigated; in total 64 % lung tumours*, above all squamous cell and adenocarcinomas (no other details)	Nickell-Brady et al. 1994
p53$^{+/-}$ knock-out mouse, 15♂, 15♀	1×, 1–3 h; 1×, 1–3 h; 2.3 h/day, 3 days; 19 months follow-up	controls; 15 µg; 60 µg	0/30; 0/29; 4/28 squamous cell or adenocarcinomas**	Finch et al. 1998
p53$^{+/-}$ wild type mouse, 15♂, 15♀	1×, 1–3 h; 2.3 h/day, 3 days; 23 months follow-up	controls; 60 µg	0/30; 0/28	Finch et al. 1998
Beryllium sulfate				
Sherman rat, 83♂, 192♀	4–8 h/day, 5–6 days/week, 180 days; 180 days follow-up	controls; 0.035 mg Be/m^3	0/139; 76/136 lung tumours (18 adenomas***, 7 mucosal tumours, 47 carcinomas, 1 pleural mesothelioma, 3 sarcomas)	Schepers et al. 1957
rat, (no other details)	33–38 h/week, 8–24 months; 11–16 months; 8–21 months; 3–22 months; 9–24 months; 3–18 months; 12 months	controls; 1.8–2.0 µg Be/m^3; 18 µg Be/m^3; 18 µg Be/m^3; 18 µg Be/m^3; 18 µg Be/m^3; 55 µg Be/m^3; 180 µg Be/m^3	not stated; 71/160 lung tumours*; 9/21 lung tumours*; 31/63 lung tumours*; 72/103 lung tumours*; 47/90 lung tumours*; 55/74 lung tumours*; 11/27 lung tumours*	WHO 1990

Table 6. continued

Animal species, number/group	Exposure	Concentration	Tumours	References
rat, ♂, ♀ (no other details)	7 h/day; 5.5 days/week, 13–18 months	controls; 33–35 µg Be/m^3	not stated; 17/17 lung tumours* (adenocarcinomas, epidermoid tumours) 1/17 metastases	Vorwald et al. 1955
Sprague-Dawley rat, 43♂ and ♀	7 h/day, 5 days/week, up to 72 weeks	controls; 34 µg Be/m^3	0/150; 43/43 lung tumours*** (adenocarcinomas)	Reeves et al. 1967
Charles River rat, 20–30 (no other details)	35 h/week; 3 months; 6 months; 9 months; 12 months; 18 months	controls; 36 µg Be/m^3	not stated; 19/22 lung tumours*; 33/33 lung tumours*; 15/15 lung tumours*; 21/21 lung tumours*; 13/15 lung tumours*	WHO 1990
rhesus monkey, 7♂, 9♀	6 h/day, 5 days/week, for life, interruption in the case of disease	controls; 35 µg Be/m^3	not stated (lung cancer was not described); 8/16 lung carcinomas* (metastases in lymph nodes, liver, adrenals); (7 animals which died prematurely did not have tumours)	Vorwald 1968
rhesus monkey, 4♀	6 h/day, 7 days, 180 days follow-up	controls; 0.198 µg Be/m^3	0/15; 0/4; (organs investigated: lungs, liver, kidneys, adrenals, pancreas, thyroid, spleen)	Schepers 1964
Beryllium chloride				
rat, (no other details)	5 h/week, 4 months	controls; 0.8 µg Be/m^3; 4 µg Be/m^3; 30 µg Be/m^3; 400 µg Be/m^3	not stated; 1/41 lung tumours; 2/42 lung tumours; 8/24 lung tumours*; 11/19 lung tumours*	WHO 1990
Beryllium hydrogen phosphate				
rhesus monkey, 4♀	6 h/day, 30 days, 270 days follow-up	controls; 0.198 µg Be/m^3; 1.130 µg Be/m^3; 8.300 µg Be/m^3	0/15; 0/4; 1/4 alveolar carcinoma; 0/4; (organs investigated: lungs, liver, kidneys, adrenals, pancreas, thyroid, spleen)	Schepers 1964

Table 6. continued

Animal species, number/group	Exposure	Concentration	Tumours	References
Beryllium fluoride				
rat (no other details)	6–15 months	controls; 9 µg Be/m^3	not stated; 10–12/200 lung cancer	WHO 1990
rhesus monkey, 4♀	6 h/day, 7 days, 28 days follow-up	controls; 0.184 µg Be/m^3	0/15; 0/4; (organs investigated: lungs, liver, kidneys, adrenals, pancreas, thyroid, spleen)	Schepers 1964
Beryllium oxide				
rat (no other details)	5 h/week, 4 months	controls; 0.8 µg Be/m^3; 4 µg Be/m^3; 30 µg Be/m^3; 400 µg Be/m^3	not stated; 3/44 lung cancer; 4/39 lung cancer; 6/26 lung cancer[*]; 8/21 lung cancer[*]	WHO 1990
rat (no other details)	33–38 h/week, 3–12 months	controls; 9000 µg Be/m^3	not stated; 22/36[*] (no other details)	WHO 1990
Bertrandite or beryl				
Charles River rat, 60♂	6 h/day, 5 days/week; up to 17 months	controls; 210 µg Be/m^3 (bertrandite)	0/60 0/60	Wagner et al. 1969
Charles River rat, 60♂	6 h/day, 5 days/week; up to 17 months	controls; 620 µg Be/m^3 (beryl)	0/60; 18/19 bronchoalveolar tumours (p < 0.05) (7 adenomas, 9 adenocarcinomas, 2 epidermoid tumours)	Wagner et al. 1969
Syrian golden hamster, 48♂	6 h/day, 5 days/week; up to 17 months	controls; 210 µg Be/m^3 (bertrandite)	0/48; 0/48	Wagner et al. 1969
Syrian golden hamster, 48♂	6 h/day, 5 days/week; up to 17 months	controls; 620 µg Be/m^3 (beryl)	0/48; 0/48	Wagner et al. 1969
squirrel monkey, 12♀	6 h/day, 5 days/week; up to 23 months	controls; 210 µg Be/m^3 (bertrandite)	0/12; 0/12	Wagner et al. 1969
squirrel monkey, 12♀	6 h/day, 5 days/week; up to 23 months	controls; 620 µg Be/m^3 (beryl)	0/12; 0/12	Wagner et al. 1969

[*] despite the lack of controls, the tumour incidence can be regarded as significantly increased;
[**] Fisher's exact test: p = 0.048; [***] tumour incidence regarded as significantly increased

Oral administration

No carcinogenic effects were observed after long-term uptake of **beryllium sulfate** (beryllium doses of about 0.025–0.75 mg/kg body weight and day, for life, given with the diet or drinking water) in rats and mice (Schroeder and Mitchener 1975a, 1975b) or in dogs (beryllium doses of 1 mg/kg body weight and day, administered for 172 weeks with the diet) (ATSDR 2002), probably as a result of poor gastrointestinal absorption (see also Section 5.2.2).

Intratracheal, intraperitoneal, intravenous, and intramedullary injection

The studies of the carcinogenic effects of beryllium and its inorganic compounds after intratracheal, intravenous and intraperitoneal administration are shown in Table 7.
Intratracheal: in different studies rats developed adenomas, adenocarcinomas, broncho-alveolar carcinomas or epidermoid carcinomas in the lungs after one or two intratracheal instillations of **beryllium metal**, **oxide** or **hydroxide** (Groth et al. 1980, Ishinishi et al. 1980).
Intraperitoneal: after eight weeks of intraperitoneal administration, three times a week, of beryllium doses of up to 0.18 mg/kg body weight as **beryllium sulfate**, a significant increase in the incidence of lung tumours (up to 38 %, control animals 15 %) was found in male A/J mice. Gross pathological examination did not reveal any tumours in the kidneys, spleen, intestine or stomach (Ashby et al. 1990).
Intravenous: the repeated administration of **beryllium oxide** induced osteosarcomas in rabbits e.g. in the humerus and femur (Dutra and Largent 1950, Hoagland et al. 1950).
Intramedullary: osteosarcomas were found in rabbits and mice after the injection of **beryllium metal**, **oxide** and **phosphate** into the bone marrow (intramedullary) in the humerus, tibia, femur, *tuber ischiadicum*, lumbar vertebrae, shoulder blade and ribs. Usually several bones were affected in an animal. The cell types of the bone tumours were described as osteoblastic, chondroblastic or fibroblastic. Metastases occurred in 40 % to 100 % of the animals, mainly in the lungs, but also in the liver, kidneys, omentum, skin and lymph nodes. The lung metastases could easily be distinguished from the primary tumours in the lungs (ASTDR 2002, US EPA 1998, WHO 1990). Although there were often no controls in these older studies, the results are of great significance, since osteosarcomas are a very rare form of tumours in rabbits.

5.7 Other effects

Beryllium acetate inhibited synthesis of DNA, RNA and proteins significantly in human gingival fibroblasts at beryllium concentrations between 0.25 and 1.0 µg/ml (Messer and Lucas 1999). On the other hand, a 100 nM **beryllium chloride** solution stimulated DNA synthesis in cultured mouse macrophages (Misra et al. 1999).

Table 7. Carcinogenicity of beryllium and its inorganic compounds after intratracheal, intravenous and intraperitoneal administration

Animal species, number/group	Exposure	Concentration	Tumours	References
Beryllium metal				
Wistar rat, 35♀	i.tr.: 1×, up to 19 months follow-up	controls; 0.5 mg Be/animal; 2.5 mg Be/animal	0/21[1]; 2/3 ($p = 0.011$); tumours[2] 6/6 ($p < 0.0001$) tumours[2]	Groth et al. 1980
Wistar rat, 35♀	i.tr.: 1×, (99 % Be, 0.26 % Cr), up to 19 months follow-up	controls; 0.5 mg Be/animal; 2.5 mg Be/animal	0/21[1]; 7/11 ($p = 0.0001$); tumours[2] 4/4 ($p < 0.0001$) tumours[2]	Groth et al. 1980
Beryllium sulfate				
A/J mouse, 20♂	i.p.: 3×/week, 8 weeks, 24 weeks follow-up	controls; 0.04 mg Be/animal; 0.09 mg Be/animal; 0.18 mg Be/animal	3/20 lung cancer; 3/18 lung cancer; 6/18 lung cancer; 7/19 lung cancer[3]	Ashby et al. 1990
Beryllium oxide				
Wistar rat, 46♂	i.tr.: 1×/week, 15 weeks, life-long follow-up	controls; 24 mg Be/animal (calcined at 900°C)	0/16; 6/30 lung tumours (malignant and benign); 6 tumours in other organs (1 tumour in control animals) (no other details)	Ishinishi et al. 1980
rabbit, 9 (no other details)	i.v.: 1 g distributed over 16 weeks, up to 14 months follow-up	controls; 22.5 mg Be/animal	not stated; 1/9 osteosarcomas; nodules in the liver and lungs	Hoagland et al. 1950
rabbit, 9♂ and ♀ (no other details)	i.v.: 3×/week, 6–8 weeks, up to 20 month follow-up	controls; 13–116 mg Be/kg body weight	0/50; 6/9 osteosarcomas (humerus, femur and other bones); metastases in lungs, liver, and other organs	Dutra and Largent 1950
Beryllium hydroxide				
Wistar rat, 35♀	i.tr.: 2× at an interval of 10 months, up to 18 months follow-up	controls; 50 µg Be/animal, after 10 months: 25 µg Be/animal	not stated[1]; 6/25 adenomas 7/25 adenocarcinomas 1/25 epidermoid carcinoma	Groth et al. 1980

[1] no tumours in a further 200 control animals, [2] the small groups were explained by mortality or cannibalism, [3] significantly increased tumour incidence in the chi square test.

6 Manifesto (MAK value/classification)

Increased mortality from lung cancer was found in various retrospective studies of workers in beryllium production plants (Mancuso 1970, 1979, 1980, Mancuso and El-Attar 1969, Wagoner et al. 1980, Ward et al. 1992). An additional evaluation of the US Beryllium Case Registry also yielded an increased SMR for lung cancer (Infante et al. 1980, Steenland and Ward 1991). A relationship between the duration of employment and lung cancer mortality was not determined. However, death from lung cancer correlated with an increasing time span since the beginning of employment and previous acute beryllium disease. Studies in which the relevant exposure periods and levels were evaluated and considered together with different potential confounding factors show that exposure to high beryllium concentrations—which is regarded as the trigger for acute beryllium disease—is associated with a significantly increased risk of lung cancer (Sanderson et al. 2001a, 2001b). Similar evidence was found in studies in which death from lung cancer occurred disproportionately frequently in workers who had previously suffered from acute beryllium disease (Infante et al. 1980, Steenland and Ward 1991, Ward et al. 1992). Thus, beryllium and its inorganic compounds must be considered as human carcinogens and classified in Carcinogenicity category 1. However, the data as a whole indicate that the carcinogenic effects described occurred mainly at high concentrations, which, as a rule, are no longer to be expected in workplaces today. As neither the basic mechanism of action nor a NOEL for the carcinogenic effects can be determined from the studies available to date, beryllium and its inorganic compounds currently cannot be classified in Carcinogenicity category 4.

The publication of Levy et al. (2002) was not taken into account in the evaluation of beryllium and its inorganic compounds for classification in Carcinogenicity category 1. The argumentation of Levy et al. was not accepted, as in this reanalysis of the study of Ward et al. (1992) the data from the exposure evaluation by Sanderson et al. (2001a) were cited, but not those from the embedded case–control study by Sanderson et al. (2001b). Therefore, the SMR values determined by Levy et al. for lung cancer must be regarded at least in some cases as being too low.

The epidemiological data for the pulmonary carcinogenicity of beryllium and its inorganic compounds are also supported by the results from animal experiments. Although some studies are only of limited value from the current point of view, they do document that after inhalation of beryllium metal, sulfate, chloride, fluoride or oxide, malignant lung tumours were more frequent in rats and monkeys (Finch et al. 1998, Nickell-Brady et al. 1994, Reeves et al. 1967, Schepers 1964, Schepers et al. 1957, Vorwald 1968, Vorwald et al. 1955, Wagner et al. 1969, WHO 1990). Lung tumours also were frequently induced after intratracheal and intraperitoneal administration of the substances (Ashby et al. 1990, Groth et al. 1980, Ishinishi et al. 1980). The intravenous and intramedullary administration of beryllium, beryllium carbonate, oxide and phosphate caused the formation of osteosarcomas in rabbits (WHO 1990).

Soluble beryllium compounds induce sister chromatid exchange, chromosomal aberrations and gene mutations in mammalian cells. There are no data available that justify classifying beryllium and its inorganic compounds in one of the categories for germ cell mutagens.

The effects of beryllium and its inorganic compounds on reproduction cannot be evaluated on the basis of the existing studies.

As the absorption of beryllium and its inorganic compounds via the skin can be viewed as insignificant, the substance is not designated with an "H".

Beryllium and beryllium compounds are capable of causing allergic contact dermatitis or a granulomatous skin reaction of immunological origin. As a result of the greatly reduced exposure to beryllium and beryllium compounds in recent times, case reports of a more recent origin are rarely available; hence, the present evaluation is based principally on earlier investigations. The results of a number of animal experiments, some of which, however, were not carried out in accordance with standardized procedures, indicate a sensitizing effect of beryllium and beryllium compounds on the skin.

In the lungs, beryllium and beryllium compounds are capable of producing a granulomatous disease. In contrast with most other known allergic reactions in the airways, this is very probably based on a cell-mediated immunological reaction of delayed type, and is generally observed, after a long latent period, only in some of the workers exposed. In addition, even an exposure period from decades ago may be the cause of this disease. In most cases, an immunological origin can be established with certainty on account of positive results in the lymphocyte transformation test. However, no definitive data are available to be able to derive a threshold value below which sensitization and subsequent disease are not expected to occur.

For this reason, beryllium and beryllium compounds have been designated with "Sah".

7 References

Amicosante M, Sanarico N, Berretta F, Arroyo J, Lombardi G, Lechler R, Colizzi V, Saltini C (2001) Beryllium binding to HLA-DP molecule carrying the marker of susceptibility to berylliosis glutamate β69. *Hum Immunol 62*: 686–693

Apostoli P, Schaller KH (2001) Urinary beryllium–a suitable tool for assessing occupational and environmental beryllium exposure? *Int Arch Occup Environ Health 74*: 162–166

Arlauskas A, Baker RSU, Bonin AM, Tandon RK, Crisp PT, Ellis J (1985) Mutagenicity of metal ions in bacteria. *Environ Res 36*: 379–388

Ashby J, Ishidate J Jr, Stoner GD, Morgan MA, Ratpan F, Callander RD (1990) Studies on the genotoxicity of beryllium sulphate *in vitro* and *in vivo*. *Mutat Res 240*: 217–225

ATSDR (2002) Toxicological profile for beryllium. Atlanta, GA. US Department for Health and Human Services. Public Health Service. Agency for Toxic Substances and Disease Registry. http://www.atsdr.cdc.gov/ toxprofiles/tp4.html

Balkissoon RC, Newman LS (1999) Beryllium copper alloy (2 %) causes chronic beryllium disease. *J Occup Environ Med 41*: 304–308

Barna BP, Deodhar SD, Chiang T, Gautam S, Edinger M (1984a) Experimental beryllium-induced lung disease. I. Differences in immunologic responses to beryllium compounds in strains 2 and 13 guinea pigs. *Int Arch Allergy Appl Immunol 73*: 42–48

Barna BP, Deodhar SD, Gautam S, Edinger M, Chiang T, McMahon JT (1984b) Experimental beryllium-induced lung disease. II. Analyses of bronchial lavage cells in strains 2 and 13 guinea pigs. *Int Arch Allergy Appl Immunol 73*: 49–55

Barnard AE, Torma-Krajewski J, Viet SM (1996) Retrospective beryllium exposure assessment at the Rocky Flats environmental technology site. *Am Ind Hyg Assoc J 57*: 804–808

Basketter DA, Lea LJ, Cooper KJ, Ryan CA, Gerberick GF, Dearman RJ, Kimber I (1999) Identification of metal allergens in the local lymph node assay. *Am J Contact Dermatitis 10*: 207–212

Belinsky SA (1998) Role of the cytosine DNA-methyltransferase and p16INK4a genes in the development of mouse lung tumors. *Exp Lung Res 24*: 463–479

Belinsky SA, Snow SS, Nikula KJ, Finch GL, Tellez CS, Palmisano WA (2002) Aberrant CpG island methylation of the p16 $_{INK4a}$ and estrogen receptor genes in rat lung tumors induced by particulate carcinogens. *Carcinogenesis 23*: 335–339

Bobka CA, Stewart LA, Engelken GJ, Golitz LE, Newman LS (1997) Comparison of *in vivo* and *in vitro* measures of beryllium sensitization. *J Occup Environ Med 39*: 540–547

Boman A, Wahlberg JE, Hagelthorn G (1979) Sensitizing potential of beryllium, copper and molybdenum compounds studied by the guinea pig maximization method. *Contact Dermatitis 5*: 332–333

Boman A, Wahlberg JE, Hagelthorn G (1980) Sensitizing potential of beryllium, copper and molybdenum compounds studied by the guinea pig maximization method. Erratum. *Contact Dermatitis 6*: 160

Brancaleone P, Weynand B, De Vuyst P, Stanescu D, Pieters T (1998) Lung granulomatosis in a dental technician. *Am J Ind Med 34*: 628–631

Brooks AL, Griffith WC, Johnson NF (1989) The induction of chromosome damage in CHO cells by beryllium and radiation given alone and in combination. *Radiat Res 120*: 494–507

Casto BC, Meyers J, DiPaolo JA (1979) Enhancement of viral transformation for evaluation of the carcinogenic or mutagenic potential of inorganic metal salts. *Cancer Res 39*: 193–198

Clary JJ, Bland LS, Stokinger HE (1975) The effect of reproduction and lactation on the onset of latent chronic beryllium disease. *Toxicol Appl Pharmacol 33*: 214–221

Curtis GH (1951) Cutaneous hypersensitivity due to beryllium. A study of thirteen cases. *Arch Dermatol Syphilol 64*: 470–482

Curtis GH (1959) The diagnosis of beryllium disease, with special reference to the patch test. *Arch Ind Health 19*: 150–153

DeNardi JM, van Ordstrand HS, Curtis GH, Zielinski J (1953) Berylliosis. *Arch Ind Hyg Occup Med 8*: 1–24

Deubner DC, Goodman M, Iannuzzi J (2001a) Variability, predictive value, and uses of the beryllium blood lymphocyte proliferation test (BLPT): preliminary analysis of the ongoing workforce survey. *Appl Occup Environ Hyg 16*: 521–526

Deubner D, Kelsh M, Shum M, Maier L, Kent M, Lau E (2001b) Beryllium sensitization, chronic beryllium disease, and exposures at a beryllium mining and extraction facility. *Appl Occup Environ Hyg 16*: 579–592

DiPaolo JA, Casto BC (1979) Quantitative studies of *in vitro* morphological transformation of Syrian hamster cells by inorganic metal salts. *Cancer Res 39*: 1008–1013

Dunkel VC, Zeiger E, Brusick D (1984) Reproducibility of microbial mutagenicity assays: I. Tests with *Salmonella typhimurium* and *Escherichia coli* using a standardized protocol. *Environ Mutagen 6*: 1–254

Dutra FR (1949) Beryllium granulomas of the skin. *Arch Dermatol Syph 60*: 1140–1147

Dutra FR, Largent EJ (1950) Osteosarcoma induced by beryllium oxide. *Am J Pathol 26*: 197–209

Eisenbud M (1998) The standard for control of chronic beryllium disease. *Appl Occup Environ Hyg 13*: 25–31

Eisenbud M, Lisson J (1983) Epidemiological aspects of beryllium-induced nonmalignant lung disease: a 30-year update. *J Occup Med 25*: 196–202

Eisenbud M, Berghout CF, Steadman LT (1948) Environmental studies in plants and laboratories using beryllium: the acute disease. *J Ind Hyg Toxicol 30*: 281–285

Eisenbud M, Wanta RC, Dustan C, Steadman LT, Harris WB, Wolf BS (1949) Non-occupational berylliosis. *J Ind Hyg Toxicol 31*: 282–294

Finch GL, March TH, Hahn FF, Barr EB, Belinsky SA, Hoover MD, Lechner JF, Nikula FJ, Hobbs CH (1998) Carcinogenic responses of transgenic heterozygous p53 knockout mice to inhaled 239PuO$_2$ or metallic beryllium. *Toxicol Pathol 26*: 484–491

Fisher AA (1953) Nonsurgical treatment of cutaneous beryllium granuloma. *Arch Dermatol 68*: 214–216

Focher F, Verri A, Maga G, Spadari S, Hübscher U (1990) Effect of divalent and monovalent cations on calf thymus PCNA-independent DNA polymerase d and its 3'–5'exonuclease. *FEBS Lett 259*: 349–352

Folesky H (1967) Bemerkungen zum Beryllium-Granulom (Comments on beryllium granuloma) (German). *Berufsdermatosen 15*: 93–103

Goering PL, Aposhian HV, Mass MJ, Cebrian M, Beck BD, Waalkes MP (1999) The enigma of arsenic carcinogenesis: role of metabolism. *Toxicol Sci 49*: 5–14

Goodman JI, Watson RE (2002) Altered DNA methylation: a secondary mechanism involved in carcinogenesis. *Annu Rev Pharmacol Toxicol 42*: 501–525

Groth DH, Kommineni C, Mackay GR (1980) Carcinogenicity of beryllium hydroxide and alloys. *Environ Res 21*: 63–84

Haberman AL, Pratt M, Storrs FJ (1993) Contact dermatitis from beryllium in dental alloys. *Contact Dermatitis 28*: 157–162

Haley PJ, Finch GL, Hoover MD, Cuddihy RG (1990) The acute toxicity of inhaled beryllium metal in rats. *Fundam Appl Toxicol 15*: 767–778

Haley PJ, Finch GL, Mewhinney JA, Harmsen AG, Hahn FF, Hoover MD, Muggenburg BA, Bice DE (1989) A canine model of beryllium-induced granulomatous lung disease. *Lab Invest 61*: 219–227

Haley PJ, Pavia KF, Swafford DS, Davila DR, Hoover MD, Finch GL (1994) The comparative pulmonary toxicity of beryllium metal and beryllium oxide in cynomolgus monkeys. *Immunopharmacol Immunotoxicol 16*: 627–644

Hall RH, Scott JK, Laskin S (1950) Acute toxicity of inhaled beryllium. *Arch Ind Hyg Occup Med 12*: 25–48

Hardy HL, Tabershaw IR (1946) Delayed chemical pneumonitis occurring in workers exposed to beryllium compounds. *J Ind Hyg Toxicol 28*: 197–211

Harris J, Bartelson BB, Barker E, Balkissoon R, Kreiss K, Newman LS (1997) Serum neopterin in chronic beryllium disease. *Am J Ind Med 32*: 21–26

Henneberger PK, Cumro D, Deubner DD, Kent MS, McCawley M, Kreiss K (2001) Beryllium sensitization and disease among long-term and short-term workers in a beryllium ceramics plant. *Int Arch Occup Environ Health 74*: 167–176

Hinds MW, Kolonel LN, Lee J (1985) Application of a job-exposure matrix to a case-control study of lung cancer. *J Natl Cancer Inst 75*: 193–197

Hoagland MB, Grier RS, Hood MB (1950) Beryllium-induced osteogenic sarcomata. *Cancer Res 10*: 629–635

Hsie AW, Johnson NP, Couch DB (1979a) Quantitative mammalian cell mutagenesis and a preliminary study of the mutagenic potential of metallic compounds. in: Kharasch N (Ed.) *Trace metals in health and disease*, Raven Press, New York, 55–69

Hsie AW, O'Neill JP, San Sebastian JR (1979b) Quantitative mammalian cell genetic toxicology: study of the cytotoxicity and mutagenicity of seventy individual environmental agents related to energy technologies and three subfractions of crude synthetic oil in the CHO/HGPRT system. *Environ Sci Res 15*: 291–315

IARC (International Agency for Research on Cancer) (1993) *Beryllium, Cadmium, Mercury, and Exposures in the Glass Manufacturing Industry. IARC Monographs on the evaluation of carcinogenic risks to humans*, Vol 58, IARC, Lyon, France

Infante PF, Wagoner JK, Sprince NL (1980) Mortality patterns from lung cancer and nonneoplastic respiratory disease among white males in the beryllium case registry. *Environ Res 21*: 35–43

Ishinishi N, Mizunoe M, Inamasu T, Hisanaga A (1980) Experimental study of the carcinogenicity of beryllium oxide and arsenic trioxide to the lung of the rat by intratracheal instillation (jpn). *Fukuoka Igaku Zasshi 71*: 19–26

Johnson JS, Foote K, McClean M, Cogbill G (2001) Beryllium exposure control program at the Cardiff atomic weapons establishment in the United Kingdom. *Appl Occup Environ Hyg 16*: 619–630

Joseph P, Muchnok T, Ong TM (2001) Gene expression profile in BALB/c-3T3 cells transformed with beryllium sulfate. *Mol Carcinog 32*: 28–35

Kada T, Hirano K, Shirasu Y (1980) Screening of environmental chemical mutagens by the rec-assay system with *Bacillus subtilis*. in: deSerres FJ, Hollander A (Eds) *Chemical mutagens. Principles and methods for their detection*, Volume 6, Plenum Press, New York, 149–173.

Kanematsu N, Hara M, Kada T (1980) Rec assay and mutagenicity studies on metal compounds. *Mutat Res 77*: 109–116

Kelleher PC, Martyny JW, Mroz MM, Maier LA, Ruttenber AJ, Young DA, Newman LS (2001) Beryllium particulate exposure and disease relations in a beryllium machining plant. *J Occup Environ Med 43*: 238–249

Kent MS, Robins TG, Madl AK (2001) Is total mass or mass of alveolar-deposited airborne particles of beryllium a better predictor of the prevalence of disease? A preliminary study of a beryllium processing facility. *Appl Occup Environ Hyg 16*: 539–558

Kesheva N, Zhou G, Spruill M, Ensell M, Ong TM (2001) Carcinogenic potential and genomic instability of beryllium sulphate in BALB/c-3T3 cells. *Mol Cell Biochem 222*: 69–76

Kligman AM (1966) The identification of contact allergens by human assay. III. The maximization test: a procedure for screening and rating contact sensitizers. *J Invest Dermatol 47*: 393–409

Kreiss K, Mroz MM, Newman LS, Martyny J, Zhen B (1996) Machining risk of beryllium disease and sensitization with median exposures below 2 µg/m^3. *Am J Ind Med 30*: 16–25

Kreiss K, Mroz MM, Zhen B, Wiedemann H, Barna B (1997) Risks of beryllium disease related to work processes at a metal, alloy, and oxide production plant. *Occup Environ Med 54*: 605–612

Kuroda K, Endo G, Okamoto A, Yoo YS, Horiguchi S'i (1991) Genotoxicity of beryllium, gallium, and antimony in short-term assays. *Mutat Res 264*: 163–170

Larramendy ML, Popescu NC, Dipaolo JA (1981) Induction by inorganic metal salts of sister chromatid exchanges and chromosome aberrations in human and Syrian hamster cell strains. *Environ Mutagen 3*: 597–606

Lehnert NM, Gary RK, Marrone BL, Lehnert BE (2001) Inhibition of normal human fibroblast growth by beryllium. *Toxicology 160*: 119–127

Levy PS, Roth HD, Hwang PMT, Powers TE (2002) Beryllium and lung cancer: a reanalysis of a NIOSH cohort mortality study. *Inhalat Toxicol 14*: 1003–1015

Luke MZ, Hamilton L, Hollocher TC (1975) Beryllium-induced misincorporation by a DNA polymerase. Possible factor in beryllium toxicity. *Biochem Biophys Res Commun 62*: 497–501

MacMahon B (1994) The epidemiological evidence on the carcinogenicity of beryllium in humans. *J Occup Med 36*: 15–26

Maier LA (2001) Beryllium health effects in the era of the beryllium lymphocyte proliferation test. *Appl Occup Environ Hyg 16*: 514–520

Mancuso TF (1970) Relation of duration of employment and prior respiratory illness to respiratory cancer among beryllium workers. *Environ Res 3*: 251–275

Mancuso TF (1979) Occupational lung cancer among beryllium workers. in: Proceedings of the conference on occupational exposure to fibrous and particulate dust and their extensions into the environment. Lemen R, Dement J (Eds) *Dust and Disease*, Park Forest South, Pathotox Publishers Inc, 463–482

Mancuso TF (1980) Mortality study of beryllium industry workers' occupational lung cancer. *Environ Res 21*: 48–55

Mancuso TF, El-Attar AA (1969) Epidemiological study of the beryllium industry. Cohort methodology and mortality studies. *J Occup Med 11*: 422–434

Mandervelt C, Clottens FL, Demedts M, Nemery B (1997) Assessment of the sensitization potential of five metal salts in the murine local lymph node assay. *Toxicology 120*: 65–73

Markham TN (1996) Screening for chronic beryllium disease using beryllium specific lymphocyte proliferation testing. *Int Arch Occup Environ Health 68*: 405–407

McCawley MA, Kent MS, Berakis MT (2001) Ultrafine beryllium number concentration as a possible metric for chronic beryllium disease risk. *Appl Occup Environ Hyg 16*: 631–638

Messer RLW, Lucas LC (1999) Evaluations of metabolic activities as biocompatibility tools: a study of individual ions' effects on fibroblasts. *Dent Mater 15*: 1–6

Misra UK, Gawdi G, Pizzo SV, Lewis JG (1999) Exposure of cultured murine peritoneal macrophages to low concentrations of beryllium induces increases in intracellular calcium concentrations and stimulates DNA synthesis. *J Leukoc Biol 65*: 786–791

Miyaki M, Akamatsu N, Ono T, Koyama H (1979) Mutagenicity of metal cations in cultured cells from Chinese hamster. *Mutat Res 68*: 259–263

Müller-Quernheim J, Zissel G, Schopf R, Vollmer E, Schlaak M (1996) Differentialdiagnose Berylliose/Sarkoidose bei einem Zahntechniker (Differential diagnosis of berylliosis/sarcoidosis in a dental technician) (German). *Dtsch Med Wochenschr 121*: 1462–1466

Newman LS (1996) Significance of the blood beryllium lymphocyte proliferation test. *Environ Health Perspect 104*, Suppl 5: 953–956

Newman LS, Lloyd J, Daniloff E (1996) The natural history of beryllium sensitization and chronic beryllium disease. *Environ Health Perspect 104*, Suppl 5: 937–943

Newman LS, Mroz MM, Maier LA, Daniloff EM, Balkissoon R (2001) Efficacy of serial medical surveillance for chronic beryllium disease in a beryllium machining plant. *J Occup Environ Med 43*: 231–237

Nichol AD, Dominguez R (1949) Cutaneous granuloma from accidental contamination with beryllium phosphors. *J Am Med Assoc 140*: 855–860

Nickell-Brady C, Hahn FF, Finch GL, Belinsky SA (1994) Analysis of K-ras, p53 and c-raf-1 mutations in beryllium-induced rat lung tumors. *Carcinogenesis 15*: 257–262

Nikula KJ, Swafford DS, Hoover MD, Tohulka MD, Finch GL (1997) Chronic granulomatous pneumonia and lymphocytic responses induced by inhaled beryllium metal in A/J and C3H/HeJ mice. *Toxicol Pathol 25*: 2–12

Nishimura M (1966) Clinical and experimental studies on acute beryllium disease. *Nagoya J Med Sci 28*: 17–44

Nishioka H (1975) Mutagenic activities of metal compounds in bacteria. *Mutat Res 31*: 185–189.

Ogawa HI, Tsuruta S, Niyitani Y, Mino H, Sakata K, Kato Y (1987) Mutagenicity of metal salts in combination with 9-aminoacridine in *Salmonella typhimurium*. *Jpn J Genet 62*: 159–162

van Ordstrand HS, Hughes R, de Nardi JM (1945) Beryllium poisoning. *J Am Med Assoc 129*: 1084–1090

Paton GR, Allison AC (1972) Chromosome damage in human cell cultures induced by metal salts. *Mutat Res 16*: 332–336

Paustenbach DJ, Madl AK, Greene JF (2001) Identifying an appropriate occupational exposure limit (OEL) for beryllium: data gaps and current research initiatives. *Appl Occup Environ Hyg 16*: 527–538

Polak L, Turk JL, Frey JR (1973) Studies on contact hypersensitivity to chromium compounds. *Progr Allergy 17*: 145–226

Reeves AL, Deitch D, Vorwald AJ (1967) Beryllium carcinogenesis. I. Inhalation exposure of rats to beryllium sulfate aerosol. *Cancer Res 27*: 439–445

Richeldi L, Sorrentino R, Saltini C (1993) HLA-DPB1 glutamate 69: A genetic marker of beryllium disease. *Science 262*: 242–244

Richeldi L, Kreiss K, Mroz MM, Zhen B, Tartoni P, Saltini C (1997) Interaction of genetic and exposure factors in the prevalence of berylliosis. *Am J Ind Med 32*: 337–340

Rietschel RL, Fowler Jr JF (1995) *Fisher's contact dermatitis*. 4th edition, Williams and Wilkins, Baltimore, 822–823

Robinson FR, Schaffner F, Trachtenberg E (1968) Ultrastructure of the lungs of dogs exposed to beryllium-containing dusts. *Arch Environ Health 17*: 193–203

Rosenkranz HS, Poirer LA (1979) Evaluation of the mutagenicity and DNA-modifying activity of carcinogens and noncarcinogens in microbial systems. *J Natl Cancer Inst 62*: 873–892

Rossman MD (1996) Chronic beryllium disease: diagnosis and management. *Environ Health Perspect 104*, Suppl 5: 945–947

Rossman MD (2001) Chronic beryllium disease: a hypersensitivity disorder. *Appl Occup Environ Hyg 16*: 615–618

Rossman TG, Molina M (1986) The genetic toxicology of metal compounds: II. Enhancement by ultraviolet light-induced mutagenesis in *Escherichia coli* WP2. *Environ Mutagen 8*: 263–271

Rossman TG, Molina M, Meyer LW (1984) The genetic toxicology of metal compounds: I. Induction of prophage in *Escherichia coli* WP2s. *Environ Mutagen 6*: 59–69

Saltini C, Amicosante M, Franchi A, Lombardi G, Richeldi L (1998) Immunogenetic basis of environmental lung disease: lessons from the berylliosis model. *Eur Respir J 12*: 1463–1475

Saltini C, Richeldi L, Amicosante M, Voorter C, van den Berg-Loonen E, Dweik RA, Wiedemann HP, Deubner DC, Tinelli C (2001) Major histocompatibility locus genetic markers of beryllium sensitization and disease. *Eur Respir J 18*: 677–684

Sanders CL, Cannon WC, Powers GJ, Adee RR, Meier DM (1975) Toxicology of high-fired beryllium oxide inhaled by rodents. *Arch Environ Health 30*: 546–551

Sanderson WT, Henneberger PK, Martyny J, Ellis K, Mroz MM, Newman LS (1999) Beryllium contamination inside vehicles of machine shop workers. *Appl Occup Environ Hyg 14*: 223–230

Sanderson WT, Petersen MR, Ward EM (2001a) Estimating historical exposures of workers in a beryllium manufacturing plant. *Am J Ind Med 39*: 145–157

Sanderson WT, Ward EM, Steenland K, Petersen MR (2001b) Lung cancer case-control study of beryllium workers. *Am J Ind Med 39*: 133–144

Savitz DA, Whelan EA, Kleckner RC (1989) Effects of parents' occupational exposures on risk of stillbirth, preterm delivery, and small-for-gestational age infants. *Am J Epidemiol 129*: 1201–1218

Schepers GWH (1961) Neoplasia experimentally induced by beryllium compounds. *Progr Exp Tumor Res 2*: 203–244

Schepers GWH (1964) Biological action of beryllium: reaction of the monkey to inhaled aerosols. *Ind Med Surg 33*: 1–16

Schepers GWH, Durkan TM, Delahant AB, Creedon FT (1957) The biological action of inhaled beryllium sulfate. *Arch Ind Health 15*: 32–58

Schönherr S, Pevny I (1985) Berylliumallergie (Allergy to beryllium) (German). *Arbeitsmed Sozialmed Praeventivmed 20*: 281–286

Schreiber J, Zissel G, Greinert U, Galle J, Schulz KH, Schlaak M, Müller-Quernheim J (1999) Diagnostik der chronischen Berylliose (Diagnosis of chronic berylliosis) (German). *Pneumologie 53*: 193–198

Schroeder HA, Mitchener M (1975a) Life-term studies in rats: effects of aluminum, barium, beryllium, and tungsten. *J Nutr 105*: 421–427

Schroeder HA, Mitchener M (1975b) Life-term effects of mercury, methyl mercury, and nine other trace metals on mice. *J Nutr 105*: 452–458

Sendelbach LE, Witschi HP (1987) Protection by parenteral iron administration against the inhalation toxicity of beryllium sulfate. *Toxicol Lett 35*: 321–325

Simmon VF (1979) In vitro mutagenicity assays of chemical carcinogens and related compounds with *Salmonella typhimurium*. *J Natl Cancer Inst 63*: 893–899

Sirover RMA, Loeb LA (1976) Metal-induced infidelity during DNA synthesis. *Proc Natl Acad Sci USA 73*: 2331–2335

Sneddon IB (1955) Berylliosis: a case report. *Br Med J I*: 1448–1450

Stange AW, Furman FJ, Hilmas DE (1996a) Rocky Flats beryllium health surveillance. *Environ Health Perspect 104*, Suppl 5: 981–986

Stange AW, Hilmas DE, Furman FJ (1996b) Possible health risks from low level exposure to beryllium. *Toxicology 111*: 213–224

Steenland K, Ward E (1991) Cancer incidence among patients with beryllium disease: a cohort mortality study. *J Natl Cancer Inst 83*: 1380–1385

Stokinger HE, Sprague GF, Hall RH (1950) Acute inhalation toxicity of beryllium. I. Four definitive studies of beryllium sulfate at exposure concentrations of 100, 50, 10 and 1 mg per cubic meter. *Arch Ind Hyg Occup Med 1*: 379–397

Stubbs J, Argyris E, Lee CW, Monos D, Rossman MD (1996) Genetic markers in beryllium hypersensitivity. *Chest 109*, Suppl 3: 45

Tinkle SS, Newman LS (1997) Beryllium-stimulated release of tumor necrosis factor-α, interleukin-6, and their soluble receptors in chronic beryllium disease. *Am J Respir Crit Care Med 156*: 1884–1891

Tinkle SS, Schwitters PW, Newman LS (1996) Cytokine production by bronchoalveolar lavage cells in chronic beryllium disease. *Environ Health Perspect 104*, Suppl 5: 969–971

Tinkle SS, Kittle LA, Schumacher BA, Newman LS (1997) Beryllium induces IL-2 and IFN-γ in berylliosis. *J Immunol 158*: 518–526

Tinkle SS, Kittle LA, Newman LS (1999) Partial IL-10 inhibition of the cell-mediated immune response in chronic beryllium disease. *J Immunol 163*: 2747–2753

Tso W-W, Fung W-P (1981) Mutagenicity of metallic cation. *Toxicol Lett 8*: 195–200

Ulitzur S, Barak M (1988) Detection of genotoxicity of metallic compounds by the bacterial bioluminescence test. *J Biolumin Chemilumin 2*: 95–99

US EPA (Environmental Protection Agency) (1998) *Toxicological Review of Beryllium and Compounds. In support of Summary Information on the Integrated Risk Information System (IRIS)*, Washington DC

Venugopal B, Luckey TD (1978) *Metal toxicity in mammals. Volume 2: Chemical toxicity of metals and metalloids.* Plenum Press, New York, London, 43–50

Vilaplana J, Romaguera C, Grimalt F (1992) Occupational and non-occupational allergic contact dermatitis from beryllium. *Contact Dermatitis 26*: 295–298

Vilaplana J, Romaguera C, Cornellana F (1994) Contact dermatitis and adverse oral mucous membrane reactions related to the use of dental prostheses. *Contact Dermatitis 30*: 80–84

Vorwald AJ (1966) Biologic manifestations of toxic inhalants in monkeys. in: Vagtborg H (Ed.) *Use of nonhuman primates in drug evaluation*, University of Texas Press, Austin & London, 222–228

Vorwald AJ (1968) Biologic manifestations of toxic inhalants in monkeys. in: Vagtborg H (Ed.) *Use of nonhuman primates in drug evaluation: a symposium.* Southwest Foundation for Research and Education, University of Texas Press, Austin, Texas, 222–228

Vorwald AJ, Pratt PC, Urban ECJ (1955) The production of pulmonary cancer in albino rats exposed by inhalation to an aerosol of beryllium sulfate (Abstract). *Acta Union Int Contra Cancrum 11*: 735

Vorwald AJ, Reeves AL (1959) Pathologic changes induced by beryllium compounds. *Arch Ind Health 19*: 190–199

Wagner WD, Groth DH, Holtz JL, Madden GE, Stokinger HE (1969) Comparative chronic inhalation toxicity of beryllium ores, bertrandite, and beryl, with production of pulmonary tumors by beryl. *Toxicol Appl Pharmacol 15*: 10–29

Wagoner JK, Infante PF, Bayliss DL (1980) Beryllium: an etiologic agent in the induction of lung cancer, nonneoplastic respiratory disease, and heart disease among industrially exposed workers. *Environ Res 21*: 15–34

Wang Z, Farris GM, Newman LS, Shou Y, Maier LA, Smith HN, Marrone BL (2001) Beryllium sensitivity is linked to HLA-DP genotype. *Toxicology 165*: 27–38

Ward E, Okun A, Ruder A, Fingerhut M, Steenland K (1992) A mortality study of workers at seven beryllium processing plants. *Am J Ind Med 22*: 885–904

Wegner R, Heinrich-Ramm R, Nowak D, Olma K, Poschadel B, Szadkowski D (2000) Lung function, biological monitoring, and biological effect monitoring of gemstone cutters exposed to beryls. *Occup Environ Med 57*: 133–139

WHO (World Health Organization) (1990) *Beryllium*. IPCS (International Programme on Chemical Safety) Environmental Health Criteria 106, WHO, Geneva, Switzerland

WHO (World Health Organization) (2001) *Beryllium and beryllium compounds.* IPCS (International Programme on Chemical Safety). Concise International Chemical Assessment Document 32, WHO, Geneva, Switzerland

Williams GM, Mori H, McQueen CA (1989) Structure-activity relationships in the rat hepatocyte DNA-repair test for 300 chemicals. *Mutat Res 221*: 263–286

Williams WJ (1996) United Kingdom beryllium registry: mortality and autopsy study. *Environ Health Perspect 104*, Suppl 5: 949–951

Williams WJ, Wallach ER (1989) Laser microprobe mass spectometry (LAMMS) analysis of beryllium, sarcoidosis and other granulomatous diseases. *Sarcoidosis 6*: 111–117

Yoshida T, Shima S, Nagaoka K, Taniwaki H, Wada A, Kurita H, Morita K (1997) A study on the beryllium lymphocyte transformation test and the beryllium levels in working environment. *Ind Health 35*: 374–379

Zakour RA, Glickman BW (1984) Metal-induced mutagenesis in the lacI gene of *Escherichia coli*. *Mutat Res 126*: 9–18

Zissu D, Binet S, Cavelier C (1996) Patch testing with beryllium alloy samples in guinea pigs. *Contact Dermatitis 34*: 196–200

Zorn H, Fischer G (1998) Leichtmetalle – Beryllium. Kapitel IV-2.2.5 (Light metals – beryllium. Chapter IV-2.2.5) (German). in: Konietzko J, Dupuis H (Eds) *Handbuch der Arbeitsmedizin*, Ecomed-Verlag, Landsberg, 1–19

Zschunke E, Folesky H (1969) Experimentelle Untersuchungen zur Frage der Sensibilisierung gegen Beryllium (Experimental studies on the problem of sensitivity to beryllium) (German). *Hautarzt 20*: 403–404

completed 28.02.2003

Butanethiol

MAK value (1969)	0.5 ml/m^3 (ppm) ≙ 1.87 mg/m^3
Peak limitation (2002)	Category II, excursion factor 2
Absorption through the skin	–
Sensitization	–
Carcinogenicity	–
Prenatal toxicity (2000)	Group C
Germ cell mutagenicity	–
BAT value	–
Synonyms	normal butyl thioalcohol thiobutyl alcohol *N*-butyl mercaptan
Chemical name (CAS)	1-butanethiol
CAS number	109-79-5
Structural formula	H$_3$C–(CH$_2$)$_3$–SH
Molecular formula	C$_4$H$_{10}$S
Molecular weight	90.18
Melting point	–116 °C
Boiling point	98 °C
Density at 20 °C	0.834 g/cm^3
Vapour pressure at 20 °C	40 hPa
log P$_{ow}$*	2.28
1 ml/m^3 (ppm) ≙ 3.742 mg/m^3	**1 mg/m^3 ≙ 0.267 ml/m^3 (ppm)**

The MAK value for butanethiol of 0.5 ml/m^3 was established in 1969 in analogy to the TLV value at the time. The present documentation is based on reviews of the toxicological data for butanethiol (Farr and Kirwin 1994, NL Health Council 1998).

---

* *n*-octanol/water distribution coefficient

Butanethiol is used as a solvent, as an intermediate in the synthesis of certain pesticides and, as a result of its odour, as an additive with warning function (Farr and Kirwin 1994).

The substance has an extremely unpleasant odour. The perception threshold is between 0.0001 and 0.01 ml/m^3 (Blinova 1965, Farr and Kirwin 1994, Ruth 1986).

1 Toxic Effects and Mode of Action

Animal studies of the acute toxicity of butanethiol indicate that the substance is absorbed via the respiratory passages and gastrointestinal tract. Mercaptans are oxidized in the liver and form, among other things, sulfides and sulfates. In man, single exposures to butanethiol concentrations of 50 to 500 ml/m^3 cause central nervous effects. Butanethiol causes mild irritation of the eyes and mucous membranes. There is no evidence the substance has sensitizing effects. The NOAEL (no observed adverse effect level) in the medium-term inhalation study with rats was found to be 9 ml/m^3. Butanethiol was not found to cause developmental toxicity at concentrations that were non-toxic or only slightly toxic to the dams. A *Salmonella* mutagenicity test, SCE test and mouse lymphoma test did not provide conclusive evidence of genotoxic effects.

2 Mechanism of Action

Butanethiol binds to the central iron ion of enzymes containing haem, e.g. cytochrome P450, chloroperoxidase (Nastainczyk *et al.* 1976, Sono *et al.* 1984) and inhibits cytochrome c oxidase *in vitro*. However, as a result of the steric hindrance caused by the butyl residue, the dissociation constant of the enzyme inhibitor complex (K_i) is much higher than that for alkyl mercaptans with shorter chain lengths (Wilms *et al.* 1980). Substance-related changes in the enzyme activities of acetyl and butyryl cholinesterases were detected in chickens (see Section 5.7).

Generally, autoxidation of thiols can yield reactive oxygen species in the presence of suitable metal ions. The disulfides formed in this way can be reduced back to thiols. This redox cycling can lead to oxidative stress (Munday 1989).

3 Toxicokinetics and Metabolism

On the basis of animal studies of the acute toxicity of butanethiol, the substance can be expected to be absorbed via the respiratory passages and gastrointestinal tract. Quantitative data are not available.

It is known that, in general, mercaptans are oxidized in the liver to form, among other things, sulfides and sulfates (Farr and Kirwin 1994).

The half-life of butanethiol in the blood plasma of hens administered single oral doses of 100 mg/kg body weight is given as eight days (Abou-Donia *et al.* 1979).

4 Effects in Man

When a polymerization autoclave in which butanethiol had evidently developed from the stabilizer *n*-octylmercaptan was opened, seven employees suffered intoxication. Butanethiol was detected by gas chromatography in an air sample taken 3 hours after the accident. Based on the strength of the odour, the exposure to butanethiol was estimated to be 50 to 500 ml/m^3 for a period of about an hour. All persons complained of indisposition and muscle weakness and most of them also of a feeling of warmth, increased sweat production, headaches, nausea and vomiting. The other symptoms, such as lethargy, excitability, dizziness, visual disorders and confusion, occurred sporadically. One of the exposed workers was unconscious for several minutes. Clinical and biochemical examinations of the exposed persons 30 to 40 days after the accident did not yield evidence of sequelae; liver and kidney values were in the normal range (Gobbato and Terribile 1968).

5 Animal Experiments and *in vitro* Studies

5.1 Acute toxicity

Independent of the administration route, high alkyl mercaptan doses cause depression of the central nervous system in rodents; death occurs as a result of respiratory arrest. Symptoms of intoxication are restlessness, an increased respiration rate, co-ordination impairment, muscle weakness and skeletal muscle paralysis, cyanosis, lethargy, sedation, respiratory depression and coma (Fairchild and Stokinger 1958).

Inhalation exposure of rats to butanethiol concentrations of 200000 mg/m^3 (about 53400 ml/m^3) led to the death of all the animals within 97 minutes (Farr and Kirwin

1994). The 4-hour LC_{50} for rats is given as 4020 ml/m^3 (Fairchild and Stokinger 1958) and 6060 ml/m^3 (Farr and Kirwin 1994) and that for mice as 2500 ml/m^3 (Fairchild and Stokinger 1958).

Oral LD_{50} values in rats are 1500 mg/kg body weight (Fairchild and Stokinger 1958) and 1800 mg/kg body weight (Farr and Kirwin 1994). After intraperitoneal administration of the substance, an LD_{50} value of 399 mg/kg body weight was determined for rats (Fairchild and Stokinger 1958). The dermal LD_{50} value for rabbits is above 34600 mg/kg body weight (Farr and Kirwin 1994).

5.2 Subacute, subchronic and chronic toxicity

In the dose-finding study for a 13-week experiment (see below), groups of 10 male and 10 female Sprague-Dawley rats were exposed to butanethiol concentrations of about 200, 1100 and 1900 ml/m^3 in a whole animal chamber 6 hours a day for 14 days. In the low exposure group the relative kidney weights were increased in both sexes and, in addition, the lung and trachea weights in the male animals. In the male and female animals of the middle exposure groups the relative weights of the kidneys, spleen, lungs, trachea and heart were significantly increased. The kidneys were darkly discoloured. The highest concentration was lethal for all the rats, the middle concentration for one rat. The body weights of the rats of these two groups were lower than those of the controls after the first week and also at the end of the experiment in the animals that survived. Autopsy of the dead animals of the high exposure group revealed congestion in the lungs. The kidneys of the dead rats from the high exposure group and those of the control animals were the only organs to be examined histologically. Tubule degeneration was detected in the kidneys of the exposed rats (Phillips Petroleum Company 1992). The kidney changes and the lung congestion may be secondary effects resulting from preterminal cardiovascular collapse.

In the subsequent experiment, groups of 15 male and 15 female Sprague-Dawley rats were exposed to average butanethiol concentrations of 9, 70 and 150 ml/m^3 for 6 hours a day on 5 days a week for 13 weeks. With the exception of one rat from the middle exposure group, all the animals survived. According to the authors, the treatment did not have a toxicologically relevant impact on body weight gains nor on haematological or clinico-chemical parameters. The slight, but significant decrease in the erythrocyte count reported in female rats (average of 7.91 in the controls down to 7.64, 7.25 and 6.97 × 10^6/mm^3 in the exposed groups), which correlated with a slight decrease in haemoglobin, and the reduced lymphocyte and neutrophil counts observed in the high exposure group are regarded by the authors as biologically not significant and as not exposure-related, as the changes are within the range of the historical control data of the laboratory. Reported as exposure-related changes were significantly increased lung weights in the male rats at concentrations of 70 ml/m^3 (1.73 in the controls, 1.77, 2.03 and 1.78 g in the three exposed groups) and slight to moderate fibrosis of the lungs in rats of both sexes at 150 ml/m^3. The biological relevance of the increased lung weights is questionable as it was not dose-dependent and was not observed in the female rats. The authors suggest the increased lung weights are connected with the histologically detected lung fibrosis in the rats treated with 150 ml/m^3. The suspected connection cannot, however, be demonstrated experimentally, as the lungs of the rats treated with 70 and 9 ml/m^3 were not histologically examined. The behaviour of the rats during exposure is not reported. It is merely stated that, in general, no signs of exposure-related toxic changes were observed in the rats at the weekly evaluation (Phillips Chemical Company 1982). The NOAEL in the study was 9 ml/m^3. However, only the animals in the highest concentration group were examined histopathologically.

Oral doses of butanethiol over four days (days 1 and 2: 200 mg/kg body weight, days 3 and 4: 400 mg/kg body weight) caused necrosis of the adrenal cortex in two of five female Sprague-Dawley rats (Szabo and Reynolds 1975).

5.3 Local effects on skin and mucous membranes

In various studies butanethiol is described as not irritative to slightly irritative to the skin (no other details; Farr and Kirwin 1994).

Various alkyl mercaptans, including butanethiol, were tested in the rabbit and found to cause weak to moderate mucosal irritation (Fairchild and Stokinger 1958). In another study, impairment of the iris within the first 24 hours and moderate to slight conjunctival irritation up to 72 hours after the application were observed (no other details; Farr and Kirwin 1994). Mucosal irritation was likewise observed in rats and mice after inhalation exposure to high concentrations of butanethiol (Fairchild and Stokinger 1958).

5.4 Allergenic effects

In a study to test the contact sensitizing properties of various thiols (which does not meet present-day requirements), 200 µl of a 20 % butanethiol solution in acetone was applied to the shaved skin of guinea pigs on ten consecutive days. No evidence was found of sensitizing effects (Cirstea 1972).

5.5 Reproductive and developmental toxicity

Groups of 5 Charles-River-COBS-CD rats were exposed to a butanethiol atmosphere of about 200, 1100 or 1900 ml/m^3 from days 6 to 20 of gestation for 6 hours a day for two weeks. While all the animals of the high exposure group died after two to three days, four of five rats survived exposure to concentrations of about 1100 ml/m^3. In the animals of these groups, a red nasal matting and discharge, a yellow-brown or red anogenital discharge, and emaciation were observed. The body weight gains of the dams of all exposure groups were reduced. In the middle exposure group, the percentage of post-implantation losses resulting from early resorptions was 100 %, although the number of *corpora lutea* and implantations was unchanged. In the low exposure group no effects were seen on the number of *corpora lutea*, implantations or resorptions (Phillips Petroleum Company 1992).

In a second study, groups of 25 pregnant COBS-CD rats were exposed in whole animal chambers to butanethiol concentrations of 10, 68 or 152 ml/m^3 from days 6 to 19 of gestation for 6 hours a day. The body weight gains of the dams were reduced at concentrations of 68 and 152 ml/m^3 in the high exposure group; in addition, a slight increase in the incidence of hair loss on the limbs was registered. Substance-related embryotoxic or foetotoxic effects were not observed (Thomas *et al.* 1987). In this study the NOEL (no observed effect level) for developmental toxicity in the rat was found to be 152 ml/m^3.

Groups of 25 CD-1 mice treated in the same way (days 6 to 16 of gestation) were found to be more sensitive. In the low exposure group (10 ml/m^3) 2 animals showed unkempt appearance. In all exposure groups body weight gains were delayed and in some cases weight loss was observed. In the two high exposure groups the concentrations were clearly lethal for the dams (8/25, 9/25) and embryos. The increased incidences of skeletal and visceral variations were not statistically significant in the low and middle exposure groups and were within the range of the historical control data. The incidence of malformations was slightly increased in the low exposure group and significantly increased in the middle exposure group. This is in particular the result of an increase in the incidence of cleft palates in several litters (0.7 % in the controls, 1.5 % and 3.4 % in the low and middle exposure groups); according to the authors the incidence in the control group is unusually high. The authors attribute the malformations to stress resulting from a lack of food and water for 6 hours and to the maternally toxic and lethal effects of the substance (Thomas *et al.* 1987). Therefore, the embryotoxic effects at a concentration of 10 ml/m^3 did not differ significantly from the data for the control animals in this study.

5.6 Genotoxicity

Various *in vitro* tests for the genotoxicity of butanethiol were carried out by a test laboratory. There is only inadequate information available about the implementation of the tests summarized below, which were carried out with material of unknown purity. They can therefore only be included in the evaluation of the genotoxicity of butanethiol with reservations.

In a *Salmonella* mutagenicity test, butanethiol dissolved in DMSO was not found to have mutagenic effects in the strains TA98, TA100, TA1535, TA1537 or TA1538 in the absence or presence of a metabolic activation system. The tested concentrations were 123.5 to 10000 µg/plate; the highest concentration represents the limit of solubility. Cytotoxic effects were observed, depending on the strain of bacteria, after concentrations of 1111 µg/plate and more (Phillips Petroleum Company 1982a).

In a test for sister chromatid exchange (SCE) with CHO cells (a cell line derived from Chinese hamster ovary), butanethiol (tested concentration range 1.2–124 µg/ml) caused a statistically significant, yet biologically probably irrelevant increase in the SCE count per cell of 1.2 and 1.3 times the control value after the cultures were exposed to concentrations of 12.4 and 41 µg/ml in the presence of a metabolic activation system. The highest tested concentration of 124 µg/ml inhibited growth (Phillips Petroleum Company 1982b).

In a mutation test with L5178Y mouse lymphoma cells, butanethiol was tested in concentrations of 12 to 300 µg/ml both with and without a metabolic activation system. The highest tested concentration reduced the survival of the cells to 40 % in the absence of a metabolic activation system and caused practically all the cells to die in the presence of such a system. Increases in the incidence of mutants of 1.7 to 2.9 times the control value were observed in the presence of the activation system; neither the relative survival of the cells nor the mutagenicity of the substance was found to be concentration-dependent (Phillips Petroleum Company 1982c). No details were given about the size distribution of the induced mutant colonies. As the effects were not found to be concentration-dependent, the results of this test cannot be regarded as clearly positive.

5.7 Other effects

In hens given single butanethiol doses of 1000 mg/kg body weight in a gelatine capsule, leg weakness and unsteadiness was observed 6 to 12 hours after administration of the substance; a watery yellowish liquid was drained from the hens' beaks. Later, the state of the hens' health deteriorated visibly: the hens showed general weakness, loss of balance, loss of co-ordination, disorientation, diarrhoea, anxiety, shortness of breath and tremor. Two days after administration of the dose all 3 hens died. A dose of 400 mg/kg body weight caused milder symptoms of intoxication, which mainly developed after 2 days; all 3 animals of this group survived the 30-day follow-up period. Partial haemolysis was observed in blood samples from both dose groups. A dose of 100 mg/kg body weight did not cause any noticeable changes in behaviour; biochemical examination revealed a slight increase in the acetylcholinesterase activity in the brain after 30 days, while the activity of plasma butyryl cholinesterase returned to the level in the controls after an initial increase. No histological changes were found in the brain, spinal cord and ischias nerve of the animals from any of the treated groups (Abou-Donia 1979, Abou-Donia et al. 1979).

A follow-up experiment with chickens (single butanethiol doses of 100–500 mg/kg body weight) confirmed the initial increase in the plasma butyryl cholinesterase activity. In addition, a reduction in the haemoglobin concentration, the haematocrit value, the erythrocyte count and the glucose-6-phosphate dehydrogenase activity was determined in the high dose group; the methaemoglobin level was increased and the erythrocytes were lysed, fragmented or contained Heinz bodies (Abdo et al. 1983).

Hens were given oral doses of butanethiol for 28 days (200 mg/kg body weight and day) and 90 days (50 mg/kg body weight and day). The treatment caused marked inhibition of the acetylcholinesterase activity in the brains of the birds, while the activity of the neuropathy target esterase in the brain and that of plasma butyryl cholinesterase increased (no other details; Abou-Donia and Abd-Elfattah 1984).

There are no data available for carcinogenic effects of butanethiol.

6 Manifesto (MAK value/classification)

After occupational exposure to butanethiol the critical effects are probably the central nervous effects. The data obtained in man are, however, insufficient for the evaluation of a threshold limit value. There is evidence the substance has irritative effects on the mucous membranes. The threshold concentration for this effect cannot be evaluated from the data in man or experiments with animals.

In a 13-week inhalation study with rats, the only target organ was the lung. At 150 ml/m^3 fibrosis was observed; the animals of the other concentration groups were not histopathologically examined. At 9 ml/m^3 lung weights were not increased, and this concentration is therefore to be regarded as the NOAEL. It cannot be excluded that the behavioural changes decisive for the setting of the threshold limit value for methyl mercaptan (which were observed in a 13-week inhalation study with rats even at the lowest tested concentration of 2 ml/m^3) did occur in a similar form in this study as well, but were not documented. Because of these uncertainties, the data for the structurally similar methyl mercaptan, which is of greater acute toxicity, are also included in the evaluation since this compound has been more extensively investigated. In view of the results of animal experiments with butanethiol as well as with methyl mercaptan (on the basis of which the MAK value for methyl mercaptan of 0.5 ml/m^3 was retained), there is no necessity to change the MAK value for butanethiol. In addition, it is unclear at which concentration the unpleasant odour of butanethiol represents an intolerable nuisance for man. Therefore, the MAK value for butanethiol of 0.5 ml/m^3 has been retained provisionally.

As a result of the odour nuisance, butanethiol is classified in Peak limitation category II with an excursion factor of 2.

The NOEL for developmental toxicity for the rat is 152 ml/m^3 and for the mouse 10 ml/m^3. A risk of prenatal toxicity is not to be expected given observance of the MAK value of 0.5 ml/m^3. Butanethiol is therefore classified in Pregnancy risk group C.

The available *in vitro* studies of the genotoxicity of butanethiol do not yield any clear evidence of mutagenic effects. There are no *in vivo* studies of the genotoxicity or the carcinogenicity of the substance. The carcinogenic and germ cell mutagenic potential of butanethiol cannot therefore be evaluated.

There is no evidence the substance has sensitizing effects. Designation with an "Sa" or "Sh" is therefore not considered to be necessary. Data for penetration of the skin are not available; therefore a decision whether designation with an "H" is necessary cannot be made.

7 References

Abdo KM, Timmons PR, Graham DG, Abou-Donia MB (1983) Heinz body production and hematological changes in the hen after administration of a single oral dose of *n*-butyl mercaptan and *n*-butyl disulfide. *Fundam Appl Toxicol 3*: 69–74

Abou-Donia MB (1979) Late acute effect of *S,S,S*-tributyl phosphorotrithioate (DEF) in hens. *Toxicol Lett 4*: 231–236

Abou-Donia MB, Abd-Elfattah AS (1984) Effects of subchronic and acute oral administration of *S,S,S*-tri-*n*-butyl phosphorotrithioate (DEF) and its metabolite *n*-butyl mercaptan (nBM) on hens neurotoxic esterase and cholinesterase activities. *Fed Proc 43*: Abstr. 346

Abou-Donia MB, Graham DG, Timmons PR, Reichert BL (1979) Delayed neurotoxic and late acute effects of *S,S,S*-tributyl phosphorotrithioate on the hen: effect of route of administration. *Neurotoxicology 1*: 425–447

Blinova EA (1965) Setting standards for the concentration of substances having a strong odour in the ambient air of industrial plants (Russian). *Gig Sanit 30*: 18–22, translation

Cirstea M (1972) Studies on the relation between chemical structure and contact sensitizing capacity of some thiol compounds. *Rev Roum Physiol 9*: 485–491

Fairchild EJ, Stokinger HE (1958) Toxicologic studies on organic sulfur compounds. 1. Acute toxicity of some aliphatic and aromatic thiols (mercaptans). *Am Ind Hyg Assoc J 19*: 171–189

Farr CH, Kirwin CJ (1994) Organic sulfur compounds. in: Clayton GD, Clayton FE (Eds) *Patty's industrial hygiene and toxicology*, Vol 2F, 4th edition, John Wiley & Sons, New York, 4311–4372

Gobbato F, Terribile PM (1968) Proprietà tossicologiche dei mercaptani (Toxicologic properties of mercaptans) (Italian). *Folia Med (Naples) 51*: 329–341

Munday R (1989) Toxicity of thiols and disulphides: involvement of free-radical species. *Free Radical Biol Med 7*: 659–673

Nastainczyk W, Ruf HH, Ullrich V (1976) Binding of thiols to microsomal cytochrome P-450. *Chem Biol Interact 14*: 251–263

NL Health Council (1998) *Butanethiol*. Committee on Updating of Occupational Exposure Limits, Public Draft, December 1998, Health Council of the Netherlands, The Hague

Phillips Chemical Company (1982) *Thirteen week inhalation toxicity study of n-butyl and t-butyl mercaptan in rats*, International Research and Development Corporation. Phillips Petroleum Company, Bartlesville, OK, USA, unpublished

Phillips Petroleum Company (1982a) *Salmonella typhimurium mammalian microsome plate incorporation assay with n-butyl mercaptan*, Hazleton Laboratories Inc. Phillips Petroleum Company, Bartlesville, OK, USA, unpublished

Phillips Petroleum Company (1982b) *In vitro sister chromatid exchange in Chinese hamster ovary cells – n-butyl mercaptan*, Hazleton Laboratories Inc. Phillips Petroleum Company, Bartlesville, OK, USA, unpublished

Phillips Petroleum Company (1982c) *Mouse lymphoma forward mutation assay – n-butyl mercaptan*, Hazleton Laboratories Inc. Phillips Petroleum Company, Bartlesville, OK, USA, unpublished

Phillips Petroleum Company (1992) *Initial Submission: two week range-finding inhalation toxicity study in rats with n-butyl mercaptan with cover letter dated 08/24/92*. International Research and Development Corporation, NTIS/OTS 0555470, EPA/OTS Doc ID 88-920009417, NTIS, Springfield, VA, USA

Ruth JH (1986) Odor thresholds and irritation levels of several chemical substances: a review. *Am Ind Hyg Assoc J 47*: A 142–A 151

Sono M, Dawson JH, Hager LP (1984) The generation of a hyperporphyrin spectrum upon thiol binding to ferric chloroperoxidase. *J Biol Chem 259*: 13209–13216

Szabo S, Reynolds ES (1975) Structure-activity relationships for ulcerogenic and adrenocorticolytic effects of alkyl nitriles, amines, and thiols. *Environ Health Perspect 11*: 135–140

Thomas WC, Seckar JA, Johnson JT, Ulrich CE, Klonne DR, Schardein JL, Kirwin CJ (1987) Inhalation teratology studies of n-butyl mercaptan in rats and mice. *Fundam Appl Toxicol 8*: 170–178

Wilms J, Lub J, Wever R (1980) Reactions of mercaptans with cytochrome c oxidase and cytochrome c. *Biochim Biophys Acta 589*: 324–335

completed 24.02.2000

Carbon disulfide

MAK value (1997)	5 ml/m³ (ppm) ≙ 16 mg/m³
Peak limitation (2001)	Category II, excursion factor 2
Absorption through the skin (1980)	H
Sensitization	–
Carcinogenicity	–
Prenatal toxicity (1985)	Pregnancy risk group B
Germ cell mutagenicity	–
BAT (1997)	4 mg 2-thio-thiazolidin-4-carboxyl acid (TTCA)/g creatinine
Synonyms	carbon bisulfide carbon sulfide dithiocarbonic anhydride
Chemical name (CAS)	carbon disulfide
CAS number	75-15-0
Structural formula	S=C=S
Molecular formula	CS_2
Molecular weight	76.14
Melting point	−112°C
Boiling point	46°C
Vapour pressure at 20°C	400 hPa
1 ml/m³ (ppm) ≙ 3.16 mg/m³	**1 mg/m³ ≙ 0.317 ml/m³ (ppm)**

The toxicity of carbon disulfide was evaluated by the Commission in 1975 and 1997 (see Henschler 1975 and the 1997 documentation for Carbon disulfide in Volume 12 of the present series). The toxic effects discussed in these MAK documentations and the report from the Beratergremium für Altstoffe (BUA 1991) are summarised.

The long-known toxic effects of carbon disulfide on the central and peripheral nervous system have been confirmed in many studies of workers in the viscose industry. The peripheral neurotoxicity of carbon disulfide manifests itself in a characteristic polyneuropathy with an irreversible slowing of the sensorimotor nerve conduction velocity. It is dependent on the cumulative exposure (product of concentration and length of exposure in years) (Henschler 1975, the 1997 documentation for Carbon disulfide in volume 12 of the present series). On this basis a NOEC (no observed effect concentration) for 40 years' exposure to carbon disulfide (8 hours/day, 5 days/week) of 4 ml/m^3 (about 12 mg/m^3) was calculated (Ruijten *et al.* 1990), which is beset with considerable uncertainties, however, since the earlier higher concentrations were not taken properly into account (see the 1997 documentation for Carbon disulfide in Volume 12 of the present series). In a cross-sectional study of 247 male workers with a median carbon disulfide concentration of 4 ml/m^3 (about 12 mg/m^3), a maximum concentration of 65.7 ml/m^3 (208 mg/m^3) and a median length of exposure of 66 months (4–220 months), no indications of neurotoxic effects compared with a group of 222 controls were found (Reinhardt *et al.* 1997a, 1997b, the 1997 documentation for Carbon disulfide in Volume 12 of the present series).

The vestibular damage, such as nystagmus and impaired hearing, observed in viscose workers after exposure to carbon disulfide concentrations of more than 30 mg/m^3 (about 10 ml/m^3) and balance disturbances are attributed to the neurotoxicity of carbon disulfide (BUA 1991). In a study of 74 workers exposed to carbon disulfide, effects of this kind were reported in workers with more than 20 years' exposure. Current exposure concentrations were 15 mg/m^3 (about 5 ml/m^3) on average, while earlier concentrations were higher (no further details) (Hirata *et al.* 1992, the 1997 documentation for Carbon disulfide in Volume 12 of the present series).

In many studies of viscose workers, vascular changes in the retina of the eye with microaneurisms, haemorrhages, vascular scleroses and pigment changes were described after several years' exposure to carbon disulfide concentrations of more than 30 mg/m^3 (10 ml/m^3) (BUA 1991, Vanhoorne *et al.* 1995, 1996). The studies carried out do not show a uniform picture in the different countries, so that ethnic differences have been postulated (BUA 1991). In the authors' opinion (Vanhoorne *et al.* 1995) it is still not clear, however, what effects on the eye are caused by carbon disulfide, by hydrogen sulfide and by a combination of the two (see the 1997 documentation for Carbon disulfide in Volume 12 of the present series).

After many years' exposure to carbon disulfide, increased blood pressure, angina pectoris, accelerated arteriosclerosis and higher mortality from myocardial infarct can be observed in viscose workers. There are no indications that hydrogen sulfide, which is always present at the same time during exposure in the viscose industry, plays any part. A NOEC for cardiotoxicity with long-term exposure cannot be derived with any certainty from the available studies, but it is probably around 12 mg/m^3 (about 4 ml/m^3) (see the 1997 documentation for Carbon disulfide in Volume 12 of the present series).

Various studies on workers exposed to carbon disulfide have been published since the 1997 documentation, necessitating a re-evaluation of the MAK value.

1 Effects in Man

1.1 Case reports

In four male patients, polyneuropathy occurred following 6–21 years' exposure to carbon disulfide (Huang *et al.* 2001). No data were given on the level of exposure.

Six male workers aged between 43 and 50 exposed to carbon disulfide for 4 to 23 years showed polyneuropathies and reduced nerve conduction velocities. The highest measured air concentrations were between 100 and 200 ml/m^3 (316 and 632 mg/m^3). Two years after the diagnosis, a biopsy of the *Nervus suralis* was carried out in one patient. This showed degeneration of axon und myelin, a loss of large myelinized fibres, but also remyelinization. Three years after the diagnosis, the carbon disulfide concentrations were only 10 to 20 ml/m^3 (32 to 63 mg/m^3). The reduced nerve conduction velocities showed a slight but insignificant improvement (Huang *et al.* 2002).

Four male workers aged between 50 and 58 exposed to carbon disulfide for 15 to 23 years (no data on level of exposure) showed signs of high blood pressure, coronary heart diseases, hypercholesterolaemia, hyperlipidaemia, atrial fibrillation and colour sense disturbances (Valic *et al.* 2001).

Thirteen workers aged between 21 and 63 were exposed to current carbon disulfide concentrations of between 20 and 50 ml/m^3 (63 and 158 mg/m^3) and peak concentrations of up to 100 or 200 ml/m^3 (316 or 632 mg/m^3). Five workers showed ECG changes (four of these workers were the most highly exposed), four workers increased cholesterol levels and a further four workers hearing impairments (Guidotti and Hoffman 1999).

Neuropsychological tests (Wechsler Adult Intelligence Scale) and magnetic resonance imaging (MRI) were carried out on patients with diagnosed carbon disulfide poisoning from a viscose factory which closed down in 1993. A total of 37 highly exposed patients (high level and long duration of exposure) and 37 slightly exposed patients were examined. There were no differences in the intelligence quotient between the groups. However, in 5 of 12 highly exposed persons (42 %), but in only 1 of 19 slightly exposed persons (5.3 %), an increased number of cerebral lacunae could be seen on the MRI images (Cho *et al.* 2002).

In another study, 91 patients with carbon disulfide poisoning were examined by MRI. In T2-weighted MRI images 70 patients (77 %) showed hyperintensities in the white cerebellar region and 27 patients lacunar infarcts (Cha *et al.* 2002).

After 27 years' exposure to carbon disulfide, a 56-year-old man developed headache, limb tremors, gait disturbance, dysarthria, memory impairment, and emotional lability. The brain MRI showed diffuse hyperintensity lesions in T2-weighted images in the subcortical white matter, basal ganglia, and brain stem. The brain computed tomography perfusion study revealed a diffusely decreased regional cerebral blood flow and prolonged regional mean transit time in the subcortical white matter and basal ganglion (Ku *et al.* 2003). Data on exposure levels are not given.

A man who worked in a viscose plant for 27 years showed, seven years after he had stopped working, coordination disturbances, dysmetria (wrong estimation of distance in target movements), speech disorders and adiadochokinesis (inability to perform antagonistic movements in rapid alternation). These were subsequently accompanied by cognitive impairment, memory disturbances, disorientation, emotional lability and paranoid-obsessive disturbances. The brain showed advanced atrophies of the cerebellum and less severe atrophies of the cortex (Fonte *et al.* 2003).

1.2 Cross-sectional studies

1.2.1 Effects on the nervous system

In a cross-sectional study carried out from 1992 to 1993 on a total of 432 Japanese viscose workers (on average 35.5 years old and exposed for 13.4 years) and 402 controls (average age 35.8), the prevalence of subjective symptoms (e.g. "feeling of heaviness" in the head, tremor in the limbs, numbness in the limbs, increased sensitivity of the skin, reduced strength of grip, lower sex drive, rough skin) was significantly higher than in the controls. The authors regard the increase as marginal and the significance for health to be unclear. In 30 workers from the sector "spinning/refining" (on average 34.9 years old and exposed for 13.8 years) the nerve conduction velocity was slightly but statistically significantly reduced compared with 123 workers from other sectors (on average 36.9 years old and exposed for 12.6 years) or with 402 controls (Takebayashi *et al.* 1998). In 1992, median carbon disulfide concentrations were 4.1 ml/m^3 (13 mg/m^3) and maximum concentrations up to 39.7 ml/m^3 (125 mg/m^3) (Omae *et al.* 1998). Prior to 1992 there were no data on exposure concentrations.

In a Chinese study, which is available as an English abstract, it was reported that a dose-dependent increase in electroneuromyographic anomalies in keeping with axonal polyneuropathy was observed among 175 workers exposed to carbon disulfide (Jiang *et al.* 1998).

Neuropsychological tests were carried out on 120 workers exposed to carbon disulfide and hydrogen sulfide and 67 controls. Depending on the activity, the carbon disulfide concentrations were between 3 mg/m^3 (1 ml/m^3) and 147 mg/m^3 (47 ml/m^3). Hydrogen sulfide exposure was lower than 14 mg/m^3. Only in the group of workers exposed to carbon disulfide concentrations of more than 90 mg/m^3 (about 28.5 ml/m^3) did psychomotor tests show significant impairments in the speed and quality of execution (De Fruyt *et al.* 1998).

There are also studies on heart rate variability, recording parameters of the vegetative nervous system, especially the sympathetic-parasympathetic equilibrium. Of 830 workers in a viscose factory with diagnosed carbon disulfide poisoning, 71 males were examined for heart rate variability and compared with 127 non-exposed persons. From the changes in the parameters the authors conclude that carbon disulfide can affect heart rate variability (Jhun *et al.* 2003).

152 male workers (24 to 66 years old and exposed for between 5 and 38 years) were exposed to a current carbon disulfide concentration of 18.1 ± 15.8 mg/m^3 (5.7 ± 5.0 ml/m^3) and a maximum concentration of 109 mg/m^3 (34 ml/m^3). Compared with 93

controls, heart rate variability parameters were changed, indicating an impairment of neurovegetative regulation. In the group with the lowest carbon disulfide exposure of 1 to 10 mg/m^3 (0.3 to 3.2 ml/m^3), only one parameter (VLF power spectrum 0.0167–0.05 Hz) was reduced. In the intermediate exposure group of 10 to 18 mg/m^3 (3.2 to 5.7 ml/m^3) and the high exposure group (> 18 mg/m^3 or > 5.7 ml/m^3), several parameters were changed. However, individual parameters did not correlate with the current exposure. A correlation could be observed only if current exposure, length of exposure and cumulative lifetime exposure were taken into account together. The authors conclude that in addition to carbon disulfide other parameters may also be of significance (Bortkiewicz et al. 1997).

Neurological effects on the eye

In 90 workers from three sectors (preparation, twisting, spinning) exposed to carbon disulfide for between 8.6 and 13.7 years on average, there was an increase in colour vision deficiency measured as the colour confusion index (CCI). The median CCI was 1.15 among viscose preparation workers, 1.11 among twisting workers and 1.24 among spinning workers. Average carbon disulfide concentrations in the air were 3.0 ml/m^3 (9.5 mg/m^3) in the preparation sector, 4.16 ml/m^3 (12.8 mg/m^3) in the twisting sector and 4.73 ml/m^3 (15.0 mg/m^3) in the spinning sector (Valic et al. 2001). It should be noted that a median CCI of 1.24 does not necessarily signify a clinical peculiarity. It is known from experience with persons exposed to toluene (HVBG 2001) that the Lanthony-D15d test used is influenced not only by the age of the persons examined but also by their level of qualification (CCI > 1.15 among workers) and that vocational practice in assessing colours has a training effect. Experience also shows that subjects with little knowledge of the language used in the test have difficulties in performing it. However, the study by Valic et al. (2001) does not give any information about the qualifications and the nationalities of persons examined.

191 male and 80 female workers at a viscose factory in China were exposed to current carbon disulfide concentrations of 13.7 to 20.05 mg/m^3 (4.3 to 6.4 ml/m^3). According to the authors, the exposure conditions had apparently not changed from the construction of the plant in the 1970s up to the time of the study (1997–1999). Male workers in particular reported increased burning of the eyes and hazy sight. The FM 100-Hue test results showed that the total error score of the exposed group, whether male or female, was higher than that of the control. Discrimination of the green and blue zones was also impaired significantly. The differences were significant in the case of men for the whole spectrum and for the green and blue zones and in the case of women only for the green zone; the lack of statistical significance might be attributable to the smaller number of women tested. Differentiated analyses, which took account of length of service, showed for controls that with increasing length of employment more errors in discriminating colours can be observed in both men and women. This would seem to indicate that colour discrimination deteriorates with age. This deterioration was most marked in the workers exposed to carbon disulfide compared with the controls in the group with the shortest length of employment (up to five years). With increasing length of employment (or age) the differences between controls and exposed persons

diminished. However, the poorest colour discrimination was observed with men exposed to carbon disulfide for 15–20 years (Wang et al. 2002). The interpretation of the results is made more difficult by the differing degrees of impairment of individual colour zones, by the fact that with increasing length of employment (or age) initial differences in colour discrimination between exposed persons and controls become smaller, and in particular by the failure to take account of possible confounders (especially age) in the analyses.

Neurological effects on hearing

40 exposed workers with symptoms of chronic carbon disulfide poisoning (group I, on average 56.0 years old and exposed for 28.4 years), 40 exposed workers without carbon disulfide-specific symptoms (group II, on average 52.3 years old and exposed for 25.3 years) and 40 controls (on average 52.0 years old and employed elsewhere for 26.7 years) were tested for hearing ability. Current carbon disulfide concentrations were reported at 25.6 mg/m^3 (8.1 ml/m^3), in a range between 10 and 35 mg/m^3 (3.2 to 11 ml/m^3). The exposure to noise was above the permissible levels for both the workers exposed to carbon disulfide, at 88 to 92 dB(A), and the controls, at 86 to 93 dB(A). In group I the proportion of retrocochlear hearing reduction (hearing components lying behind the inner ear) was 97.5 %. In group II 45% of the workers showed retrocochlear hearing reduction, 32.5 % cochlear hearing reduction and 22.5 % no hearing reduction. In the control group there was a predominance of hearing reduction caused by cochlear damage (Kowalska et al. 2000). The authors do not give any precise data on the hearing ability of the control group. Nor does the study contain a statistical comparison between the measured hearing thresholds of group II and the controls. The authors interpret the results rather qualitatively and from the location of the damage conclude for group II that carbon disulfide exposure has a predominantly toxic effect. This study makes it clear that exposure periods of over 25 years and current carbon disulfide exposures of up to 35 mg/m^3 (11 ml/m^3) are associated with hearing impairments for which a toxic cause seems physiologically plausible. However, earlier carbon disulfide concentrations may have been higher.

Hearing loss was investigated in 131 men with exposure to noise (80–91 dB(A)) and carbon disulfide (1.6–20.1 ml/m^3) in a viscose rayon plant. The prevalence of > 25 dB hearing loss (dBHL) in rayon workers (67.9 %) was much higher than that in 110 administrative workers exposed to low noise (23.6 %) and in workers exposed to noise only (32.4 %). Hearing loss occurred mainly for speech frequencies of 0.5, 1, and 2 kHz. When the carbon disulfide exposure was measured by the product of carbon disulfide exposure level and employment years, the adjusted odds ratios (OR) of hearing loss of > 25 dBHL in rayon workers, compared with administrative workers, were 3.8 (95 % confidence interval (CI) 1.5–9.4) for those with the exposure of 37–214 year-ppm, 14.2 (95 % CI 4.4–45.9) for those with exposure of 215–453 year-ppm, and 70.3 (95 % CI 8.7–569.7) for those with exposure of > 453 year-ppm. The study suggests that carbon disulfide exposure enhances human hearing loss in a noisy environment and mainly affects hearing in lower frequencies (Chang et al. 2003).

1.2.2 Effects on the cardiovascular system

Various cross-sectional studies on cardiovascular effects in workers exposed to carbon disulfide have been published since the MAK documentation of 1997.

In 118 workers at a viscose factory in Taiwan exposed to current carbon disulfide concentrations of 0.1 to 54.6 ml/m^3 (0.3–173 mg/m^3), there were ECG changes compared with 44 controls (Kuo et al. 1997).

22 women (average age 42.5 years) chronically exposed (no further details) to carbon disulfide concentrations between 9.36 and 23.4 mg/m^3 (3.0–7.4 ml/m^3) showed in comparison with 20 women (average age 44.5 years) significant effects on the plasma lipid fraction and changes in the coagulation system (Stanosz et al. 1998).

In a group of 33 workers exposed to high levels of carbon disulfide of 50.6 ± 25.6 ml/m^3 (159.8 ± 76.8 mg/m^3) from two viscose factories in Taiwan, increased triglyceride and reduced HDL concentrations (after adjustment) were detected. In 64 moderately exposed workers (12.9 ± 5 ml/m^3; 40.7 ± 15.8 mg/m^3) and 35 low-exposed workers (3.5 ± 1.2 ml/m^3; 11.6 ± 3.7 mg/m^3), no increases were detected. The average length of exposure for all the groups was more than 21 years (Luo et al. 2003).

In order to test the hypothesis whether carbon disulfide has a weak negative inotropic effect on the heart muscle (Drexler et al. 1996), the diameter of the left ventricle of 325 workers exposed to carbon disulfide (on average 37 years old and exposed for 10 years) and 179 controls (average age 35 years) was determined by echocardiograph. The current carbon disulfide concentration in the air was given as 6 ml/m^3 (19 mg/m^3), with a range of 0.03–91.08 ml/m^3 (0.09–288 mg/m^3). The diameter of the left ventricle tended to be larger in the exposed workers than in the controls, but not to a significant extent. An evaluation by multiple linear regression showed that the diameter of the left ventricle correlates with the body mass index, body fat, alcohol consumption and physical fitness, but not with internal or external carbon disulfide exposure (Korinth et al. 2003).

In a group of 177 workers (on average 44 years old and exposed for between 5 and 38 years) and 93 controls (average age 41 years), the at-rest or 24-hour ECG, blood pressure and heart rate variability were recorded. Changes in the at-rest or 24-hour ECG were particularly apparent in workers with more than 20 years' exposure (Bortkiewicz et al. 2001). A further differentiation of the workers due to the level of exposure was not made.

252 workers in a Bulgarian viscose factory (111 men and 141 women, average age 42.4 years and exposed for at least 1 year) were compared with 252 controls of the same age and sex. Current carbon disulfide concentrations were between 10 and 64 mg/m^3 (3.2–20.3 ml/m^3). By multiplying the workplace-specific exposure concentration (in mg/m^3) by the length of exposure (in years) the authors calculated cumulative exposure indices. The odds ratios for elevated cholesterol were significantly higher with both a cumulative exposure index below 300 mg/m^3 at 2.19 (95 % CI 1.79–4.75; n=134) and a cumulative exposure index over 300 mg/m^3 at 3.05 (95 % CI 1.84–5.09; n=118). In the high-exposure group the odds ratio for coronary heart diseases was also significantly higher at 4.32 (95 % CI 1.84–10.11) (Kotseva and De Bacquer 2000).

In a later publication 141 workers (64 men and 77 women, on average 42.6 years old and exposed for at least 1 year) were compared with 141 similar controls. This is presumably a subgroup of the workers described by Kotseva and De Bacquer (2000). Current carbon disulfide concentrations were given as 1 to 30 mg/m^3 (about 0.3–9.5 ml/m^3) for this group. With a cumulative exposure index over 100 mg/m^3 the odds ratio for elevated cholesterol was significantly higher at 5.52 (95 % CI 2.81–10.83; n=70), but not with a cumulative exposure index under 100 mg/m^3. The author concludes that carbon disulfide concentrations under 31 mg/m^3 (about 10 ml/m^3) and cumulative exposure indices over 100 mg/m^3 can lead to higher cholesterol levels (Kotseva 2001). There are no data on earlier carbon disulfide concentrations.

Several studies were carried out in a Belgian viscose factory. One of them was published in 1992 (Vanhoorne et al. 1992, the 1997 documentation for Carbon disulfide in Volume 12 of the present series). For workers exposed to current carbon disulfide concentrations of 4 to 113 mg/m^3 (1.2–36 ml/m^3) increased blood pressure, reduced HDL cholesterol and higher apolipoproteins A1 and B were detected.

In more recent studies the current carbon disulfide concentrations in viscose processing were 1.3 to 20.8 mg/m^3 (0.4–6.6 ml/m^3) and in spinning 3.9 to 32.4 mg/m^3 (1.2–10.2 ml/m^3). In breathing masks the carbon disulfide concentrations in viscose processing were 6.0 to 10.09 mg/m^3 (about 1.9–3.2 ml/m^3) and in spinning 2.2 to 13.02 mg/m^3 (0.7–4.1 ml/m^3). The results for 91 exposed male workers (average age 41.1 years) were compared with those for 81 male controls (average age 39.5 years). There were no cardiovascular effects in workers with cumulative exposure indices under 150 mg/m^3 (n=47), but with cumulative exposure indices over 150 mg/m^3 (n=43) there were higher adjusted odds ratios for coronary heart diseases (OR_{adj}: 18.20; 95 % CI 2.44–135), ischaemic ECG (OR_{adj}: 8.48; 95 % CI 1.25–57.26) and ischaemia (OR_{adj}: 4.65; 95 % CI 1.60–13.54) (Kotseva et al. 2001a).

In order to determine earlier changes in the elasticity of arteries and functional changes in the vessel walls, 85 male workers (average age 37.1 years) and 37 controls (average age 40.5 years) from the same viscose plant were examined. In the exposed workers the distensibility of the vessel walls and the heart rate were increased (Braeckman et al. 2001, Kotseva et al. 2001b).

In 367 workers exposed to carbon disulfide concentrations of 15.47 ± 2.34 mg/m^3 (4.9 ± 0.74 ml/m^3) no significant increase in clinical complaints and abnormal electrocardiograms was noted compared to 125 reference workers. Nor were significant effects of carbon disulfide on blood pressure, cholesterol, HDL and LDL cholesterol or triglycerides detected (Tan et al. 2004).

In a study of a total of 432 Japanese viscose workers the prevalence of microaneurisms of the retinal artery at 8.1 % was significantly higher than for 402 controls at 3.4 %. The prevalence of arteriosclerosis was not increased (Takebayashi et al. 1998). The median carbon disulfide concentration was 4.1 ml/m^3 (13 mg/m^3) and the maximum concentration 39.7 ml/m^3 (125 mg/m^3) (Omae et al. 1998).

In an overview study the published literature on cardiovascular effects of carbon disulfide following occupational exposure was critically assessed (Sulsky et al. 2002). The evaluation covered 37 publications; 12 publications were excluded on account of methodological shortcomings (unsatisfactory presentation of exposure or the results, manifest selection bias without correction, no consideration of important confounders). Following a critical appraisal of the individual studies, specific hypotheses were tested on the basis of the evidence and consistency of the findings. A meta-analysis was not carried out. An important objective of this work was to describe the risks of cardiovascular diseases at various exposure concentrations. Concentrations over 20 ml/m^3 (63 mg/m^3) were regarded as "high-level" exposure, concentrations below that as "low-level" exposure. In 15 studies risk factors for cardiovascular diseases were examined at concentrations below 20 ml/m^3, with 11 of them showing at least one weak point. The four most significant studies were identified as Drexler et al. (1995), Kotseva (2001), Kotseva and De Bacquer (2000) and Kotseva et al. (2001a). In these studies a multivariant adjustment for important confounders (such as sex, smoking, alcohol, stress, shift work, noise, training), a dose-effect analysis and a classification of exposure were carried out. The authors conclude that there are no strong or consistent associations between carbon disulfide exposure and coronary heart disease mortality or relevant risk factors. Even at higher exposure concentrations the associations with coronary heart diseases or other clinical indicators were different, not very consistent and to some extent contradictory. At low exposure concentrations the only relatively consistent finding, which was also backed up by findings at high concentrations, was an increase in total or LDL cholesterol. As this effect was shown in only half of the studies (of comparable quality), it would seem that other factors than carbon disulfide exposure alone may play a part. It is unclear whether a slight increase in these parameters, even if they were associated with exposure, contributes to a disease or whether it is a physiological response to the exposure (Sulsky et al. 2002).

1.2.3 Other effects

In 29 men exposed to carbon disulfide for more than 20 years and in 24 patients with peripheral arteriosclerosis, but not in 30 non-exposed healthy subjects, there were indications of lipid peroxidation in the plasma (increased concentration of thiobarbituric reactive substances, reduced concentrations of α-tocopherol, glutathione-peroxidase, catalase). The carbon disulfide concentrations were between 30 and 80 mg/m^3 (9.5 to 25.4 ml/m^3) in the 1960s, between 22 and 45 mg/m^3 (7.0 to 14.3 ml/m^3) in the 1970s and over 18 mg/m^3 (5.7 ml/m^3) in later years (Wronska-Nofer et al. 2002). The increased incidence of coronary heart diseases observed in workers exposed to carbon disulfide is attributed by the authors to oxidative stress, leading to arteriosclerosis.

In a group of 67 workers and 88 controls biochemical indications of oxidative stress were obtained both in the case of workers exposed to carbon disulfide concentrations below 10 mg/m^3 (3.2 ml/m^3) (CuZn superoxide-dismutase increased, malondialdehyde reduced) and at concentrations over 10 mg/m^3 (3.2 ml/m^3) (CuZn superoxide-dismutase and malondialdehyde increased) (Jian and Hu 2000).

In workers exposed to carbon disulfide, the NO concentration in the plasma was significantly lower at 43.28 µmol/l (high-exposure group) and 50.00 µmol/l (low-exposure group) compared with controls at 70.66 µmol/l. The superoxide-dismutase activities in the erythrocytes were significantly higher in the high and low-exposure groups at 4832 U/g Hb and 3520 U/g Hb, respectively, than in the control group at 2425 U/g Hb. The concentrations of lipid peroxides in the plasma were also significantly higher in the high and low-exposure groups at 19.38 and 17.09 µmol/l, respectively, compared with the controls (4.37 µmol/l). The authors conclude that occupational long-term exposure to carbon disulfide reduces NO concentrations and increases those of the superoxide radicals (Jiang et al. 2000).

50 workers exposed to carbon disulfide at a concentration of 14.4 ± 4.62 mg/m^3 (4.6 ± 1.5 ml/m^3) in a viscose factory showed in comparison with the control group reduced serum FSH (7.50 ± 7.07 IU/l to 10.4 ± 7.35 IU/l) and prolactin (75.72 ± 4.18 ng/l to 6.89 ± 4.64 ng/l). With increasing length of exposure serum LH decreased, as did serum FSH with increasing level of exposure (measured as TTCA elimination in urine) (Wang et al. 1999).

1.3 Cohort studies

There is a mortality study of workers in a Polish viscose factory. Male workers were engaged in production or the maintenance of equipment for at least one year in the period between 1950 and 1985 and lived in the vicinity of the factory. 43.7 % of the workers were employed for more than 21 years, while the average length of employment was 17.3 years. Over 50 % of the workers were considered to be severely exposed. Of 2878 workers in the cohort, 2762 were monitored, giving a total of 76465 man-years. The male population of Poland was taken as the reference. For the cohort there was a significantly increased SMR of 108 (95 % CI 101–115) for all fatalities. There was a significant increase in the risks of mortality from cardiovascular (SMR 114, 95 % CI 104–125) or cerebrovascular (SMR 208, 95 % CI 170–252) diseases. For the groups of highly exposed workers and workers in the spinning plant who were first employed before 1974, the risks of dying from a disease of the circulatory system were significantly increased, with SMRs of 209 (95 % CI 157–273) and 204 (95 % CI 138–291). There was also a statistically significant trend in mortality from all cardiovascular diseases in relation to the level of exposure. No clear relationship was apparent between the length of exposure and increased mortality (Pepłońska et al. 2001).

A number of publications are available on a cohort of workers from 11 Japanese viscose rayon factories. In an initial survey in 1992 and 1993, a total of 432 viscose workers and 402 control workers were covered (see cross-sectional studies by Omae et al. (1998) and Takebayashi et al. (1998)). The median carbon disulfide concentration in 1992 was 4.1 ml/m^3 (13 mg/m^3) and the maximum concentration 39.7 ml/m^3 (125 mg/m^3) (Omae et al. 1998). Following these cross-sectional studies, longitudinal studies (cohort studies) were carried out. In 1994 and 1995 some viscose factories were shut down, so that 140 viscose workers were no longer exposed (ex-exposed workers). The other workers in the cohort study were exposed until the end of the study in 1998 or 1999 (exposed workers).

Cerebrovascular effects of carbon disulfide exposures were studied by brain magnetic resonance imaging to determine hyperintense spots (HIS, "silent cerebral infarctions"). The six-year follow-up study comprised 217 exposed workers, 125 ex-exposed workers and 324 referent workers. Mean duration of exposure was 19.6 years. The geometric mean exposure to carbon disulfide was 4.9 ml/m^3 (15.5 mg/m^3). The exposure level was

2.5 ml/m^3 (7.9 mg/m^3) in the lowest quartile and 8.1 ml/m^3 (25.6 mg/m^3) in the highest quartile. No significant difference in the prevalence of HIS was observed in the exposed or ex-exposed workers compared to reference workers. The proportion of increases in HIS in the follow-up period was 24, 15, and 12 % in the exposed, ex-exposed and referent workers, respectively. The multivariate adjusted odds ratio (95 % CL) was significantly increased in exposed (2.27 (1.37–3.76)), but not in the ex-exposed workers (1.33 (0.70–2.54)) (Nishiwaki et al. 2004). The authors of the study conclude that the results should be interpreted cautiously due to lack of an exposure-response relation and due to several potential limitations of the study. In addition, the relative newness of the application of MRI scans in the identification of toxic responses with the lack of criteria for the definition of adverse effects indicates that the results cannot be taken into account for setting an OEL at present.

Cardiovascular effects of carbon disulfide exposures were examined in 251 exposed, 140 ex-exposed and 359 referent workers. The incidence of coronary artery ischaemia, defined as Minnesota electrocardiographic codes I, IV$_{1-3}$, V$_{1-3}$ (tested at rest and after step test), or treatment received for ischaemia, was 6.4 % in referent workers and 11.8 % in exposed workers. When exposures were divided into quartiles based on the urinary TTCA concentration, the confounder-adjusted odds ratio for ischaemic findings was significantly increased in the highest quartile (4.2 (1.8–9.7)) corresponding to an exposure level of 8.7 ml/m^3 (27.5 mg/m^3). In the age restricted groups (< 35 years), the incidence of ischaemic findings was 3.8 % in the non-exposed group and 12.4 % in the exposed group. This indicates that recent exposure contributed to the development of the ischaemic findings. However, with more rigorous criteria for defining ischaemia (ST depression ≥ 2 mm) both incidence and prevalence of ischaemic findings were comparable among the exposed and non-exposed groups. In addition to ischaemic findings, the incidence of retinal microaneurysm was increased in exposed workers (9.2 %) compared to referent workers (5.0 %). In the highest quartile the confounder adjusted odds ratio for microaneurysms was increased significantly (3.8 (1.2–11.4)) (Takebayashi et al. 2004). The value of exposure-related findings with the Minnesota code is questioned, as with the application of generally accepted rigorous ECG criteria (ST depression ≥ 2 mm) no significant effects of carbon disulfide exposure were found. The incidence of retinal microaneurysm was increased with marginal significance in the high exposure group (8.7 ml/m^3) and has been reduced over time. In 1998 peak exposure was up to 30 ml/m^3 and may be responsible for the effects observed. In addition, the clinical relevance of this marginally statistically significant increase seems to be low.

In order to ascertain endocrinal changes, 259 workers exposed to carbon disulfide, 133 ex-exposed workers, and 352 controls were examined. The parameters measured with regard to glucose metabolism (glucose after fasting, insulin, HbA), the pituitary hormones (luteinising hormone, LH; follicle-stimulating hormone, FSH; adrenocorticotropic hormone, ACTH), the male sex organs (testosterone), the thyroid (triiodothyronine, T$_3$; tetraiodothyronine, T$_4$; thyroid-stimulating hormone, TSH; thyroxine-binding α-globulin, TBG) and subjective data on reduced sexual activity showed no biologically relevant changes between the groups (Takebayashi et al. 2003).

The cohort studies of workers in the Japanese viscose rayon industry thus show only in the group of workers exposed to carbon disulfide concentrations of more than 8 ml/m^3

(25 mg/m^3) initial indications of possible carbon disulfide-induced subclinical changes such as increased ischaemic findings in the ECG using the Minnesota code, retinal microaneurisms and hyperintense spots in T2-weighted MRI images of the brain. The study indicates no clinically relevant effects at mean exposure concentrations of 5 ml/m^3 (16 ml/m^3) carbon disulfide.

A meta-analysis of cohort studies on cardiovascular effects (Tan et al. 2002) considers only older studies, which are already described in the earlier MAK documentation.

2 Manifesto (MAK value/classification)

In 1997 the MAK value was lowered from 10 to 5 ml/m^3 (30 to 16 mg/m^3), as there had been indications that with several decades' exposure at the workplace neurotoxic and cardiotoxic effects can occur in a concentration range below 10 ml/m^3 (30 mg/m^3). For 40 years' exposure one working group estimated a NOEC for neurotoxicity of 4 ml/m^3 (12 mg/m^3) (Ruijten et al. 1990). Another working group found at 4 ml/m^3 (12 mg/m^3) no increase in blood lipids, no indication of an increase in the prevalence of coronary heart disease or arteriosclerotic findings, no difference in the distribution of the performance of ergometric tests (Drexler et al. 1995, 1996) and also no signs of effects on the nervous system (Reinhardt et al. 1997a, 1997b). A provisional MAK value was therefore set at 5 ml/m^3 (16 mg/m^3). The main problem in setting a NOEC was and still is that, although the current concentrations are relatively well documented at the time of each study, earlier exposure concentrations at the workplace, which are of significance on account of the cumulative effect of carbon disulfide, are not available and can only be estimated. Previously, they were appreciably higher for the most part. Moreover, the usual method of measuring carbon disulfide in the air has considerable shortcomings, and it used to be quite common that the values measured were lower than actual exposure.

In cohort studies of workers in 11 Japanese viscose rayon factories, carried out from 1992 to 1999, signs of possible carbon disulfide-induced subclinical changes, such as ischaemic ECG findings, retinal microaneurisms (Takebayashi et al. 2004) and hyperintense spots in T2-weighted MRT brain images (Nishiwaki et al. 2004) were observed in workers of the highest exposure category of more than 8 ml/m^3 (25 mg/m^3). No effects were observed in groups of workers exposed to 5 ml/m^3 (16 mg/m^3) of carbon disulfide. These cohort studies support the MAK value established in 1997 of 5 ml/m^3 (16 mg/m^3).

Well-performed cross-sectional studies did not show relevant cardiovascular effects at carbon disulfide concentrations of 5 ml/m^3 (16 mg/m^3) (Tan et al. 2004) or 6 ml/m^3 (19 mg/m^3) (Korinth et al. 2003). Cardiovascular or neurotoxic effects were mostly observed after long-term exposure to carbon disulfide at higher concentrations. However, some cardiovascular or neurological effects were reported at carbon disulfide concentrations of 3 to 10 ml/m^3 (9.5 to 32 mg/m^3). Due to methodological limitations, questionable clinical relevance of the findings, or higher exposure concentrations in previous years, these results cannot be used for lowering the MAK value below 5 ml/m^3 (16 mg/m^3). The provisional MAK value of 5 ml/m^3 (16 mg/m^3) will therefore be retained. Further research is required into effects at concentrations in the range of the MAK value.

For the limitation of exposure peaks carbon disulfide is classified in Peak limitation category II with an excursion factor of 2.

As prenatal toxic effects cannot be excluded in the concentration range of 5 ml/m^3 (16 mg/m^3), classification of carbon disulfide in Pregnancy risk group B is retained.

Designation with "H" has been retained because of the good dermal absorption of carbon disulfide.

There are no reports of sensitization of the skin or respiratory tract, nor have such results of animal experiments been published. Therefore carbon disulfide has not been designated with either "Sh" or "Sa".

3 References

Bortkiewicz A, Gadzicka E, Szymczak W (1997) Heart rate variability in workers exposed to carbon disulfide. *J Auton Nerv Syst 66*: 62–68
Bortkiewicz A, Gadzicka E, Szymczak W (2001) Cardiovascular disturbances in workers exposed to carbon disulfide. *Appl Occup Environ Hyg 16*: 455–463
Braeckman L, Kotseva K, Duprez D, De Bacquer D, De Buyzere M, Van De Veire N, Vanhoorne M (2001) Vascular changes in workers exposed to carbon disulfide. *Ann Acad Med Singapore 30*: 475–480
BUA (Beratergremium für umweltrelevante Altstoffe der Gesellschaft deutscher Chemiker) (1991) *Carbon disulfide, BUA report 83*. S. Hirzel, Wissenschaftliche Verlagsgesellschaft, Stuttgart
Cha JH, Kim SS, Han H, Kim RH, Yim SH, Kim MJ (2002) Brain MRI findings of carbon disulfide poisoning. *Korean J Radiol 3*: 158–162
Chang SJ, Shih TS, Chou TC, Chen CJ, Chang HY, Sung FC (2003) Hearing loss in workers exposed to carbon disulfide and noise. *Environ Health Perspect 111*: 1620–1624
Cho SK, Kim RH, Yim SH, Tak SW, Lee YK, Son MA (2002) Long-term neuropsychological effects and MRI findings in patients with CS$_2$ poisoning. *Acta Neurol Scand 106*: 269–275
De Fruyt F, Thiery E, De Bacquer D, Vanhoorne M (1998) Neuropsychological effects of occupational exposures to carbon disulfide and hydrogen sulfide. *Int J Occup Environ Health 4*: 139–146
Drexler H, Ulm K, Hubmann M, Hardt R, Göen T, Mondorf W, Lang E, Angerer J, Lehnert G (1995) Carbon disulphide. III. Risk factors for coronary heart diseases in workers in the viscose industry. *Int Arch Occup Environ Health 67*: 243–252
Drexler H, Ulm K, Hardt R, Hubmann M, Göen T, Lang E, Angerer J, Lehnert G (1996) Carbon disulphide. IV. Cardiovascular function in workers in the viscose industry. *Int Arch Occup Environ Health 69*: 27–32
Fonte R, Edallo A, Candura SM (2003) Cerebellar atrophy as a delayed manifestation of chronic carbon disulfide poisoning. *Ind Health 41*: 43–47
Guidotti TL, Hoffman H (1999) Indicators of cardiovascular risk among workers exposed to high intermittent levels of carbon disulphide. *Occup Med 49*: 507–515
Henschler D (Ed.) (1975) Schwefelkohlenstoffe (Carbon disulfide) (German). in: *Toxikologisch-arbeitsmedizinische Begründungen von MAK-Werten*, 4th issue. VCH-Verlagsgesellschaft, Weinheim

Hirata M, Ogawa Y, Okayama A, Goto S (1992) A cross-sectional study on the brainstem auditory evoked potential among workers exposed to carbon disulfide. *Int Arch Occup Environ Health* 64: 321–324

Huang CC, Chu CC, Chu NS, Wu TN (2001) Carbon disulfide vasculopathy: a small vessel disease. *Cerebrovasc Dis 11*: 245–250

Huang CC, Chu CC, Wu TN, Shih TS, Chu NS (2002) Clinical course in patients with chronic carbon disulfide polyneuropathy. *Clin Neurol Neurosurg 104*: 115–120

HVBG (Hauptverband der gewerblichen Berufsgenossenschaften) (2001) *Toluol in Tiefdruckereien. Abschlussbericht zu einem Forschungsprojekt.* (Toluene in rotogravure printing operations. Final report on a research project) (German). HVBG, Sankt Augustin

Jhun H-J, Yim S-H, Kim R, Paek D (2003) Heart-rate variability of carbon disulfide-poisoned subjects in Korea. *Int Arch Occup Environ Health 76*: 156–160

Jian L, Hu D (2000) Antioxidative stress response in workers exposed to carbon disulfide. *Int Arch Occup Environ Health 73*: 503–506

Jiang B, Zhang S, Huang J, Lu J (1998) Study on electroneuromyography of 175 workers exposed to carbon disulfide (Chinese). *Wei Sheng Yan Jiu 27*: 84–86

Jiang X, Jiann L, Hu D (2000) A preliminary study on body level of nitric oxide in workers exposed to carbon disulfide (Chinese). *Zhonghua Yu Fang Yi Xue Za Zhi 34*: 348–350

Korinth G, Göen T, Ulm K, Hardt R, Hubmann M, Drexler H (2003) Cardiovascular function of workers exposed to carbon disulphide. *Int Arch Occup Environ Health 76*: 81–85

Kotseva K (2001) Occupational exposure to low concentrations of carbon disulfide as a risk factor for hypercholesterolaemia. *Int Arch Occup Environ Health 74*: 38–42

Kotseva KP, De Bacquer D (2000) Cardiovascular effects of occupational exposure to carbon disulphide. Occup Med 50: 43–47

Kotseva K, Braeckman L, De Bacquer D, Bulat P, Vanhoorne M (2001a) Cardiovascular effects in viscose rayon workers exposed to carbon disulfide. *Int J Occup Environ Health 7*: 7–13

Kotseva K, Braeckman L, Duprez D, De Bacquer D, De Buyzere M, Van De Veire N, Vanhoorne M (2001b) Decreased carotid artery distensibility as a sign of early atherosclerosis in viscose rayon workers. *Occup Med 51*: 223–229

Kowalska S, Sulkowski W, Sinczuk-Walczak H (2000) Assessment of the hearing system in workers chronically exposed to carbon disulfide and noise (Polish). *Med Pra 51*: 123–138

Ku MC, Huang CC, Kuo HC, Yen TC, Chen CJ, Shih TS, Chang HY (2003) Diffuse white matter lesions in carbon disulfide intoxication: microangiopathy or demyelination. *Eur Neurol 50*: 220–224

Kuo HW, Lai JS, Lin M, Su ES (1997) Effects of exposure to carbon disulfide (CS_2) on electrocardiographic features of ischemic heart disease among viscose rayon factory workers. *Int Arch Occup Environ Health 70*: 61–66

Luo J-CJ, Chang H-Y, Chang S-J, Chou T-C, Chen C-J, Shih T-S, Huang C-C (2003) Elevated triglyceride and decreased high density lipoprotein level in carbon disulfide workers in Taiwan. *J Ocuup Environ Med 45*: 73–78

Nishiwaki Y, Takebayashi T, O'Uchi T, Nomiyama T, Uemura T, Sakurai H, Omae K (2004) Six year observational cohort study of the effect of carbon disulphide on brain MRI in rayon manufacturing workers. *Occup Environ Med 61*: 225–232

Omae K, Takebayashi T, Nomiyama T, Ishizuka C, Nakashima H, Uemura T, Tanaka S, Yamauchi T, O'Uchi T, Horichi Y, Sakurai H (1998) Cross sectional observation of the effects of carbon disulphide on arteriosclerosis in rayon manufacturing workers. *Occup Environ Med 55*: 468–472

Peplońska B, Sobala W, Szeszenia-Dabrowska N (2001) Mortality pattern in the cohort of workers exposed to carbon disulfide. *Int J Occup Med Environ Health 14*: 267–274

Reinhardt F, Drexler H, Bickel A, Claus D, Angerer J, Ulm K, Lehnert G, Neundorfer B (1997a) Neurotoxicity of long-term low-level exposure to carbon disulphide: results of questionnaire, clinical neurological examination and neuropsychological testing. *Int Arch Occup Environ Health 69*: 332–338

Reinhardt F, Drexler H, Bickel A, Claus D, Ulm K, Angerer J, Lehnert G, Neundorfer B (1997b) Electrophysiological investigation of central, peripheral and autonomic nerve function in workers with long-term low-level exposure to carbon disulphide in the viscose industry. *Int Arch Occup Environ Health 70*: 249–256

Ruijten MWMM, Sallé HJA, Verbek MM, Muijser H (1990) Special nerve function and colour discrimination in workers with long term low level exposure to carbon disulphide. *Br J Ind Med 47*: 589–595

Stanosz S, Kuligowska E, Kuligowski D (1998) Coefficient of linear correlation between levels of fibrinogen, antithrombin III, thrombin-antithrombin complex and lipid fractions in women exposed chronically to carbon disulfide (Polish). *Med Pra 49*: 51–57

Sulsky SI, Hooven FH, Burch MT, Mundt KA (2002) Critical review of the epidemiological literature on the potential cardiovascular effects of occupational carbon disulfide exposure. *Int Arch Occup Environ Health 75*: 365–380

Takebayashi T, Omae K, Ishizuka C, Nomiyama T, Sakurai H (1998) Cross sectional observation of the effects of carbon disulphide on the nervous system, endocrine system, and subjective symptoms in rayon manufacturing workers. *Occup Environ Med 55*: 473–479

Takebayashi T, Nishiwaki Y, Nomiyama T, Uemura T, Yamauchi T, Tanaka S, Sakurai H, Omae K (2003) Lack of relationship between occupational exposure to carbon disulfide and endocrine dysfunction: a six-year cohort study of the Japanese rayon workers. *J Occup Health 45*: 111–118

Takebayashi T, Nishiwaki Y, Uemura T, Nakashima H, Nomiyama T, Sakurai H, Omae K (2004) A six year follow up study of the subclinical effects of carbon disulphide exposure on the cardiovascular system. *Occup Environ Med 61*: 127–134

Tan X, Peng X, Wang F, Joyeux M, Hartemann P (2002) Cardiovascular effects of carbon disulfide: meta-analysis of cohort studies. *Int J Hyg Environ Health 205*: 473–477

Tan X, Chen G, Peng X, Wang F, Bi Y, Tao N, Wang C, Yan J, Ma S, Cao Z, He J, Yi P, Braeckman L, Vanhoorne M (2004) Cross-sectional study of cardiovascular effects of carbon disulfide among Chinese workers of a viscose factory. *Int J Hyg Environ Health 207*: 217–225

Valic E, Pilger A, Pirsch P, Pospischil E, Waldhör T, Rüdiger HW, Wolf C (2001) Nachweis erworbener Farbsinnstörungen bei CS_2 exponierten Arbeitern in der Viskoseproduktion (The detection of acquired colour vision deficiency in workers exposed to CS_2 during the production of viscose) (German). *Arbeitsmed Sozialmed Umweltmed 36*: 59–63

Vanhoorne M, De Bacquer D, De Backer G (1992) Epidemiological study of the cardiovascular effects of carbon disulphide. *Int J Epidemiol 21*: 745–752

Vanhoorne M, De Rouck A, De Bacquer D (1995) Epidemiological study of eye irritation by hydrogen sulphide and/or carbon disulphide exposure in viscose rayon workers. *Ann Occup Hyg 39*: 307–315

Vanhoorne M, De Rouck A, Bacquer D (1996) Epidemiological study of the systemic ophthalmological effects of carbon disulfide. *Arch Environ Health 51*: 181–188

Wang C, Bi Y, Tan X, Yan J (1999) Serum sex hormone and urinary metabolites of male workers exposed to carbon disulfide (Chinese). *Wei Sheng Yan Jiu 30*: 132–133

Wang C, Tan X, Bi Y, Su Y, Yan J, Ma S, He J, Braeckman L, De Bacquer D, Wang F, Vanhoorne D (2002) Cross-sectional study of the ophthalmological effects of carbon disulfide in Chinese viscose workers. *Int J Hyg Environ Health 205*: 367–372

Wronska-Nofer T, Chojnowska-Jezierska J, Nofer JR, Halatek T, Wisniewska-Knypl J (2002) Increased oxidative stress in subjects exposed to carbon disulfide (CS_2) - an occupational coronary risk factor. *Arch Toxicol 76*: 152–157

completed 28.11.2002; updated 10.11.2004

Diisopropyl ether

MAK value (2000)	200 ml/m^3 (ppm) ≙ 850 mg/m^3
Peak limitation (2000)	Category I, excursion factor 2
Absorption through the skin	–
Sensitization	–
Carcinogenicity	–
Prenatal toxicity (2000)	Pregnancy risk group D
Germ cell mutagenicity	–
BAT value	–
Synonyms	2-isopropoxypropane isopropyl ether 2,2´-oxybispropane
Chemical name (CAS)	diisopropyl ether
CAS number	108-20-3
Structural formula	(CH$_3$)$_2$CH–O–CH(CH$_3$)$_2$
Molecular formula	C$_6$H$_{14}$O
Molecular weight	102.17
Melting point	−85.9°C
Boiling point	68.5°C
Density at 20°C	0.73 g/cm^3
Vapour pressure at 20°C	180 hPa
log P$_{ow}$*	1.52
1 ml/m^3 (ppm ≙ 4.24 mg/m^3	1 mg/m^3 ≙ 0.236 ml/m^3 (ppm)

The MAK value of 500 ml/m^3 for diisopropyl ether which was valid until 2000 was established in 1958 in accordance with the threshold limit value (TLV) at the time.

---

* *n*-octanol/water distribution coefficient

1 Toxic Effects and Mode of Action

Diisopropyl ether has an anaesthetic effect, but is not used in the clinical field because of its intensive effects and its unpleasant odour. Concentrations close to anaesthetic concentrations cause rapid fall in blood pressure and respiratory failure. For the situation at the workplace the critical effects are irritative effects on the mucous membranes and adverse effects on the central nervous system.

Animal study results showed a very low acute toxicity with all routes of administration. The substance had a slight irritative effect on the skin and caused irritation to the eye of the rabbit. Repeated exposure of rats yielded no effects up to 480 ml/m^3. Concentrations above 3000 ml/m^3 led to increases in liver and kidney weights. At concentrations not toxic to the dams diisopropyl ether had no prenatal toxic effects on the offspring of pregnant rats; concentrations above 3000 ml/m^3 can lead to skeletal changes.

The substance was not found to be genotoxic *in vitro* in prokaryotic and eukaryotic test systems.

2 Mechanism of Action

The irritative effects of diisopropyl ether are probably the result of non-covalent interaction with the receptors of the sensory nerve endings in the mucous membranes of the respiratory tract and are a function of the physico-chemical properties (e.g. vapour pressure, lipophilia) of the substance (Alarie *et al*. 1998). The central nervous effects are probably also the result of the lipophilia of the substance and a corresponding interaction with neuronal membranes.

3 Toxicokinetics and Metabolism

There are no studies available. It may be assumed that, after absorption, the substance is exhaled largely unchanged. After inhalation of small amounts, elimination of the substance via the lungs or saliva glands could be detected for up to 24 hours, as the specific odour revealed (Browning 1965).

4 Effects in Man

After exposure of 24 volunteers of both sexes to diisopropyl ether at concentrations of 300 ml/m^3 for 15 minutes, 35 % of them complained about the unpleasant odour. No irritation of the eyes, nose and throat occurred under these exposure conditions. Consequently, the test persons regarded a concentration above 300 ml/m^3 as an acceptable concentration for 8-hour exposure (Silverman et al. 1946). In reviews of the toxicity profile of diisopropyl ether, it is also mentioned that 5-minute exposure to 800 ml/m^3 led to irritation of the eyes and nose as well as to breathing difficulties (Browning 1965, Kirwin and Sandmeyer 1981). The derivation of these findings is not apparent, however. This is also true of the statement that exposure to 500 ml/m^3 for 15 minutes did not lead to irritative effects (Browning 1965).

The odour threshold for diisopropyl ether has been cited as 0.17 ml/m^3 (Ruth 1986).

5 Animal Experiments and *in vitro* Studies

There are no data available for the carcinogenicity of diisopropyl ether in animals.

5.1 Acute toxicity

5.1.1 Inhalation

For rats, exposure to a diisopropyl ether concentration of 16000 ml/m^3 for 4 hours was lethal (no further details) (Kirwin and Sandmeyer 1981). For the mouse, an LC$_{50}$ of 36000 ml/m^3 after 15-minute exposure was reported (ECB 1995).

The exposure of one rhesus monkey, rabbit and guinea pig to a diisopropyl ether concentration of 60000 ml/m^3 led to death of the animals due to respiratory arrest within a maximum of 78 minutes. The most marked signs of intoxication were irritative effects on the respiratory tract with increased salivation and anaesthesia. Apart from visceral congestion and oedema in the brain, lethal concentrations led to only a few specific lesions of the internal organs. Toxic changes were established in the heart muscle, liver and kidneys. The blood count of exposed animals revealed, especially for the monkey, a reduction in erythrocytes and leukocytes as well as a decreased haemoglobin content; the causal relationship, however, is not certain. This species survived exposures to 30000 ml/m^3 (1 hour) or 3000 ml/m^3 (2 hours) with signs of an anaesthetising effect being observed at the higher concentration (Machle et al. 1939).

5.1.2 Ingestion

The oral LD_{50} value was approximately 4600 mg/kg body weight for 14-day-old rats and approximately 12000 mg/kg body weight for young and older adult animals (Kimura et al. 1971). Other authors described for rats an oral LD_{50} value of 8470 mg/kg body weight (Nishimura et al. 1994). For rabbits, the minimum oral lethal dose was 6000 mg/kg body weight. The cause of death was respiratory failure due to the narcotic effect (Machle et al. 1939).

5.1.3 Dermal absorption

Dropwise application of a total of 150 ml diisopropyl ether (about 50 g/kg body weight) over one hour under open conditions onto a clipped skin area of about 80 cm^2 did not lead to externally visible signs of intoxication in rabbits (Machle et al. 1939), so that no toxicity is to be expected after the dermal application of high doses.

5.2 Subacute, subchronic and chronic toxicity

Fourteen female and 14 male Sprague Dawley rats per group were exposed to diisopropyl ether at concentrations of 480, 3300 or 7100 ml/m^3 for 6 hours a day, on 5 days a week for a period of 90 days. A group of sham-exposed animals and a group of untreated ones served as controls. Body weight development, clinico-chemical parameters as well as the number of sperms and spermatids were not changed in any group compared with the control values. Neither were any clinical signs observed. Exposure to 3300 ml/m^3 or more led to increased liver and kidney weights in males. Females in the same concentration group also showed higher liver weights, but only compared with the sham-exposed animals. Histopathological changes were found only in males of the 7100 ml/m^3 group, such as hypertrophic liver cells and an increased number of hyaline droplets in the proximal renal tubules (Dalbey and Feuston 1996). According to these investigations, 480 ml/m^3 can be considered as being the NOEL (no observed effect level) for medium-term exposure of rats.

In addition to the above-mentioned 90-day study with its classical study design, the same group of authors conducted another 90-day inhalation study with the specific aim of detecting potential neurotoxic effects. A test battery was used involving repeated observation of sensorial, motor and autonomic functions as well as repeated automated analysis of motor activity and an extensive neuropathological examination at the end of the study. The animals were exposed to concentrations of 450, 3250 and 7060 ml/m^3. Although some values were significantly different from the respective control values, they were nevertheless sex-specific, not concentration-dependent and not reproducible at all examination times. The pinna reflex, for example, was reduced only in the 2nd study week and only in males of the 450 ml/m^3 group compared with the control group. Females showed a decreased motor activity in the 4th week of exposure in the low and high, but not in the middle concentration group, while increased motor activity was observed during week 8 of exposure, but only in the 450 ml/m³ group. Sensorial and autonomic functions as well as the clinical picture and the results of the neuropathological examinations were normal. The authors concluded that exposures to diisopropyl ether concentrations up to 7060 ml/m³ did not lead to significant neurotoxic effects in rats (Rodriguez and Dalbey 1997), so that these findings are not contradictory to the NOEL of 480 ml/m³ obtained above.

The results of an earlier study involving repeated inhalation exposure of guinea pigs, rabbits and monkeys to high diisopropyl ether concentrations confirm the low toxicity of the substance, but they are not relevant for the derivation of a threshold limit value: Exposures to 30000 ml/m³ for one-hour periods at 4-day intervals until 10 exposures led to deep anaesthesia with signs of depression of the medulla and impending failure of respiration especially in the monkey. After one-hour exposures to 10000 ml/m³ on 20 days, there were signs of a central nervous depression, especially in the monkey. There were no visible signs of adverse effects after the 2-hour exposure to 3000 ml/m³ and the 3-hour exposure to 1000 ml/m³ on 20 days in each case (Machle et al. 1939).

5.3 Effects on skin and mucous membranes

Dropwise application of 150 ml diisopropyl ether for one hour a day over 10 days onto the clipped skin (area of about 80 cm²) of the rabbit under open conditions led to erythema and dermatitis, which were reversible within two weeks after the end of application. Single treatment was without effect (Machle et al. 1939).

In the rabbit eye, diisopropyl ether was irritating and caused minimal injuries (no further details) (Kirwin and Sandmeyer 1981).

5.4 Allergenic effects

The fact that diisopropyl ether does not react with lysine-containing peptides *in vitro* shows that the substance cannot act as a hapten. This finding, however, does not allow us to exclude a sensitizing potential (Wass and Belin 1990).

5.5 Reproductive toxicity

Twenty-two Sprague Dawley rats per group were exposed to diisopropyl ether concentrations of 430, 3095 or 6745 ml/m³ for 6 hours daily from day 6 to day 15 of gestation. A group of sham-exposed animals and a group of untreated animals served as controls. In the two high concentration groups, food consumption and body weight gain of the dams were reduced, although the differences to the sham-exposed group only reached a significant level at 6745 ml/m³. Some animals showed transient lacrimation and salivation during or immediately after exposure. Reproductive parameters and foetal development (size, weight, fraction of surviving foetuses) were not affected by the treatment. The only unusual finding was a significant increase in the incidence of rudimentary or shortened 14th ribs in the offspring of the two high concentration groups, which is generally not considered to be a teratogenic effect but a skeletal variation. The low concentration of 430 ml/m^3 is thus the NOEL for both maternal and developmental toxicity (Dalbey and Feuston 1996).

5.6 Genotoxicity

Diisopropyl ether was not mutagenic in the *Salmonella* mutagenicity test (pre-incubation assay) with and without the addition of a metabolic activation system. The strains TA98, TA100, TA1535, TA1537, and TA1538 were used. A differential killing test with the *E. coli* strains WP$_2$ and WP$_2$ uvrA also yielded negative results. Incubation of *Saccharomyces cerevisiae* with diisopropyl ether did not lead to mitotic gene conversions. The substance did not induce chromosomal aberrations in the epithelial rat liver cell line RL4 and in CHO cells (a cell line derived from Chinese hamster ovary). In each of the various study designs, concentrations up to the cytotoxic range or to the limit of solubility were tested. (Brooks *et al.* 1988).

6 Manifesto (MAK value, classification)

The few data available for man and the information obtained from animal studies indicate that irritation of the mucous membranes and, at higher concentrations, central nervous depression ought to be the critical effects in persons exposed to diisopropyl ether at the workplace. With respect to the corresponding threshold concentration in man data are available only from one short-term exposure study in test persons who, at 300 ml/m^3, complained about the unpleasant odour, but showed no other acute effects. Although this investigation does not meet present-day methodical requirements, it is nevertheless the only source of information on this subject. Derivation of a MAK value can additionally be based on modern 90-day inhalation studies on rats, yielding a NOEL of 480 ml/m^3 for local and for systemic effects.

On the basis of this NOEL, the MAK value for diisopropyl ether, which was valid up to 2000, has been reduced to 200 ml/m^3. Peak limitation category I with a permissable excursion factor of 2 has been set as a result of the irritative effects. With regard to the thus permissible exposure peaks of 400 ml/m^3, no signs of irritative effects in man are known. This value is still lower than the NOEL of 480 ml/m^3 determined in the animal study, and the effects observed at 3300 ml/m^3, the next highest concentration, are still very weak.

On the basis of the data for acute dermal toxicity, designation of the substance with "H" is not necessary.

The carcinogenic and germ cell mutagenic potential of the substance has not been investigated. As no genotoxicity has been found, a germ cell mutagenic effect is at any rate not to be expected either.

No reproductive toxicity was observed in rats exposed to diisopropyl ether in a concentration range non-toxic to dams. Concentrations above 3000 ml/m^3 led to skeletal variations. Owing to a lack of studies testing a second species, the substance has been classified in Pregnancy risk group D (tending towards C).

No decision can be made whether designation of the substance with "Sa" or "Sh" is necessary, because no data are available.

7 References

Alarie Y, Schaper M, Nielsen GD, Abraham MH (1998) Structure-activity relationships of volatile organic chemicals as sensory irritants. *Arch Toxicol 72*: 125–140

Brooks TM, Meyer AL, Hutson DH (1988) The genetic toxicology of some hydrocarbon and oxygenated solvents. *Mutagenesis 3*: 227–232

Browning E (1965) *Toxicity and metabolism of industrial solvents*. Elsevier, Amsterdam, 502–505

Dalbey W, Feuston M (1996) Subchronic and developmental toxicity studies of vaporized diisopropyl ether in rats. *J Toxicol Environ Health 49*: 29–43

ECB (European Chemicals Bureau) (1995) *IUCLID data sheet, diisopropyl ether*. Ispra, Italy

Kimura ET, Ebert DM, Dodge PW (1971) Acute toxicity and limits of solvent residue for sixteen organic solvents. *Toxicol Appl Pharmacol 19*: 699–704

Kirwin Jr CJ, Sandmeyer EE (1981) Ethers. in: Clayton GD, Clayton FE (Eds) *Patty's industrial hygiene and toxicology*, Volume 2A, John Wiley & Sons, New York

Machle W, Scott EW, Treon J (1939) The physiological response to isopropyl ether and to a mixture of isopropyl ether and gasoline. *J Ind Hyg Toxicol 21*: 72–96

Nishimura H, Saito S, Kishida F, Matsuo M (1994) Analysis of acute toxicity (LD50-value) of organic chemicals to mammals by solubility parameter (d). (1) Acute oral toxicity to rats. *Jpn J Ind Health 36*: 421–427

Rodriguez SC, Dalbey WE (1997) Subchronic neurotoxicity of vaporized diisopropyl ether in rats. *Int J Toxicol 16*: 599–610

Ruth JH (1986) Odor thresholds and irritation levels of several chemical substances: A review. *Am Ind Hyg Assoc J 47*: 142–151

Silverman L, Schulte HF, First MW (1946) Further studies on sensory response to certain industrial solvent vapors. *J Ind Hyg Toxicol 28*: 262–266

Wass U, Belin L (1990) An *in vitro* method for predicting sensitizing properties of inhaled chemicals. *Scand J Work Environ Health 16*: 208–214

completed 24.02.2000

Ethanethiol

MAK value (1969)	0.5 ml/m^3 (ppm) ≙ 1.29 mg/m^3
Peak limitation (2002)	Category II, excursion factor 2
Absorption through the skin	–
Sensitization	–
Carcinogenicity	–
Prenatal toxicity (2000)	see Section IIc of the *List of MAK and BAT Values*
Germ cell mutagenicity	–
BAT value	–
Synonyms	ethyl mercaptan mercaptoethane ethyl sulfhydrate thioethyl alcohol
Chemical name (CAS)	ethanethiol
CAS number	75-08-1
Structural formula	H$_3$C–CH$_2$SH
Molecular formula	C$_2$H$_6$S
Molecular weight	62.13
Melting point	–148°C
Boiling point	35°C
Density at 20°C	0.839 g/cm^3
Vapour pressure at 20°C	589 hPa
log P$_{ow}$*	not specified
1 ml/m^3 (ppm) ≙ 2.578 mg/m^3	**1 mg/m^3 ≙ 0.388 ml/m^3 (ppm)**

---

* *n*-octanol/water distribution coefficient

The MAK value for ethanethiol of 0.5 ml/m^3 was established in 1969 in accordance with the TLV value at the time. The present documentation is based on reviews of the toxicological data for the substance (Farr and Kirwin 1994, NL Health Council 2000).

Ethanethiol is used as a starting or intermediate product in the manufacture of certain plastics, insecticides and antioxidants, and as a result of its odour as an additive with warning function. It is naturally present in some kinds of petroleum and natural gas (Farr and Kirwin 1994).

The substance has an extremely unpleasant odour. The perception threshold is between 2.6×10^{-10} and 0.002 ml/m^3 (Amoore and Hautala 1983, Farr and Kirwin 1994, Ruth 1986).

1 Toxic Effects and Mode of Action

Ethanethiol is absorbed via the respiratory and gastrointestinal tracts and some of the absorbed amount is exhaled unchanged. The end product of metabolism is, among other things, sulfate. Metabolic products of ethanethiol can enter biosynthetic pathways as a source of sulfur or carbon. Ethanethiol causes mild irritation of the skin and mucous membranes. In an inadequately documented study with 3 test persons, central nervous effects and irritation of the mucous membranes were observed after repeated exposure to concentrations of 3.9 ml/m^3; concentrations of 0.39 ml/m^3 did not produce effects. The short-term effects of ethanethiol observed in animal experiments included CNS depression and cyanosis. Long-term subcutaneous injection of ethanethiol caused haemolysis in rats and rabbits. In the *Salmonella* mutagenicity test, sister chromatid exchange (SCE) test and mouse lymphoma test, only high ethanethiol concentrations in the cytotoxic range produced occasional positive results; these investigations do not, therefore, yield any clear evidence for genotoxic effects of the substance.

2 Mechanism of Action

Ethanethiol binds to the central iron ion of enzymes containing haem, e.g. cytochrome P450, cytochrome c and chloroperoxidase (Nastainczyk et al. 1976, Sono et al. 1984, Wilms et al. 1980); it inhibits cytochrome c oxidase *in vitro*, but, as a result of the steric hindrance caused by the ethyl residue, to a lesser extent than hydrogen sulfide or methyl mercaptan (Wilms et al. 1980). In mitochondria isolated from rat liver and rat brain, impairment of the electron transfer in the respiratory chain was detected (Vahlkamp et al. 1979). Na$^+$, K$^+$-ATPase is also inhibited by ethanethiol, but to a lesser extent than by methyl mercaptan (Foster et al. 1974). The described inhibition of gluconeogenesis and ureogenesis in rat hepatocytes exposed to ethanethiol is attributed to the suppression of the phosphorylation of the respiratory chain; inhibition of cytochrome c oxidase is thought to be of decisive importance in this process (Vahlkamp et al. 1979).

Generally, autoxidation of thiols can yield reactive oxygen species in the presence of suitable metal ions. The disulfides formed in this way can be reduced back to thiols. This redox cycling can lead to oxidative stress; this would explain the haemolytic effects sometimes observed in animal experiments and the occasional positive results in genotoxicity tests with high concentrations. In the case of ethanethiol, formation of the superoxide radical was detected (Munday 1989).

3 Toxicokinetics and Metabolism

During exposure to ethanethiol concentrations of 50 ml/m^3 for 35 to 60 minutes, between 60 % and 80 % of the inhaled ethanethiol was absorbed by three test persons (Shibata 1966b). After rabbits inhaled ethanethiol concentrations of 30 ml/m^3 for 25 minutes, traces of the substance were detected in their blood. When the animals were exposed to 10000 ml/m^3 for an hour, significant concentrations of the substance were determined in blood. After the end of exposure, the levels in blood rapidly decreased. In addition to the substance itself, ethyl methyl sulfide and diethyl sulfide were also detected in blood and exhaled air (no other details; Farr and Kirwin 1994).

The penetration rate of a saturated aqueous solution of ethanethiol, calculated from the physico-chemical data, is 0.57 mg/cm^2 and hour (Fiserova-Bergerova et al. 1990). Assuming contact of both hands and forearms with the substance (about 2000 cm^2 of skin), percutaneous absorption of ethanethiol thus amounts to 1140 mg per hour. In comparison, with observance of the MAK value of 1.29 mg/m^3, 9.03 mg ethanethiol is taken up during exposure for 8 hours (10 m^3 inhaled air volume, 70 % pulmonary retention). Thus ethanethiol may be assumed to penetrate the skin very readily. Metabolic products of ethanethiol can enter biosynthetic pathways as a source of sulfur or carbon in the mammalian organism (Farr and Kirwin 1994). In rabbit liver microsomes an S-adenosylmethyltransferase was detected for which ethanethiol and other short-chain mercaptans are a substrate (Holloway et al. 1979).

The results available from experiments and conclusions by analogy with homologous mercaptans support the hypothesis that in addition to the direct oxidation of the sulfhydryl group, which ultimately leads to the formation of inorganic sulfate (Farr and Kirwin 1994), an alternative metabolic pathway is methylation to ethyl methyl sulfide. The metabolite ethyl methyl sulfide can be oxidized via ethyl methyl sulfoxide to ethyl methyl sulfone (Snow 1957).

4 Effects in Man

Over a period of at least four and a half days, about 3 g ethanethiol escaped from a defective capillary tube into the preparation room and the neighbouring classroom of a school (in total 325 m^3). On the first day, lessons had to be terminated in the affected classroom after an hour as a result of the intolerable odour. A third of the thirty pupils, aged 16 to 18, developed indisposition, headaches and stomach aches. Three of those exposed suffered vomiting and diarrhoea. The symptoms disappeared after a short time. On the second day, the pupils remained in the classroom, which had been continuously ventilated, for 3 hours. The same pupils who had shown symptoms on the previous day again suffered headaches, but less severely. One pupil examined by the author was found to have haematuria and proteinuria, but the author questioned that there was a causal relationship with the exposure to ethanethiol (Pichler 1918).

Two studies with volunteers from the 1960s are, as a result of the low number of test persons and shortcomings in the documentation, only of limited use.

In an inhalation chamber experiment, test persons were exposed to ethanethiol concentrations of 10 mg/m^3 (3.9 ml/m^3) for 3 hours a day for 5 (n = 2) and 10 days (n = 1). Odour perception decreased after one and a half to two hours. The exposed persons complained of nausea, irritation of the oral and nasal mucosa, tiredness and a sensation of heaviness in the head. Ethanethiol concentrations of 1 mg/m^3 (0.39 ml/m^3) did not produce any symptoms (Blinova 1965).

Exposure of male volunteers to ethanethiol concentrations of 50 (n = 2) and 112 ml/m^3 (n = 3) for twenty minutes caused a decrease in the respiration rate in two of three persons and an increase in the respiratory volume. The test persons did not report any symptoms, even when the duration of exposure for the low concentration was extended to 35 to 60 minutes (Shibata 1966b).

5 Animal Experiments and *in vitro* Studies

5.1 Acute toxicity

The 4-hour LC$_{50}$ for rats was found to be 4420 ml/m^3 and for mice 2770 ml/m^3. Characteristic symptoms of intoxication in rodents after exposure to single sublethal and lethal doses were increased respiration rate, initial restlessness (hyperactivity in mice), later sedation, uncoordinated movements, staggering gait, muscle weakness, in some cases paralysis of the skeletal muscles, cyanosis and tolerance of prone position. The use of very high concentrations led to death from respiratory arrest. With minimal lethal concentrations, death was preceded by a prolonged period of torpor (Fairchild and Stokinger 1958). All five male rats survived inhalation exposure to a concentration of

28400 ml/m^3 for one hour. Five female rats survived exposure to 15000 ml/m^3 for one hour. Three of five female rats died, however, after exposure to 27000 ml/m^3 for one hour (Vernot et al. 1977). Interpolation of dose–response relationships obtained with groups of 3 to 5 rats given different doses revealed that ethanethiol concentrations of about 3300 ml/m^3 led to coma in 50 % of the animals after exposure for 15 minutes. This corresponded to an ethanethiol concentration in blood of 200 nmol/ml (Zieve et al. 1974). Inhalation of ethanethiol concentrations of 1000 ml/m^3 for twenty minutes caused a decrease in the respiration rate in rabbits. The effect was observed in a weaker form also after 100 ml/m^3; 10 ml/m^3 still caused slight destabilization of the respiratory function (Shibata 1966a).

In rats, the LD$_{50}$ values were 682 mg/kg body weight after oral administration and 226 mg/kg body weight after intraperitoneal injection (Fairchild and Stokinger 1958). The dermal LD$_{50}$ value for rabbits (24 hours under occlusive conditions) is above 2000 mg/kg body weight (Farr and Kirwin 1994).

5.2 Subacute, subchronic and chronic toxicity

Mice exposed daily to ethanethiol concentrations of about 1972 ml/m^3 for four hours died 4 to 9 days after the beginning of exposure. Severe CNS depression was thought to be the cause of death. Autopsy revealed inflammatory changes in the lungs, degeneration of the liver and increased heart weights. In mice exposed to concentrations of about 394 ml/m^3 for 4 hours a day for 3 weeks, body weights were reduced. Autopsy of the animals revealed no abnormalities (no other details; Farr and Kirwin 1994). Exposure of rabbits, rats and mice to ethanethiol concentrations of about 40 ml/m^3 for five months caused cardiovascular effects in rabbits, increased excitability in rats and increased body weights in mice. At the end of the study the exposed animals were not found to have any morphological organ changes (no other details; Blinova 1965).

In rabbits exposed to concentrations of 1000 ml/m^3 on nine consecutive days, there was a transient decrease in the leukocyte count on the seventh day. The erythrocyte counts and body weights remained unchanged (Shibata 1966a).

Subcutaneous injection of ethanethiol doses of 10 and 90 mg/kg body weight once a day or every second day for a year caused necrosis at the site of injection in rats and rabbits and clear signs of haemolysis in the blood count and spleen (e.g. haemosiderin deposits, fibrosis, sinusoid dilation, hyperaemia and increased haematopoiesis). However, some of the histopathological findings were observed in control animals as well and may be unspecific consequences of a general weakness resulting from the frequent injections. Also, the slight changes in the liver, lungs, kidneys and testes detected during gross pathological examination may not have been substance-related (no other details; Farr and Kirwin 1994).

5.3 Local effects on skin and mucous membranes

The instillation of 0.1 ml of the undiluted substance in the conjunctival sac caused slight irritation in rabbits (Fairchild and Stokinger 1958). Dermal application of ethanethiol caused discoloration of the skin and pain reactions in rats, but not persistent irritation. Moderate erythema was observed on rabbit skin after contact with the substance for 4 hours; this disappeared after less than 24 hours.

Inhalation exposure to 35 ml/m^3 twice for one minute did not cause any signs of irritation of the upper respiratory passages in mice (whole animal plethysmography) (no other details; Farr and Kirwin 1994).

5.4 Genotoxicity

Various *in vitro* tests for the genotoxicity of ethanethiol were carried out by a test laboratory. There is only inadequate information available about the implementation of the tests, which were carried out with material of unknown purity. They can therefore only be included in the evaluation of the genotoxicity of ethanethiol with reservations.

In the *Salmonella* mutagenicity test, ethanethiol dissolved in DMSO in concentrations of 123.5 to 10000 µg/plate was investigated with the strains TA98, TA100, TA1535, TA1537 and TA1538 in the absence and presence of a metabolic activation system. The highest tested concentration produced cytotoxic effects. Ethanethiol was not found to be mutagenic with the strains TA100, TA1535 and TA1538. With the strains TA98 and TA1537 only the second highest concentration of 3333 µg/plate caused a doubling of the mutant frequency in the presence of the activation system (Phillips Petroleum Company 1992). In the SCE test with CHO cells (a cell line derived from Chinese hamster ovary) in the presence and absence of a metabolic activation system, ethanethiol concentrations of 840 µg/ml caused a statistically significant but biologically probably irrelevant increase in the frequency of SCE of 1.2 times the control value. At the next highest concentration of 2500 µg/ml, which caused the inhibition of cell growth, all cells were in the metaphase of the first cell cycle and could not, therefore, be evaluated. In a repeat experiment, in which only this concentration was used and chromosome analysis was not carried out until 43 hours after the treatment as a result of the great delay in growth, there was a statistically significant increase in the frequency of SCE of 2.4 and 2.5 times the control value in the presence and absence of the activation system (Phillips Petroleum Company 1983). In a mutation test with L5178Y mouse lymphoma cells, ethanethiol was tested in concentrations of 60.6 to 1000 µg/ml both in the presence and absence of a metabolic activation system. Even the lowest tested concentration reduced the survival of the cells to 10 % and 31 % of the untreated controls, and concentrations of 670 µg/ml and above led to the death of practically all cells. Increases in the incidence of mutants of 1.6 to 3.7 times the control value were observed in the absence of the activation system, which, however, were not concentration-dependent and probably the result of the pronounced cytotoxicity (Phillips Petroleum Company 1992). No details were given about the size distribution of the induced mutant colonies. The *in vitro* genotoxicity tests described above yielded occasional positive results only at high concentrations. These studies do not, therefore, show any clear evidence of a genotoxic potential of ethanethiol.

There are no data available concerning allergenic or carcinogenic effects of ethanethiol, or the toxic effects of the substance on reproduction.

6 Manifesto (MAK value/classification)

The critical effects after occupational exposure to ethanethiol are presumably the central nervous effects and irritation of the mucous membranes. The data obtained in man are, however, insufficient for the derivation of a threshold limit value. In a study with 3 volunteers, effects were observed after repeated exposure to ethanethiol concentrations of 3.9 ml/m^3, but not after 0.39 ml/m^3. The study is, however, insufficiently documented. Ethanethiol is an odour nuisance even in low concentrations; it is unclear at which concentration this odour is intolerable for man.

There are no valid studies with animals with medium-term and long-term exposure. The data for the structurally similar methyl mercaptan, which is of greater acute toxicity and has been more extensively investigated, are therefore also included in the evaluation. In a valid medium-term inhalation study with rats, the first systemic-toxic effects (reduced body weight gains) were observed at methyl mercaptan concentrations of 57 ml/m^3. The LOAEL (lowest observed adverse effect level) in this study was 2 ml/m^3 and was based on slight behavioural changes in the exposed animals. Thus, the existing MAK value for methyl mercaptan of 0.5 ml/m^3 was not changed. As a result of the findings in man with ethanethiol and those in animal experiments with methyl mercaptan, the MAK value for ethanethiol has also been provisionally left at 0.5 ml/m^3.

As a result of the odour nuisance, the substance is classified in Peak limitation category II with an excursion factor of 2.

As there are no data available concerning toxic effects of the substance on reproduction, ethanethiol is listed in Section IIc of the *List of MAK and BAT Values*.

The available *in vitro* studies of the genotoxicity of ethanethiol, which produced occasional positive results only at high concentrations in the cytotoxic range, yield no clear evidence that the substance is mutagenic. There are no *in vivo* studies of the genotoxicity or data about the carcinogenicity of the substance. The carcinogenic and germ cell mutagenic potential of ethanethiol cannot therefore be evaluated.

Despite the high percutaneous penetration calculated for ethanethiol, designation with an "H" is not necessary because of the low acute toxicity of the substance after dermal application. As there are no data available for possible sensitizing effects of ethanethiol, it cannot be decided whether the substance should be designated with an "Sa" or "Sh".

7 References

Amoore JE, Hautala E (1983) Odor as an aid to chemical safety: odor thresholds compared with threshold limit values and volatilities for 214 industrial chemicals in air and water dilution. *J Appl Toxicol 3*: 272–290

Blinova EA (1965) Setting standards for the concentration of substances having a strong odour in the ambient air of industrial plants (Russian). *Gig Sanit 30*: 18–22, translation

Fairchild EJ, Stokinger HE (1958) Toxicologic studies on organic sulfur compounds. 1. Acute toxicity of some aliphatic and aromatic thiols (mercaptans). *Am Ind Hyg Assoc J 19*: 171–189

Farr CH, Kirwin CJ (1994) Organic sulfur compounds. in: Clayton GD, Clayton FE (Eds) *Patty's industrial hygiene and toxicology*, Vol 2F, 4th edition, John Wiley & Sons, New York, 4311–4372

Fiserova-Bergerova V, Pierce JT, Droz PO (1990) Dermal absorption potential of industrial chemicals: criteria for skin notation. *Am J Ind Med 17*: 617–635

Foster D, Ahmed K, Zieve L (1974) Action of methanethiol on Na+, K+ ATPase: implications for hepatic coma. *Ann NY Acad Sci 242*: 573–576

Holloway CJ, Husmann-Holloway S, Brunner G (1979) Enzymatic methylation of alkane thiols. *Enzyme 24*: 307–312

Munday R (1989) Toxicity of thiols and disulphides: involvement of free-radical species. *Free Radical Biol Med 7*: 659–673

Nastainczyk W, Ruf HH, Ullrich V (1976) Binding of thiols to microsomal cytochrome P-450. *Chem Biol Interact 14*: 251–263

NL Health Council (2000) *Ethanethiol*. Committee on Updating of Occupational Exposure Limits, Public draft, July 2000, Health Council of the Netherlands, The Hague

Phillips Petroleum Company (1992) *Initial Submission: mouse lymphoma forward mutation assay and in vitro sister chromatid exchange in Chinese hamster ovary cells*. Hazelton Laboratories America Inc., NTIS/OTS 0571884, EPA/OTS Doc ID 88-920010738, NTIS, Springfield, VA, USA

Phillips Petroleum Company (1983) *Salmonella typhimurium mammalian microsome plate incorporation assay with ethyl mercaptan*. Hazleton Laboratories Inc. Phillips Petroleum Company, Bartlesville, OK, USA, unpublished

Pichler K (1918) Vergiftung durch Einatmen von Äthylmerkaptan (Poisoning by inhalation of ethyl mercaptan) (German). *Zentralbl Inn Med 39*: 689–693

Ruth JH (1986) Odor thresholds and irritation levels of several chemical substances: a review. *Am Ind Hyg Assoc J 47*: A 142–A 151

Shibata Y (1966a) Studies on the influence of ethyl mercaptan upon living body, 2. On the respiratory function and clinical findings in rabbits inhaled ethyl mercaptan gas (Japanese). *Shikoku Acta Med 22*: 834–843

Shibata Y (1966b) Studies on the influence of ethyl mercaptan upon living body, 3. Inhalation experiment of ethyl mercaptan gas in human body (Japanese). *Shikoku Acta Med 22*: 844–850

Snow GA (1957) The metabolism of compounds related to ethanethiol. *Biochem J 65*: 77–82

Sono M, Dawson JH, Hager LP (1984) The generation of a hyperporphyrin spectrum upon thiol binding to ferric chloroperoxidase. *J Biol Chem 259*: 13209–13216

Vahlkamp T, Meijer AJ, Wilms J, Chamuleau RAFM (1979) Inhibition of mitochondrial electron transfer in rats by ethanethiol and methanethiol. *Clin Sci 56*: 147–156

Vernot EH, MacEwen JD, Haun CC, Kinkead ER (1977) Acute toxicity and skin corrosion data for some organic and inorganic compounds and aqueous solutions. *Toxicol Appl Pharmacol 42*: 417–423

Wilms J, Lub J, Wever R (1980) Reactions of mercaptans with cytochrome c oxidase and cytochrome c. *Biochim Biophys Acta 589*: 324–335

Zieve L, Doizaki WM, Zieve FJ (1974) Synergism between mercaptans and ammonia or fatty acids in the production of coma: a possible role for mercaptans in the pathogenesis of hepatic coma. *J Lab Clin Med 83*: 16–28

completed 24.02.2000

Nitrogen dioxide

MAK value	–
Peak limitation	–
Absorption through the skin	–
Sensitization	–
Carcinogenicity (2003)	Category 3B
Prenatal toxicity	–
Germ cell mutagenicity	–
BAT value	–
Synonyms	nitrogen peroxide
Chemical name (CAS)	nitrogen dioxide
CAS number	10102-44-0
Molecular formula	NO_2
Molecular weight	46.01
Melting point	−11.2°C
Boiling point	21.2°C
Density at 20°C	1.45 g/cm^3
Relative gas density	1.59
Vapour pressure at 20°C	960 hPa
log P_{ow}*	−0.58
1 ml/m^3 (ppm) ≙ 1.88 mg/m^3	**1 mg/m^3 ≙ 0.53 ml/m^3 (ppm)**

The MAK value for nitrogen dioxide (NO_2) of 5 ml/m^3 was established in 1958 in analogy to the American TLV value at the time. The present documentation is based in particular on an IPCS report of the WHO (1997) and earlier and more recent literature relevant to the evaluation.

---

* *n*-octanol/water distribution coefficient

1 Toxic Effects and Mode of Action

NO_2 enters the terminal airways via the nose and the upper respiratory tract; nitrous acid, nitric acid and nitrogen monoxide (NO) are formed, which can cause irritative effects. Of greater importance, however, is the damage to the tissue in the terminal airways, which is caused by radical reactions of NO_2 with components of the alveolar fluid and the epithelial cells. Damage to type I pneumocytes and cilia-bearing epithelial cells occurs; these are replaced by less sensitive cells, such as type II pneumocytes and Clara cells. Also inflammatory symptoms are observed. After long-term exposure emphysema-like changes develop.

Studies with healthy volunteers and short-term exposure yielded inconsistent evidence of inflammatory symptoms and reduced viral resistance after NO_2 concentrations of 0.6 ml/m^3 and above. After concentrations of 1.5 ml/m^3 and above, increased bronchial reactivity is observed. Lung function changes (increased airway resistance) are seen after 2.0 ml/m^3 and above.

Exposure to 25 to 75 ml/m^3 leads to bronchitis or bronchopneumonia, 50 to 100 ml/m^3 to reversible bronchiolitis and focal pneumonitis, 150 to 200 ml/m^3 to lethal *bronchiolitis fibrosa obliterans* and more than 300 ml/m^3 to lethal pulmonary oedema and asphyxia (as a result of a lack of oxygen, due to methaemoglobinaemia).

In studies with animals, continuous long-term exposure (23 to 24 hours a day) to NO_2 concentrations of 5 ml/m^3 and above caused emphysema, as is also observed in the human lung. In rats exposed to NO_2 concentrations of 0.04 ml/m^3 continuously for 27 months, evidence of lipid peroxidation was observed, and after 0.4 ml/m^3 and above, of histopathological effects on the lungs. Increased sensitivity of mice to bacterial and viral lung infections is observed after NO_2 concentrations of 0.2 and 0.3 ml/m^3 and above.

In vitro NO_2 produces clear mutagenic and clastogenic effects. Evidence of genotoxic effects *in vivo* was found in a study with the rat lung, which, however, is only of limited use.

Valid long-term studies of the carcinogenic effects of the substance are not available. Initiation promotion experiments and a short-term test with the mouse lung yielded evidence of promoting and carcinogenic effects of NO_2 at concentrations of 4 to 10 ml/m^3.

There are no valid studies of the toxic effects of NO_2 on development. The studies carried out, which are only of very limited use, indicate toxic effects on development in a range as low as 0.5 ml/m^3 but need verification.

2 Mechanism of Action

NO_2 is an irritant gas. In water NO_2 slowly hydrolyses to nitric acid and nitrous acid, both of which have irritative to caustic effects, and to NO (Elsayed 1994, WHO 1997).

Molecular reactions

NO_2 is a moderate oxidation agent and an uncharged, stable free radical capable of reacting in various ways (Kirsch et al. 2002):

1) Abstraction of hydrogen from organic molecules with the formation of nitrous acid and a molecule radical (•NO_2 + R – H → HONO + R•). This reaction is irreversible. With unsaturated fatty acids the hydrogen is abstracted from the allyl position (–HC=CH–CH•–). The alkenyl radicals formed produce toxic effects, as they further react with molecular oxygen to form peroxyl radicals which break down lipids by initiating the radical chain reactions of lipid peroxidation (see below).

2) Addition of organic molecules to double bonds (•NO_2 + –HC=CH–CH_2– ⇌ –(NO_2)HC–•CH–CH_2–) in a reversible reaction competing with the cleavage of hydrogen (see above).

3) Electron transfer of organic molecules to •NO_2. In this reaction a nitrite ion, the radical of the biomolecule and a proton are formed. This reaction is just as destructive as that described under 1), as the biomolecule radical reacts with molecular oxygen to form a peroxyl radical and as such can set off oxidizing chain reactions. •NO_2 itself has only moderate oxidation capability, but there are numerous biomolecules which are changed by electron transfer. •NO_2 can thus oxidize thiolates, for example.

4) Recombination with another radical. Both the nitrogen atom and one of the oxygen atoms of NO_2 can form the covalent bond. As a result, nitro compounds and alkyl nitrites are formed. The relative amounts of the two reaction products are determined by the spin density of the unpaired electron in the nitrogen (50 %) and the two oxygen atoms (both 25 %) of the •NO_2 molecule. The pathobiological importance of the alkyl nitrites is based on their strong nitrosation capability with amines of the organism (see below), e.g. to form 3-nitrotyrosine. Nitrated biomolecules are unphysiological and are therefore degraded rapidly. This is true in particular for proteins with nitrotyrosine residues.

When an organism exposed to •NO_2 was previously exposed to polycyclic aromatic hydrocarbons, the corresponding (often carcinogenic) nitro aromatics (see below) can be formed.

A biologically important exception among radical recombinations is the coupling of •NO_2 with the •NO radical (which takes place, for example, in inflamed tissue) to form N_2O_3. The latter is a strongly nitrosating oxide. It reacts with secondary amines of the organism to form (carcinogenic) nitrosamines (see below), transforms terminal amino groups of peptides into mutagenic diazopeptide groups and cleaves amino groups from DNA bases, so that transition mutations take place (e.g. GC → AT during the deamination of cytosine to uracil; further deamination processes: adenine → hypoxanthine; guanine → xanthine) (Halliwell 1999). The same deamination in DNA, followed by mutations, can also be triggered by N_2O_4, the recombination product of •NO_2 with itself.

Combination with the superoxide radical (O_2•) also has far-reaching biological consequences; the strongly oxidizing peroxynitrate (O_2NOO^-) is formed, which oxidatively attacks NADH, aromatic compounds and even Cl^- and Br^-.

5) It is possible for •NO$_2$ to enter •NO metabolism to a certain extent, insofar as it causes a disproportion in the equation between nitrate and •NO:

$$2 \cdot NO_2 \rightleftharpoons N_2O_4$$
$$N_2O_4 + H_2O \rightarrow NO_2^- + NO_3^- + 2 H^+$$
$$2 NO_2^- + 2 H^+ \rightleftharpoons N_2O_3 + H_2O$$
$$N_2O_3 \rightleftharpoons \cdot NO + \cdot NO_2$$

•NO is a highly active signal molecule which influences complex biological processes in cells by regulating genes (Kröncke et al. 2001) and modulating the activity of numerous protein kinases (Park et al. 2000, Schindler and Bogdan 2001). Among the complex processes governed by •NO are e.g. inflammation, stress, apoptosis and necrosis. •NO also reacts with haemoglobin and forms methaemoglobin (MetHb) (Kim et al. 2001).

As a result of the reactions listed •NO$_2$ is regarded as highly toxic.

The peroxidation of membrane lipids is an integral part of cell damage and toxic processes. It leads among other things to changes in the composition of phospholipids, which alter the membrane fluidity. The healthy organism, however, possesses antioxidants which can counteract •NO$_2$ and its reactive intermediates (N$_2$O$_3$, N$_2$O$_4$, RONO). Antioxidants can also reverse the first radical reaction into which a biomolecule enters; the intact (bio-)molecular structure can be restituted (Kirsch et al. 2002). Ascorbic acid, glutathione and its thiolate anion, vitamin E (α-tocopherol), γ-tocopherol, β-carotene are here the principal antioxidants worth mentioning.

Physiological reactions

One working group investigated lipid peroxidation and the activities of antioxidative enzymes in the lungs of rats after short, medium and long-term exposure to NO$_2$. During the first two days of exposure to NO$_2$ (0.04, 0.4, 4 ml/m^3) a transient reduction in the amount of ethane exhaled, a measure of lipid peroxidation, was observed at 0.4 and 4 ml/m^3. The authors attributed this reduction to the activity of the antioxidants. After this period, a steep increase in the amount of ethane exhaled (depending on the exposure concentration) was observed and, following a maximum level after about 7 days, another steep decrease. The authors saw a relationship with the initial damage to type I cells (up to day 1), the commencement of the repair process and the proliferation of type II cells (up to the maximum level), and the subsequent decrease in the proliferation of type II cells. After about 4 weeks, with increasing exposure duration, a slow increase in the amount of ethane exhaled was again observed, parallel to the increase in the control group (Sagai and Ichinose 1987). This indicates a reduced antioxidative capacity in the animals with increasing age.

Increased cell proliferation was observed in particular in the bronchiolar epithelium of male Sprague-Dawley rats after exposure to NO$_2$ concentrations of 5, 10 or 20 ml/m^3 air for 3 days. After exposure for 25 days, only very slight proliferation was found at the 5 ml/m^3 concentration, while marked proliferation was still found at NO$_2$ concentrations of 10 and 20 ml/m^3 (Barth and Müller 1999). In a follow-up study by the same working group, increased proliferation of epithelial cells of the bronchioli was determined after the animals were exposed to 0.8 ml/m^3 for only one hour. Proliferation of epithelial cells of the bronchi and type II cells was observed only after exposure of the animals to NO_2 concentrations of 5 ml/m^3 for 1 and 5 hours, respectively (Barth *et al.* 1994).

Effects on the immune defence of the lungs were also observed, and inflammatory symptoms with the development of activated macrophages in the alveolar fluid film, which thus increase the oxidative stress, were seen. In addition, collagen synthesis in the lungs is induced by NO_2. After long-term exposure, the alveolar capillary membrane becomes thicker and emphysema-like changes are observed as a result of the formation of collagen (Blomberg 2000).

As a result of the reactivity of NO_2, reactions and formation of NO_2 products in the living cell are numerous. It is hardly possible to detect the many various reaction products and the mechanisms involved.

The formation of nitrosamines

Both NO and NO_2 can react in aqueous solution with amines to form nitrosamines and nitramines (Challis and Shuker 1982). Nitrosation reactions can take place in the air, in exhaust gases, on the surfaces of dust particles, on filters and on the surfaces of walls (Pitts *et al.* 1978, Tokiwa *et al.* 1983).

Nitrosamines have also been detected in animals given amino compounds and exposed via inhalation to NO_2, while after exposure to NO the formation of nitrosamines was not observed (Uozumi *et al.* 1982). In rats and rabbits given oral doses of aminopyrine and exposed via inhalation to NO_2 concentrations of 25 to 100 ml/m^3 for one hour, the dose-dependent formation of *N*-nitrosodimethylamine was detected (Uozumi *et al.* 1982). In mice given intravenous or oral doses of 2 mg morpholine per animal and exposed to NO_2 concentrations of 45 ml/m^3 for 2 hours, the formation of *N*-nitrosomorpholine was observed (Norkus *et al.* 1984). In mice given oral doses of dimethylamine and subsequently exposed to NO_2 concentrations of 0.04 to 44.5 ml/m^3 for 4 hours, an increase in the formation of dimethylnitrosamine was observed in particular after concentrations of 10 ml/m^3 and more (Iqbal *et al.* 1981).

After exposure of mice to NO_2 concentrations of 50 ml/m^3 for 4 hours and application to the skin of 25 mg morpholine per animal, no *N*-nitrosomorpholine was detected on the skin of the mice 20 hours later. If, however, mice whose skin had already been painted with morpholine were then exposed to NO_2 concentrations of 55 ml/m^3 for 30 minutes, 19 nmol *N*-nitrosomorpholine per animal was found, which was probably the result of direct nitrosation (Mirvish *et al.* 1988).

The formation of nitroaromatics

The formation of genotoxic nitropyrene from NO_2 and pyrene, a polycyclic aromatic hydrocarbon, has also been detected. Nitropyrenes were found in the urine of mice exposed to NO_2 concentrations of 5, 10 or 20 ml/m^3 for 3 days, then injected with pyrene in doses of 10 to 20 mg/kg body weight and finally exposed to NO_2 for another 24 hours (Kanoh *et al.* 1987, 1990).

Changes in neuroendocrine (APUD) cells

Exposure to NO_2 changed the number and function of neuroendocrine lung cells (also called APUD cells, from 'amine precursor uptake and decarboxylation') in experimental animals. These cells are involved in the regulation of pulmonary blood pressure; they secrete vasoactive substances (Witschi 1988). The quantitative identification of these cells requires the use of special methods such as fluorescence techniques or silver staining (Palisano and Kleinerman 1980, Pan et al. 2000). Neuroendocrine cells have been detected e.g. in hamsters and rats in the walls of the bronchioli, bronchi and trachea; this is also the case with the so-called neuroendocrine bodies (Kleinerman et al. 1981, Palisano and Kleinerman 1980). After almost continuous exposure to NO_2 concentrations of 30 ml/m^3 for 29 days, rats were found to have a two-fold increase in tracheal neuroendocrine cells (Kleinerman et al. 1981). In Syrian hamsters, a numerical decrease in these cells was observed under the same exposure conditions (Palisano and Kleinerman 1980).

Other irritant gases and substances hazardous to the lungs can also change the number of neuroendocrine cells, in particular when oxidative stress is involved. In hamsters treated with diethylnitrosamine or other nitrosamines, persistent hyperplasia of the neuroendocrine cells in the bronchioli was observed. After longer exposure, signs of cell metaplasia were detected (Reznick-Schuller 1976, 1977a, 1977b). Neuroendocrine lung tumours were found in hamsters after inhalation of nitrosamine under hyperoxic conditions (Schuller et al. 1990). Rhesus monkeys developed marked foci of neuroendocrine cells after exposure to ozone (Castleman et al. 1980).

It was discussed whether both the presence of carcinogenic substances and an imbalance in the oxygen equilibrium are necessary for the development of neuroendocrine lung tumours as are induced e.g. in the hamster after inhalation exposure to nitrosamines under hyperoxic conditions (Schuller et al. 1990). NO_2 and ozone could be important because they contribute towards creating conditions in the lungs able to stimulate the proliferation of neuroendocrine cells (Witschi 1988).

In small cell bronchial carcinoma in man, which accounts for about 15 % to 20 % of lung tumours, neuroendocrine cells are found particularly frequently, so that their detection can be used as a diagnostic criterion (Sattler and Salgia 2003, Wistuba et al. 2001). Whether these cells are directly involved in the formation of the tumour is, however, unclear in experimental animals.

3 Toxicokinetics and Metabolism

NO_2 is a reddish-brown gas, which is in equilibrium with the colourless dinitrogen tetroxide (N_2O_4). At 25°C NO_2 makes up 25 % of the mixture, at 37°C 30 %. At greater dilution, e.g. with air, the amount of NO_2 is greater. As a result of its poor solubility in water, NO_2 reaches deep areas of the lung. The target organs are thus the terminal airways (Kirsch et al. 2002).

NO_2 reacts only slowly in water to form nitric acid and nitrous acid, both of which have irritative to caustic effects (Mücke and Wagner 1998). It was concluded from a study with monkeys with radioactively labelled NO_2 that nitric acid and nitrous acid are formed also in the respiratory tract. The acids or their salts were detected in blood and urine. Experimental studies also showed that NO_2 or its products can remain in the lungs for a longer period (WHO 1997). In man 80 % to 90 % NO_2 is taken up via the respiratory tract during normal breathing, and over 90 % at maximum breathing. Dosimetric model calculations showed that NO_2 is absorbed mainly in the lower respiratory tract, where NO_2 accumulates particularly in the area between the conductive and respiratory airways (pulmonary acinus), in which the morphological changes are observed (WHO 1997).

4 Effects in Man

NO_2 has a penetrating odour. Depending on the study conditions, the perception threshold is between 0.1 and 0.2 ml/m^3 (Feldman 1974, Shalamberidze 1967). With slowly increasing concentrations the odour is not perceived until much higher concentrations have been reached (Henschler et al. 1960), so that the warning effect of the gas is poor under this condition.

Background exposure

The main source of atmospheric NO_2 is the combustion of fossil fuels for heating, power plants and vehicles. NO_2 concentrations in towns depend on the time of day and time of year, meteorological conditions and human activities. Normally peak concentrations occur in the mornings and afternoons during traffic rush hours. As NO_2 concentrations indoors may be considerably higher than the environmental concentrations, indoor exposure, in particular from gas fires and ovens, provides the main source of an individual's exposure to NO_2. Indoor NO_2 concentrations can reach an average value of 0.5 ml/m^3 over 24 hours with peak concentrations of up to 1 ml/m^3 (Blomberg 2000).

4.1 Single exposures

Studies with volunteers

In the majority of studies with **healthy volunteers** (Table 1), no effects on lung function were observed after short-term exposure (a maximum of 4 hours) to NO_2 concentrations in the air of 0.1 to about 2 ml/m^3. Effects on lung function were described even at lower concentrations in only two studies (Bylin *et al.* 1985, Kulle 1982). Increased airway resistance was reported in several studies only after 2 ml/m^3 and above (Beil and Ulmer 1976, Blomberg *et al.* 1999, von Nieding and Wagner 1975, von Nieding *et al.* 1970, Stresemann and von Nieding 1970). Evidence of increased bronchial reactivity was observed after NO_2 concentrations of 1.5 ml/m^3 and above (Frampton *et al.* 1989b, 1991, Mohsenin 1987, 1988).

In some studies with volunteers bronchoalveolar lavage (BAL) was also carried out. Reduced inactivation of influenza viruses *in vitro* and increased interleukin-1 activity were detected in 4/9 persons 3.5 hours after exposure to NO_2 concentrations of 0.6 ml/m^3 for 3 hours with physical exercise. In the other 5 persons, interleukin-1 activity was decreased (Frampton *et al.* 1989a). The protein concentration in the BAL fluid was slightly, but not significantly increased, the concentrations of α-2-macroglobulin significantly increased. The observed effects were no longer in evidence in the BAL carried out 18 hours after exposure (Frampton *et al.* 1989b). In a further study with 21 volunteers exposed to NO_2 concentrations of 0.6 or 1.5 ml/m^3 for 3 hours during physical exercise, dose-dependent effects were observed in some cases: these included reduced haemoglobin and a decreased mean haemoglobin concentration in the individual erythrocytes (MCHC), a reduced lymphocyte count, an increased number of polymorphonuclear leukocytes in men and an increase in the proportion of CD4$^+$ lymphocytes (Frampton *et al.* 2002). However, the effects were relatively slight, and it is not clear from the publication whether the effects mentioned were already significant at concentrations of 0.6 ml/m^3, as the data for significance is given only generally as "NO_2 effects". Overall, the BAL investigations are to be considered relevant as a result of the cell-damaging effects of NO_2. The findings at 0.6 ml/m^3 are, however, slight and not consistent, but indicate possible first effects. Clear effects in the BAL were obtained at NO_2 concentrations of 1.5 ml/m^3 (Sandström *et al.* 1992) and 2 ml/m^3 (Blomberg *et al.* 1999, Devlin *et al.* 1999, Solomon *et al.* 2000).

Table 1. Controlled inhalation studies with healthy volunteers

Test persons	Duration of physical activity (min)	Exposure duration (min)	Exposure concentration (ml/m³)	Parameters investigated, findings	References
15♂	–	60	0.1	lung function, bronchial reactivity (METH): no effects	Hazucha et al. 1982, 1983
20♂, 20♀	–	60	0.1	lung function: no effects; bronchial reactivity (CARB): variable effects	Ahmed et al. 1982
4♂, 6♀	–	60	0.12	lung function: no effects	Koenig et al. 1985
7♂	4×15	120	0.15	lung function: no effects	Kagawa 1983
10 (♂ + ♀)	10	40	0.12, 0.18	lung function: no effects	Koenig et al. 1987
19♂ (15♂ controls)	–	120	0.2	blood: GSH significantly increased (no significant change in glutathione reductase, 2,3-diphosphoglycerate, methaemoglobin, IgA, vitamin E, complement C3)	Chaney et al. 1981
9♂	26	30	0.18, 0.30	lung function: no effects	Kim et al. 1991
5♂, 7♀	2×15	60	0.3	lung function: no effects	Koenig et al. 1988
7 (no other details)	4×10	60	0.3	lung function, bronchial reactivity (METH): no effects	Vagaggini et al. 1994
5♂, 3♀	–	20	0.12, 0.24, 0.48	lung function: airway resistance increased at 0.24 ml/m³ and reduced at 0.48 ml/m³; bronchial reactivity (HIST): no effects	Bylin et al. 1985
10♂ (3 S)	4×15	120	0.5	lung function: individual effects (quasistatic compliance significantly reduced, closing volume significantly increased)	Kerr et al. 1979, Kulle 1982
10♂	2×15	240	0.5	lung function: no effects	Stacy et al. 1983
20♂, 20♀	60	60	0.6	lung function: no effects (significant effects with exposure to O_3)	Adams et al. 1987
21♀	4×15	120	0.6	lung function, bronchial reactivity (METH): no significant effects	Hazucha et al. 1994
8♂, 8♀	3×20	120	0.6	lung function: no significant effects	Drechsler-Parks et al. 1987

Table 1. continued

Test persons	Duration of physical activity (min)	Exposure duration (min)	Exposure concentration (ml/m3)	Parameters investigated, findings	References
groups of 5♂	1×15, 1×30, 4×15	120 in each case	0.62	lung function: no effects	Folinsbee et al. 1978
5 (no other details)	"intermittent"	4 days for 120 minutes	0.6	blood, BAL: no significant effects on the phenotype of the lymphocytes in blood and BAL; slight (significant) increase in the ratio of NK cells	Rubinstein et al. 1990b
7♂, 2♀	6×10	180	0.6	lung function, bronchial reactivity (CARB): no effects; BAL: in 4/9 persons reduced inactivation of influenza viruses *in vitro* (not significant) and increased IL-1 production; in the other 5/9 persons IL-1 reduced, antiprotease α-2-macroglobulin increased (BAL sampled 3.5 hours after exposure); no changes in macrophages, lymphocytes, neutrophils, protein, albumin	Frampton et al. 1989a, 1989b, 1991
9♀, 12♂	6×10	180	0.6, 1.5	lung function: no effects; blood: ≥ 0.6 ml/m^3: haematocrit, lymphocyte count reduced, ratio of T helper cells to T suppressor cells increased (♂) or reduced (♀); BAL: ≥ 0.6 ml/m^3: ratio of T helper cells to T suppressor cells increased (♂) or reduced (♀), 1.5 ml/m^3: number of polymorphonuclear leukocytes increased (♂); evidence of slight inflammation. The statistical significance is not stated.	Frampton et al. 2002
16♂	4×15	120	1.0	lung function: no relevant effects (marginal reduction in the FVC)	Hackney et al. 1978
10♂	5–6×15	150–180	1.0, 2.0	blood: ≥ 1 ml/m^3: acetylcholinesterase activity (erythrocyte membrane) reduced; haemoglobin, haematocrit slightly reduced; 2 ml/m^3: glucose-6-phosphate dehydrogenase increased	Posin et al. 1978
5♀, 3♂	6×15	180	1.0	lung function, symptoms, bronchoscopy: no changes; BAL: thromboxan B$_2$ slightly increased	Jörres et al. 1995

Table 1. continued

Test persons	Duration of physical activity (min)	Exposure duration (min)	Exposure concentration (ml/m^3)	Parameters investigated, findings	References
11♂, 4♀	6×10	180 with 3×15 min peaks	0.05 + 2.0 peaks	lung function, bronchial reactivity (CARB), BAL: no effects	Frampton et al. 1989a, 1989b, 1991, Johnson et al. 1990b
12♂, 3♀	6×10	180	1.5	lung function, BAL: no effects; bronchial reactivity (CARB): increased	Frampton et al. 1989b, 1991, Johnson et al. 1990b
8♂	–	6 days for 20 minutes	1.5	BAL: cytotoxic suppressor T cells reduced (ratio of T helper cells to T suppressor cells increased), NK cells reduced	Sandström et al. 1992
13♂, 5♀	–	60	2.0	bronchial reactivity (METH): increased; lung function: no effects	Mohsenin 1988
8♂, 3♀	–	60	2.0	bronchial reactivity (METH): increased (prevented by the administration of ascorbic acid)	Mohsenin 1987
7♂	–	240	2.0	nasal epithelium: minimal changes (ultrastructural membrane changes of ciliary cells in 6/7 samples; morphometric statistical analysis not significant)	Carson et al. 1993
8♂	8×15	240	2.0	lung function: no changes (FEV$_1$, sR$_{aw}$), but evidence of slight obstructive changes; BAL: reduced capacity for phagocytosis of alveolar macrophages, reduced superoxide production; in the bronchial fraction PMNs, IL-6, IL-8, α1-antitrypsin and tissue plasminogen activator increased, number of epithelial cells reduced	Devlin et al. 1999
11♂, 4♀	4×30 in each case	3 days for 240 minutes	2.0	bronchoscopy, blood: no effects; BAL: CD4$^+$ cells reduced, neutrophils in bronchial fraction increased; evidence of inflammation	Solomon et al. 2000

Table 1. continued

Test persons	Duration of physical activity (min)	Exposure duration (min)	Exposure concentration (ml/m³)	Parameters investigated, findings	References
8♂, 4♀	8×15 in each case	4 days for 240 minutes	2.0	bronchoscopy with endobronchial biopsy: neutrophils in epithelium reduced; lung function: FEV_1 and FVC reduced (only after first exposure); BAL: $CD25^+$ lymphocytes and $HLA-DR^+$ macrophages increased; in bronchial fraction neutrophils and myeloperoxidase increased, albumin reduced; antioxidants (GSH, GSSG, ascorbic acid, uric acid) in BW and BAL unchanged; evidence of inflammation	Blomberg et al. 1999
5♂, 2♀	–	300	2.3	lung function: no effects; alveolar permeability: reduced; serum glutathione peroxidase: reduced	Rasmussen et al. 1992
groups of 21–23 (no other details)	–	3 days for 120 minutes in each case	1.0, 2.0, 3.0	lung function; bronchial reactivity (METH): no effects; antiviral resistance after influenza infection: 1–2 ml/m³: increased antibody formation; 3 ml/m³: infections (5/22)	Goings et al. 1989
14♂, 10♀	–	20	1.5, 3.5	mucociliary clearance (45 min after exposure): no effects; mucociliary clearance (24 h after exposure): increased	Helleday et al. 1995
	1×15	240	3.5		
10 exposed, 7 controls (16♂, 1♀)	–	180	3.0, 4.0	BAL: functional activity of α-1 proteinase inhibitor reduced	Mohsenin and Gee 1987
16♂, 9♀	2×15	75	4.0	lung function, symptoms, heart frequency, permeability of the skin, subjective emotional status: no effects (systolic blood pressure slightly increased)	Linn et al. 1985a
32♂, 18♂	1×15	20	2.25, 4.0, 5.5	BAL: changes (evidence of inflammation) ≥ 2.25 ml/m³: inflammatory symptoms; mastoid air cells increased; ≥ 4 ml/m³: lymphocytes increased; 5.5 ml/m³: lysozyme-positive alveolar macrophages increased	Sandström et al. 1990, 1991

Table 1. continued

Test persons	Duration of physical activity (min)	Exposure duration (min)	Exposure concentration (ml/m3)	Parameters investigated, findings	References
15♂ (4 S)	–	(20 breaths)	4.6–5.4	lung function: airway resistance increased (in 11/11 NS and 2/4 S)	Stresemann and von Nieding 1970
298 (no other details)	–	15	< 1.0, 1.0–1.5, 1.6–2.0, 2.1–2.5, 3.0–5.0	lung function: < 1.5 ml/m^3: no significant changes; ≥ 1.6–2.0 ml/m^3: increased airway resistances; ≥ 4 ml/m^3: arterial O$_2$ partial pressure decreased	von Nieding and Wagner 1975
12♂, 1♀ (10 S)	–	15	5.0	lung function: lung function changes (no differences between S and NS)	von Nieding et al. 1970
16♂ (11 S)	–	120	1.0, 2.5, 5.0, 7.5	lung function, bronchial reactivity (ACH): changes at > 2.5 ml/m^3	Beil and Ulmer 1976

ACH: acetylcholine; AL: alveolar lavage; BAL: bronchoalveolar lavage; BL: bronchial lavage; BW: bronchial washing; CARB: carbachol; closing volume: lung volume at which the large airways close, before the residual volume with maximum breathing is reached; FEV$_1$: forced expiratory volume in 1 sec; FVC: forced vital capacity; GSH: glutathione; GSSG: oxidized GSH; HIST: histamine; IL: interleukin; LDH: lactate dehydrogenase; METH: methacholine; NK cells: natural killer cells; NS: non-smoker; PMN: polymorphonuclear leukocytes; S: smoker; sR$_{aw}$: specific airway resistance

In some cases an increase in bronchial reactivity or lung function changes was reported in **asthmatic patients** (Table 2) at NO$_2$ concentrations of 0.1 to 0.2 ml/m^3 (Orehek et al. 1976) and 0.25 to 0.3 ml/m^3 (Avol et al. 1989, Bauer et al. 1986, Jörres and Magnussen 1990, Koenig et al. 1988), but there was great individual variability. After 0.4 ml/m^3 and concentrations above this the effects found were consistent between the studies (Bylin et al. 1985, Jenkins et al. 1999, Kulle 1982). Studies are, however, available in which no changes in bronchial reactivity were described at concentrations of 4 ml/m^3 (Linn et al. 1985a).

Table 2. Controlled inhalation studies with asthmatic patients

Test persons	Duration of physical activity (min)	Exposure duration (min)	Exposure concentration (ml/m^3)	Parameters investigated, findings	References
24♂, 10♀ (8–16 years old)	9×10	180	0.09 (0.01–0.26)	lung function, bronchial reactivity (cold): no effects	Avol et al. 1989
9 (no other details)	–	60	0.1	lung function, specific bronchial reactivity (ragweed): no effects	Ahmed et al. 1983
15♂	–	60	0.1	lung function, bronchial reactivity (METH): no effects	Hazucha et al. 1982, 1983
13♂, 7♀	–	60	0.1	lung function: airway resistance increased (13/20); bronchial reactivity (CARB): in some cases increased (3/20)	Orehek et al. 1976
4♂, 6♀	–	60	0.12	lung function: no effects	Koenig et al. 1985
4♂, 6♀ and 7♂, 3♀	10	40	0.12 and 0.18	lung function: no effects	Koenig et al. 1987
12♂, 19♀	4×15	120	0.20	lung function: no effects; bronchial reactivity (METH): minimally increased (great individual variability)	Kleinman et al. 1983
9♂, 2♀	10	30	0.25	bronchial reactivity (METH): no effects	Jörres and Magnussen 1991
10♂, 4♀	–	30	0.25	bronchial reactivity (hyperventilation and SO$_2$): increased	Jörres and Magnussen 1990
15 (no other details)	10	30	0.30	lung function, bronchial reactivity (cold air): no effects without physical activity; with physical activity and after provocation with cold air FEV$_1$ and partial expiratory flow rate reduced at 60 % of the total lung capacity	Bauer et al. 1986
5♂, 4♀	20	30	0.30	lung function, bronchial reactivity (SO): no effects	Rubinstein et al. 1990a
9♂, 3♀	2×15	60	0.30	lung function: FVC significantly reduced (but no reduction after combined exposure with O$_3$)	Koenig et al. 1988

Table 2. continued

Test persons	Duration of physical activity (min)	Exposure duration (min)	Exposure concentration (ml/m^3)	Parameters investigated, findings	References
15♂, 6♀	3×10	60	0.30, 1.0, 3.0	bronchial reactivity (cold): no effects	Linn et al. 1986
24♂, 10♀ (8–16 years old)	9×10	180	0.30	lung function, bronchial reactivity (cold): slightly different effects	Avol et al. 1989
9♂, 2♀	7×10,	360	0.2	bronchial reactivity (mite allergen): no effects;	Jenkins et al. 1999
	3×10	180	0.4	bronchial reactivity (mite allergen): increased	
6♂, 2♀	–	4×20	0.12, 0.24, 0.48	lung function: at 0.48 ml/m^3 reduced (not significant); bronchial reactivity (HIST): at 0.48 ml/m^3 increased	Bylin et al. 1985
9♂, 4♀	15	120	0.5	lung function: individual effects (quasistatic compliance significantly reduced)	Kulle 1982
8♂, 12♀	–	4×30	0.14, 0.27, 0.53	lung function: no effects; bronchial reactivity (HIST): only at 0.27 ml/m^3 significantly increased	Bylin et al. 1988
21♂	3×10	75	0.15, 0.30, 0.60	lung function, bronchial reactivity (METH): no effects	Roger et al. 1990
27♂, 32♀	6×10	120	0.30, 0.60	lung function, bronchial reactivity (cold): no effects	Avol et al. 1988
13♂, 9♀	2×15	60	0.5, 1.0, 2.0	lung function: no effects	Linn et al. 1985b
15♂, 6♀	3×10	60	0.3, 1.0, 3.0	lung function, bronchial reactivity (cold): no effects	Linn et al. 1986
12♂, 11♀	2×15	75	4.0	lung function, symptoms, heart frequency, permeability of the skin, subjective emotional status: no effects (systolic blood pressure slightly increased)	Linn et al. 1985a
88–111 (no other details)	–	15–60	0.5–5	lung function: ≥ 1.5 ml/m^3: airway resistance increased; ≥ 4 ml/m^3: arterial O$_2$ partial pressure reduced	von Nieding and Wagner 1975, 1979, von Nieding et al. 1971

Abbreviations: see Table 1

Workplace exposures

Numerous reports are available of short-term exposure to nitrous gases, which occurred as a result of accidents during welding (Norwood *et al.* 1966), the industrial use of nitrates and nitric acid (Touze *et al.* 1983, Tse and Bockman 1970, Wagner 1917), during blasting or explosions in mines or tunnels (Becklake *et al.* 1957, Müller 1969, Wantz and Dechoux 1982), during the combustion of nitrocellulose (Larcan *et al.* 1970), or during work in silos containing fermentation gases (Pavelchak *et al.* 1999, Ramirez and Dowell 1971). Short-term exposure induces coughing, shortness of breath and retching and leads, after a latency period of up to several hours, to acute bronchitis and pulmonary oedema. After a further 2 to 3 weeks symptoms of *bronchiolitis fibrosa obliterans* can develop once the acute symptoms have regressed. The biphasic course of the illness is typical for cases in which persons were exposed to high concentrations of nitrous gases (Milne 1969).

Exposure to 25 to 75 ml/m^3 leads to bronchitis or bronchopneumonia with complete restitution, 50 to 100 ml/m^3 to reversible bronchiolitis and focal pneumonitis, 150 to 200 ml/m^3 to lethal *bronchiolitis fibrosa obliterans* and more than 300 ml/m^3 to lethal pulmonary oedema and asphyxia (combined with methaemoglobinaemia) (Grayson 1956).

4.2 Repeated exposures

Long-term damage after exposure to nitrous gases has been frequently described. The severity of such effects varies from slight dyspnoea to severe chronic bronchitis. Often increased sputum production and changed lung function parameters, in particular an increased residual volume and increased expiratory resistance, can still be detected years later (Becklake *et al.* 1957, Müller 1969, Ramirez and Dowell 1971).

Workplace exposures

In studies with occupational exposure to NO_2 the exposure is always to a mixture of substances e.g. with diesel motor emissions, NO, sulfur dioxide, smoke or mineral dust, which also impair the respiratory tract. Diesel motors are an important source of exposure to NO_2. As a result of the exposure to a mixture of substances these studies are, however, unsuitable for the adequate evaluation of NO_2-related effects. For the sake of completeness, all those studies in which the exposure to NO_2 was at least determined are discussed below.

In a questionnaire 232 workers from 4 diesel bus garages were asked about acute respiratory diseases. Lung function tests were carried out before and after the shift, and the NO_2 concentration was determined during the shift. Concentrations of other irritant gases were below the national standards. The highest NO_2 concentrations in the air were 0.56 ± 0.38 ml/m^3, the lowest 0.13 ± 0.06 ml/m^3. Short-term personal air sampling often yielded exposure concentrations of over 1 ml/m^3. The authors reported that the prevalence of acute respiratory symptoms was higher than expected only in the high exposure group (> 0.3 ml/m^3) (Gamble *et al.* 1987).

In 259 workers of a salt mine who were exposed to diesel motor emissions, chronic respiratory effects were reported in a questionnaire and X-rays and lung function tests were carried out. The cumulative NO_2 exposure was determined by personal air sampling and was on average between 0.2 ± 0.1 ml/m^3 and 2.5 ± 1.3 ml/m^3. The authors reported that coughing was associated with age and smoking, and dyspnoea with age. There was no association, however, with the NO_2 exposure (Gamble *et al.* 1983).

In a 4-year study with 560 coal mining workers exposed to diesel soot, no impairment in lung function (FEV_1) and no respiratory symptoms were observed. The concentrations of NO_2 were in the range between 0.02 and 0.06 ml/m^3 air, the concentrations of NO were much higher and between 0.13 and 1.19 ml/m^3 (Robertson *et al.* 1984).

In a study with 20000 coal mining workers exposed on average to NO_2 concentrations of 0.03 ml/m^3 and NO concentrations of 0.2 ml/m^3, workers were not absent due to illness more frequently because of airway infections. There were, however, problems in this study as regards the classification of exposure (Jacobsen *et al.* 1988).

Population studies and studies with children

Other studies investigated e.g. correlations between the use of gas for cooking or heating and the frequency of illness, in particular in children. The studies are also unsuitable for evaluating NO_2-related effects because possible confounders, such as the smoking habits of the parents and the time spent indoors, were not taken into account. In a review (Schwela 2000) the author concludes that after long-term exposure from cooking with gas an increase in respiratory symptoms, impaired lung function, an increase in the incidence of chronic coughing, bronchitis and conjunctivitis are observed in children, but not in adults. A causal relationship between these adverse effects on health and exposure to NO_2 could not, however, be determined to date (Schwela 2000). Studies of morbidity resulting from air pollution (Abbey *et al.* 1999) are also not suitable for evaluating NO_2-related effects because exposure to a mixture of substances is involved.

4.3 Local effects on skin and mucous membranes

The irritation threshold for NO_2 in air was given as 20 to 30 ml/m^3 (Henschler *et al.* 1960).

4.4 Reproductive and developmental toxicity

51 pregnant women who were exposed to nitrogen oxides in the air (average NO_2 concentrations of 0.023 mg/m^3 with peak levels up to 0.239 mg/m^3), drinking water (nitrate concentrations in well water of up to 400 mg/l) and food, were examined during the birth. Methaemoglobin in the blood of the mother and in the umbilical cord blood was determined as the effect marker, and blood lipids and glutathione as markers for oxidative stress. In the newborn babies the birth weight, the APGAR index evaluating the most important body functions of the baby and clinical diagnoses at birth were recorded. Methaemoglobin concentrations (MetHb) of a maximum of 2 % in maternal blood and of a maximum of 2.8 % in cord blood were regarded as normal. Around 55 % of the maternal blood samples were above 2 % MetHb and about 20 % above 5 %.

Around 80% of the cord blood samples were above 2 % MetHb and about 45 % above 5 %. In the case of increased MetHb concentrations in maternal or cord blood the values for glutathione (total and reduced) were decreased and those for lipid peroxides increased. A strong association was found between increased lipid peroxides in cord blood and adverse birth outcome. With premature births in particular the levels of MetHb in cord blood were increased (Tabacova *et al.* 1998). The study suggests that increased concentrations of MetHb and lipid peroxides in cord blood can lead to premature births and other impairments at birth. A correlation of the findings with NO_2 concentrations in the air is, however, not possible from this study. As nitrite in particular (formed from nitrate) is responsible for the formation of MetHb, the oral uptake of nitrate and nitrite via water and food is regarded in the study of Tabacova *et al.* (1998) as having a decisive influence. In an earlier study, after volunteers were exposed to NO_2 concentrations of 20 ml/m^3 for 2 hours the MetHb levels in blood were increased by 1 % (Henschler and Lüdtke 1963). A significant increase in the MetHb concentration is therefore not to be expected where NO_2 concentrations are below 1 ml/m^3 in air. In determining the MetHb in cord blood the possibility of overestimating its value due to the presence of foetal haemoglobin (Lynch *et al.* 1998) must also be considered.

The influence of air pollution with regard to birth defects was investigated in a study. Associations between individual defects and carbon monoxide and ozone were found; no such associations were found with the other substances (including NO_2) (Ritz *et al.* 2002).

4.5 Genotoxicity

There are no relevant studies available.

In an investigation of 23 workers of a fertilizer factory who were exposed to NO_2 among other substances at concentrations below 5 ml/m^3, no increase in chromosomal aberrations of peripheral lymphocytes was found (Rojas 1992).

4.6 Carcinogenicity

Numerous population-based studies investigating the influence of urban air pollution on the formation of tumours (in particular lung tumours) are available.

In particular NO_2 contributes to air pollution in addition to diesel motor emissions, ozone and SO_2. The studies available mainly investigated the influence of diesel motor emissions, particle fractions and ozone. As a result of exposure to mixtures of substances and many possible influencing factors, these studies are not suitable for the evaluation of the carcinogenic effects of NO_2 on man.

In a more recent study, for example, exposure to average NO_2 concentrations of over 0.03 mg/m^3 (0.016 ml/m^3) for 30 years (as a measure of air pollution caused by traffic) was associated with a relative risk of 1.2 (95 % confidence interval, CI 0.8–1.6) for lung tumours; this value was adjusted for smoking, socio-economic status, domestic radon concentrations and workplace-related exposures. When exposure to NO_2 over a 10-year period 20 years previously was considered, the adjusted relative risk unexpectedly increased to 1.44 (95 % CI 1.05–1.99). For non-smokers and for smokers the relative risks for lung cancer after exposure to NO_2 were higher than without exposure to NO_2 and increased with increasing cigarette consumption (Nyberg et al. 2000). The exposure concentrations given serve merely as a surrogate for the level of air pollution caused by traffic and therefore do not allow a causal relationship between NO_2 exposure and an increased risk of lung tumours to be evaluated.

No data are available for the allergenic effects of nitrogen dioxide.

5 Animal Experiments and *in vitro* Studies

5.1 Acute toxicity

For rats the LC_{50} after exposure for 15 minutes was given as about 200 ml/m³ and after 60 minutes as 115 ml/m³ (Carson et al. 1962). The LC_{50} values for various strains of rat after exposure to NO_2 for 16 hours were given as 39 to 56 ml/m³. Mice, depending on the strain, were found to have LC_{50} values between 33 and 67 ml/m³ after exposure for 16 hours, golden hamsters values of 22 ml/m³ (♀) and 28 ml/m³ (♂) and Hartley guinea pigs values of 50 ml/m³ (♀) and 62 ml/m³ (♂) (Sagai and Ichinose 1987). For rabbits the LC_{50} after exposure for 15 minutes was found to be 315 ml/m³. For dogs, 53 ml/m³ was given as the 50 % value of the LC_{50} (corresponding to an LC_{50} value of about 105 ml/m³) after exposure to NO_2 for 60 minutes. In rats and rabbits, severely impaired breathing, eye irritation and reduced body weights were observed. In dogs, only slight symptoms of toxicity were found after NO_2 concentrations which led to pulmonary oedema (Carson et al. 1962).

Other animal experiments with single exposures are described in WHO (1997).

5.2 Subacute, subchronic and chronic toxicity

Numerous animal studies with repeated exposure are described in WHO (1997). Many animal studies have shown that long-term (continuous) exposure to NO_2 concentrations of 5 ml/m³ air can cause emphysema, which has also been observed in man (WHO 1997).

In particular studies with mice, rats, golden hamsters, guinea pigs, dogs and monkeys with long-term exposure in the range of 5 ml/m³ or below for more than 4 weeks are discussed. These studies are shown in Table 3.

Mice

One working group carried out several studies of immunological changes after exposure to NO_2 (Kuraitis and Richters 1989, Richters and Damji 1988, 1990, Sherwin and Richters 1982, 1995a, 1995b) in which mice of various strains (Swiss-Webster, AKR,

C57BL/6J) were exposed to NO_2 concentrations of 0.25 to 0.35 ml/m^3 for 6 to 8 hours a day on 5 days a week for 6 to 12 weeks. In all studies immunological changes such as reduced T cell populations or increased spleen weights were observed, and in one study also histopathological changes (hypertrophy of type II cells in the lungs). In BALB/c mice exposed to NO_2 concentrations of 0.4 ml/m^3 a reduced primary immune response *in vitro* was found, and at 1.6 ml/m^3 an increased secondary immune response *in vitro*. B and T lymphocytes were unchanged (Fujimaki *et al.* 1982). After mice were exposed for 3 weeks (Coffin *et al.* 1976, 1977, Gardner *et al.* 1977, 1979) or for 6 and 12 months intermittently (3, 6 or 18 hours a day) or continuously (Ehrlich and Henry 1968, Ehrlich *et al.* 1979) to NO_2 concentrations of 0.5 ml/m^3 the mortality of the animals was increased after bacterial infection. From the mortality of the mice after bacterial infection it was shown that, with the same product resulting from the concentration and duration, exposure concentration has a greater influence than exposure duration (Coffin *et al.* 1976, 1977, Gardner *et al.* 1977, 1979). With exposure at the same level (e.g. 3.5 ml/m^3) there was an inverse linear relationship between survival and the exposure duration (Coffin *et al.* 1976, Gardner *et al.* 1979). In mice exposed continuously to NO_2 concentrations of 2 ml/m^3 for 12 weeks, the levels of serum IgA were decreased and those of serum IgG_1, IgM and IgG_2 increased. After injection of the animals with an influenza vaccine, the IgM levels increased (Ehrlich *et al.* 1975). The studies with mice serve as a model for NO_2-induced immunological changes. It must be taken into account, however, that different strains of mice differ greatly in their immunological defence (Gardner 1982) and in their behaviour towards oxidants (Kleeberger *et al.* 1997). There are also great quantitative differences between different animal species. In a comparative study mortality after bacterial infection increased in mice after NO_2 concentrations of 3.5 ml/m^3 air and more, in hamsters after 35 ml/m^3 and more and in monkeys only after 50 ml/m^3 and more. A quantitative transfer of the findings of these immunological studies to man does not, therefore, seem justified (Gardner 1982).

Histopathological changes in the lungs of mice were reported after intermittent exposure (6 hours a day, 5 days a week) to NO_2 concentrations of 0.34 ml/m^3 for 6 weeks (hypertrophy, hyperplasia) (Sherwin and Richters 1982) and after continuous exposure to 0.3 to 0.5 ml/m^3 for 6 months (hyperplasia, degeneration) (Nakajima *et al.* 1980). In mice exposed to NO_2 concentrations of 0.5 ml/m^3 for 6, 18 or 24 hours for up to 12 months, a loss of cilia and inflammatory changes were found, which were most pronounced after continuous exposure (Blair *et al.* 1969). Reduced survival with reduced body weight gains was observed in mice after continuous exposure to NO_2 concentrations of 0.5 ml/m^3 for 18 months (Csallany and Ayaz 1978). After exposure of mice to NO_2 concentrations of 1.0 to 1.5 ml/m^3 for 1 month and a follow-up period of 1 month, desquamation of the mucous membranes, obstruction of the alveolar passages and minimal enlargement of the alveolar volume were observed; most of the changes in the bronchioli were reversible (Nakajima *et al.* 1980).

Table 3. Animal studies with inhalation exposure to NO_2 concentrations of up to 5 ml/m^3 and long-term exposure of over 4 weeks

Species, strain, number of animals per sex and group	Exposure conditions	Results	References
mouse, Swiss-Webster, groups of at least 58♂, 3 weeks old	6 weeks, 0.25 ml/m^3 (**7 h/day, 5 days/week**), follow-up period 0, 10, 32 weeks	**0.25 ml/m^3**: in 9-week-old offspring effects on the development and maturation of the lungs (type II cell hypertrophy and hyperplasia, changes in the elastic tissue of the alveolar walls), which were not completely reversible	Sherwin and Richters 1995a, 1995b
mouse, AKR/cum, 18♀, 6 weeks old	7 weeks, 0.25 ml/m^3 (**7 h/day, 5 days/week**)	**0.25 ml/m^3**: total T cell population (mature T cells, T helper cells, T suppressor cells) reduced	Richters and Damji 1988
mouse, C57BL/6J, groups of 5♂, 4 weeks old	6, 8, 9, 12, 14, 16 weeks, 0.25–0.35 ml/m^3 (**8 h/day, 5 days/week**)	**0.25–0.35 ml/m^3**: spleen weights increased, spleen cells, erythrocytes and IgM-positive lymphocytes in the spleen reduced, lymphocytes and neutrophils in blood decreased	Kuraitis and Richters 1989
mouse, AKR/cum, 50♀, 5 weeks old	36 weeks, 0.25 ml/m^3 (**7 h/day, 5 days/week**)	**0.25 ml/m^3**: survival reduced, total T cell population (in particular T helper cells) reduced	Richters and Damji 1990
mouse, Swiss Webster, 60♂, young adult animals	6 weeks, 0.34 ml/m^3 (**6 h/day, 5 days/week**)	**0.34 ml/m^3**: type II cell hypertrophy and hyperplasia	Sherwin and Richters 1982
mouse, C57BL/6J, 10♂, 6 weeks old	12 weeks, 0.35 ml/m^3 (**7 h/day, 5 days/week**)	**0.35 ml/m^3**: total T cell population reduced; no effects on individual T cell sub-populations	Richters and Damji 1988
mouse, BALB/c, groups of 6♂, 7 weeks old	4 weeks, 0.4, 1.6 ml/m^3 (**continuous**)	**0.4 ml/m^3**: reduced primary immune response (*in vitro*) **1.6 ml/m^3**: increased secondary immune response (*in vitro*); no difference in T and B lymphocyte activity; questionable relevance of the *in vitro* determination	Fujimaki *et al.* 1982
mouse, Swiss Webster, 15♂, young adult animals	4 weeks, 0.45 ml/m^3 (**7 h/day**)	**0.45 ml/m^3**: serotonin and 5-hydroxyindolacetic acid concentration in the lungs unchanged, in the brain increased	Sherwin *et al.* 1986

Table 3. continued

Species, strain, number of animals per sex and group	Exposure conditions	Results	References
mouse, CD$_2$F$_1$, groups of 46–169♀, 6–8 weeks old	**1, 2, 3, 6 months**, 0.5 ml/m^3 **(3 h/day, 5 days/week)**	**0.5 ml/m^3**: after infection with *Streptococcus* increased mortality (after exposure for 3 and 6 months)	Ehrlich *et al.* 1979
mouse, Swiss, groups of at least 30♀ (bacterial infection) and groups of 4♀ (histology), age not stated (animals initially 21 g in weight)	**3, 6, 9, 12 months**, 0.5 ml/m^3 **(6, 18 or 24 h/day, 7 days/week)**	**0.5 ml/m^3**: bacterial infection: after infection with *Klebsiella pneumoniae* increased mortality (with intermittent exposure after 6 months, with continuous exposure after 3 months), reduced clearance; histopathology: pneumonitis, loss of cilia in the respiratory bronchioli, alveoli enlarged with thin walls and septal breaks, bronchiolar inflammation; effects detectable after exposure for 3 months; after exposure for 12 months moderate effects (exposure 6 h/day) and severe (exposure 18 and 24 h/day)	Blair *et al.* 1969, Ehrlich and Henry 1968
mouse, C57BL6/J, groups of 10♀, 3 weeks old	**17 months**, 0.5, 1.0 ml/m^3 **(continuous)**	**≥ 0.5 ml/m^3**: body weights and survival reduced, absolute liver and heart weights increased; no effect on GSH peroxidase activity **1.0 ml/m^3**: GSH peroxidase activity reduced	Ayaz and Csallany 1978, Csallany and Ayaz 1978
mouse, JCL:ICR, ♂, ♀ (number of animals not stated), 3–4 weeks old	**6 months**, 0.3–0.5 ml/m^3 **39 days**, 0.5–1 ml/m^3 **1 month**, 0.5–0.8 ml/m^3 **1 month**, 0.7–0.8 ml/m^3 **1 month**, 1–1.5 ml/m^3, 1 month follow-up period; **continuous exposure**	**0.3–0.5 ml/m^3 (6 months), 0.5–1 ml/m^3 (39 days), 0.5–0.8 and 1–1.5 ml/m^3 (1 month)**: hypersecretion of mucous, mucous membrane focally degenerated and desquamous hyperplasia of the terminal bronchiolar epithelium; cilia shortened; after infection with influenza virus advanced interstitial pneumonia and adenomatous proliferation of the bronichial epithelium **1–1.5 ml/m^3 (1 month) + follow-up period** (1 month): most bronchioli almost normal but still desquamation of the mucous membrane, obstruction of the alveolar passages, minimal enlargement of the alveolar lumen	Nakajima *et al.* 1980

Table 3. continued

Species, strain, number of animals per sex and group	Exposure conditions	Results	References
mouse, Swiss (CD-1), groups of 20♀, age not stated (animals initially 20–25 g in weight)	6 min–1 year, 0.5–28 ml/m³ (**continuous**) about 20 days 1.5 ml/m³ (**7 h/day, 7 days/week**) about 14 days 3.5 ml/m³ (**7 h/day, 7 days/week**)	after infection with *Streptococcus* mortality increases with exposure duration and concentration; concentration a more important factor than duration: **0.5 ml/m³**: 20 % mortality after 256 days exposure (24 h/day); **1.5 ml/m³**: 20 % mortality after about 2 days exposure (24 h/day) and 9 days exposure (7 h/day, 7 days/week; extrapolated); **3.5 ml/m³**: 20 % mortality after about 2–3 hours exposure	Coffin *et al.* 1976, 1977, Gardner *et al.* 1977, 1979
mouse, CD-1, groups of 18–21♀, 4–6 weeks old	1 year, 0.2 ml/m³ (**23 h/day, 7 days/week**) or 0.2 ml/m³ (**23 h/day, 7 days/week**) + 0.8 ml/m³ (**2 h/day, 5 days/week**)	**0.2 ml/m³**: no effects **0.2 + 0.8 ml/m³**: after infection with *Streptococcus* significantly increased mortality; lung function changed (vital capacity reduced, trend towards reduced respiratory compliance)	Miller *et al.* 1987
mouse, probably CD-1, 6 (no other details)	6 months, 0.1 ml/m³ (**continuous**) + 1.0 ml/m³ (**2 h/day**)	**0.1 + 1 ml/m³**: lung changes (dilated airways and destruction of the alveolar walls)	Port *et al.* 1977
mouse, no other details	4, 12, 24 weeks, 0.5, 2 ml/m³ (**continuous**); 0.1 or 0.5 ml/m³ (**24 h/day, 5 days/week**) + 1.0 or 2.0 ml/m³ (**3 h/day, 5 days/week**)	**0.5, 0.1 + 1.0 ml/m³**: no effects on morphology of the AM **0.5 + 2.0 ml/m³**: morphological changes in the AM (electron microscopy)	Aranyi *et al.* 1976
mouse, CD-1, groups of 9♂, 10–11 weeks old	1, 3, 6, 9, 12 months, 0.5 ml/m³ (**continuous**); 0.1 ml/m³ (**24 h/day, 5 days/week**) + 0.25, 0.5, 1.0 ml/m³ (**3 h/day**)	**0.1 ml/m³ + 0.25–1.0 ml/m³**: suppression of the response of T and B cells of the spleen to mitogens not dependent on the concentration or duration of exposure **0.5 ml/m³**: linear reduction in phytohaemagglutinin-induced mitogenesis with increasing exposure duration	Maigetter *et al.* 1978
mouse, Swiss (CD-1), groups of 14–20♂, 6 weeks old	12 weeks, 2.0 ml/m³ (**continuous**); 0.5 ml/m³ (**continuous**) + 2.0 ml/m³ (**1 h/day, 5 days/week**)	**0.5 + 2.0; 2.0 ml/m³**: serum IgA reduced, serum IgM, IgG$_1$, IgG$_2$ increased; after virus infection serum IgM increased	Ehrlich *et al.* 1975

Table 3. continued

Species, strain, number of animals per sex and group	Exposure conditions	Results	References
rat, JCL-SD, groups of 16♂ + ♀, 4 weeks old	**7 months**, 0.5, 1.0, 4.0 ml/m^3 **(continuous)**	**≥ 0.5 ml/m^3**: swelling of the terminal bronchiolar cilia, hyperplasia of the type II cells **≥ 1.0 ml/m^3**: loss of cilia in the terminal bronchioli; interstitial oedema	Yamamoto and Takahashi 1984
rat, Wistar, groups of 4–6♂, 2 months old	**4, 9, 18, 27 months**, 0.04, 0.4, 4.0 ml/m^3 **(continuous)**	**≥ 0.04 ml/m^3**: increased exhalation of ethane **≥ 0.4 ml/m^3**: increased thiobarbituric acid, NPSH, reduced GSH peroxidase activity **4.0 ml/m^3**: increased GSH reductase activity and G6PD	Sagai et al. 1984
rat, Wistar, groups of 3–4♂, 2 months old	**9, 18, 27 months**, 0.04, 0.4, 4.0 ml/m^3 **(continuous)**	**≥ 0.04 ml/m^3**: no significant changes; tendency towards an increased median thickness of the air–blood barrier, increased exhalation of ethane **≥ 0.4 ml/m^3**: slightly increased median thickness of the air–blood barrier after 18 months; significant after 27 months; some interstitial oedema and slight changes in the bronchiolar and alveolar epithelium **4.0 ml/m^3**: hypertrophy and hyperplasia of the bronchiolar epithelium; hyperplasia of Clara cells; interstitial fibrosis and hypertrophy of type I and type II cells; lipid peroxidation increased; activity of anti-oxidative enzymes decreased	Kubota et al. 1987, Sagai and Ichinose 1987
rat, Wistar, groups of 6♂, 13 weeks old	**1, 2, 4, 8, 12, 16 weeks**, 0.4, 1.2, 4.0 ml/m^3 **(continuous)**	**0.4 ml/m^3**: minimal increase in GSH reductase and G6PD (not significant) **1.2 ml/m^3**: increased exhalation of ethane in the first 4 weeks; slight increase in GSH reductase and G6PD, significant increase in the NPSH level in the lung **4 ml/m^3**: increased exhalation of ethane until the end of exposure, significant increase in GSH reductase, G6PD and SOD	Ichinose and Sagai 1982

Table 3. continued

Species, strain, number of animals per sex and group	Exposure conditions	Results	References
rat, SD, groups of 7♀, 1, 3, 12, 21 months old	1 month, 0.1, 0.5, 3, 10 ml/m³ (**continuous**)	**≥ 0.1 ml/m³**: increased density of tissue of alveolar wall (only in 1 and 3-month-old animals), increase in type II cells, interstitial cells, capillary endothelial cells and interstitial matrix **≥ 0.5 ml/m³**: increased density of the tissue of the alveolar wall (also in adult animals) **≥ 3 ml/m³**: increased thickness of the air–blood barrier (in particular in 1-month-old animals)	Kyono and Kawai 1982
rat, Wistar, a total of 86♂, (200–300 g in weight)	7, 14 days, 1, 2, 4, 6, 12, 19 months, 0.5 ml/m³ (**continuous**)	**0.5 ml/m³**: type II cell hypertrophy and interstitial oedema after 4 months; increase in the thickness of alveolar septa after 6 months; fibrous pleural thickening after 19 months	Hayashi et al. 1987
rat, SD, 4 animals, 4 weeks old, no other details	lifetime (up to 33 months), 0.8 ml/m³ (**continuous**)	**0.8 ml/m³**: increased respiration rate, minimal morphological changes (slight enlargement of the alveoli and respiratory duct, rounding of bronchial and bronchiolar epithelial cells, increase in elastic fibres around the alveolar duct)	Freeman et al. 1966, Haydon et al. 1965
rat, F344, groups of 15–20♂, newly born or 7 weeks old	1, 3, 6 weeks, 0.5, 1.0, 2.0 ml/m³ (**21.5 h/day, 7 days/week**) + 1.5, 3.0, 6.0 ml/m³ (**2× 1 h/day, 5 days/week**)	**0.5 + 1.5 ml/m³**: no changes in lung function **≥ 1 + 3 ml/m³**: newly born animals only: increased lung volumes and increased compliance only after 3 weeks exposure **2 + 6 ml/m³**: adults: reduced body weights and reduced lung compliance	Stevens et al. 1988
rat, F344, groups of 4–12♂, newly born or 6 weeks old	6 weeks, 0.5, 2.0 ml/m³ (**23 h/day, 7 days/week**) + 1.5, 6.0 ml/m³ (**2× 0.5 h/day, 5 days/week**)	**0.5 + 1.5 ml/m³** proximal alveolar region: type II cells widespread (newly born and adult animals) and hypertrophied (adults); increased AM (adults), type II cells thinner (newly born); terminal bronchiolar region: changes in Clara cells (only newly born animals) **2 + 6 ml/m³** proximal alveolar region: type I cells increased (only adults investigated); terminal bronchiolar region: changes in Clara cells (only adults investigated)	Chang et al. 1986, 1988, Crapo et al. 1984

Table 3. continued

Species, strain, number of animals per sex and group	Exposure conditions	Results	References
rat, F344, groups of 5♂, 7 weeks old	9 weeks, 0.5 ml/m^3 (**22 h/day**) + 1.5 ml/m^3 (**2× 1 h/day**)	**0.5 + 1.5 ml/m^3**: increased number of "fenestrae" in the alveolar septa of the lungs	Mercer et al. 1995
rat, F344, groups of 4–12♂, 60 days old	**1, 3, 13, 52, 78 weeks**, 0.5 ml/m^3 (**22 h/day, 7 days/week**) + 1.5 ml/m^3 (**2 h/day, 5 days/week**)	**0.5 + 1.5 ml/m^3**: body weight gains unchanged, reduced ΔFEF$_{25\%}$, increased expiratory airway resistance and reduced respiration rate after 78 weeks; immunotoxicity: no effects on B or T cell response (spleen, blood) to mitogens	Selgrade et al. 1991, Tepper et al. 1993
rat, F344, groups of 8–12 no other details	**12, 18 months**, 0.5 ml/m^3 (**20 h/day**) + 1.5 ml/m^3 (**2 h/day**)	**0.5 + 1.5 ml/m^3**: no effects on proteinase inhibitors in BAL	Johnson et al. 1990a
rat, Sherman, groups of 40–60♂, no other details	**16, 18 months**, 1, 5, 25 ml/m^3 (**6 h/day, 5 days/week**)	**0, 1, 5, 25 ml/m^3**: interstitial pneumonia in all control animals and exposed animals with increasing age	Wagner et al. 1965
rat, Wistar, 24♂, 2 months old	6 weeks, 2.0 ml/m^3 (**continuous**)	**2 ml/m^3**: minimal effects (slight emphysematous modifications), no effects on blood parameters (e.g. MetHb)	Azoulay et al. 1978
rat, Wistar, ♂, 1 month old, no other details	**up to 12 months**, 2.0 ml/m^3 (**continuous**)	**2 ml/m^3**: initial hypertrophy and hyperplasia of the epithelium of the terminal bronchioli, reduced number of cilia-bearing cells, loss of cilia; later (21 days to 12 months exposure) cilia-bearing cells normal, crystalloid bodies in the non-cilia-bearing cells	Stephens et al. 1972
rat, Wistar, a total of 36♂ exposed, 24♂ controls, 1 month old	**12, 14 months**, 2.0 ml/m^3 (**continuous**)	**2 ml/m^3**: hypertrophy of the terminal bronchiolar epithelium; increased transformation of type II cells in peripheral alveoli (day 1 to 3; normal from day 7), haematocrit and erythrocyte count increased	Evans et al. 1972, Furiosi et al. 1973
rat, Wistar, 10♂, 1 month old	**2 years**, 2.0 ml/m^3 (**continuous**)	**2 ml/m^3**: loss of cilia in the terminal bronchioli; abnormal ciliogenesis; crystalloid bodies in the bronchiolar epithelial cells, increased thickness of collagen-fibrils and of the basal membrane in the terminal bronchioli	Stephens et al. 1971a, 1971b

Table 3. continued

Species, strain, number of animals per sex and group	Exposure conditions	Results	References
rat, SD, 18♂ exposed, 14♂ controls, 2 months old	lifetime (up to 763 days), 0.8 ml/m³ (**continuous**) for the first 69 days, then 2.0 ml/m³ (**continuous**)	**0.8 + 2 ml/m³**: respiration rate increased, airway resistance and compliance not increased, lung weights increased, alveolar widening in particular in the area of the alveolar duct; loss of cilia and hypertrophy of the cells in the terminal bronchioli	Freeman et al. 1968a
rat, Long-Evans, 14♂ no other details	9 months, 2.9 ml/m³ (**24 h/day, 5 days/week**)	**2.9 ml/m³**: increased lung weights, increased total lipids, reduced level of saturated fatty acids and increased surface tension of pulmonary lavage, reduced pulmonary compliance	Arner and Rhoades 1973
rat, Wistar, groups of 3♂, 6 weeks old	1, 2, 4, 8, 16, 28 days, 0.53, 1.33, 2.66 ml/m³ (**continuous**)	**up to 1.33 ml/m³**: no morphological changes **2.66 ml/m³**: focal thickening of the septa in the pulmonary acinus; progressive loss of cilia and abnormal cilia in the trachea and main bronchi, hypertrophy of the bronchial epithelium, hypertrophy of epithelial cells	Rombout et al. 1986
rat, F344, 5–6♂, ♀, 14–16 weeks old	15 weeks, 1.0, 5.0 ml/m³ (**continuous**); 1.0 ml/m³ (**4 h/day, 5 days/week**) + 5.0 ml/m³ (**3 h/day**)	**≥ 1 ml/m³**: alkaline phosphatase, LDH increased (only after 2.7 weeks, no longer after 6 weeks), GSH reductase increased (only after 1.7 weeks, no longer after 2.7 weeks) **1 + 5 ml/m³**: accumulation of AM, changes in enzyme activity in the BAL and in the lung fluid **5 ml/m³**: accumulation of AM	Gregory et al. 1983
rat, SD, 9♂ 1 month old	16 weeks, 4.0 ml/m³ (**continuous**)	**4 ml/m³**: early histopathological changes, hyperplasia of the bronchial epithelium	Haydon et al. 1965
rat, black and white, groups of 3, 5 weeks old	3, 5, 7, 9, 11 weeks, 5 ml/m³ (**continuous**)	**5 ml/m³**: changes in epithelium of terminal bronchioli (e.g. loss of cilia and cell differentiation, hyperplasia)	Rejthar and Rejthar 1975
Syrian hamster, groups of 20–40♂, no other details	15 months, 1, 5, 25 ml/m³ (**6 h/day, 5 days/week**)	**≥ 1 ml/m³**: slight changes of unclear origin (congestion, mild interstitial pneumonia)	Wagner et al. 1965

Table 3. continued

Species, strain, number of animals per sex and group	Exposure conditions	Results	References
Syrian hamster, 7♂, 2 months old	8 weeks, 2.0 ml/m^3 (**8 h/day, 5 days/week**)	**2 ml/m^3**: lung function (vital capacity or compliance of the lungs) unchanged; lung changes (e.g. slight alveolar widening primarily at the convergence of the bronchiolar and alveolar ducts)	Lafuma et al. 1987
guinea pig, Hartley, 250–300 g, no other details	7 days, 4 months, 0.5 ml/m^3 (**8 h/day, 7 days/week**)	**0.5 ml/m^3** 7 days: reduced erythrocyte count and GSH peroxidase, increased lysozyme (plasma) and acetylcholinesterase (erythrocytes); 4 months: reduced erythrocyte count, GSH peroxidase, lysozyme (plasma), acidic phosphatase (plasma) and acetylcholinesterase (erythrocytes), increased acidic phosphatase (lungs)	Menzel et al. 1977
guinea pig, 22♂, 350–500 g, no other details	6 months, 1 ml/m^3 (**8 h/day**)	**1 ml/m^3**: increased respiration rate, reduced body weight gains, reduced haemolytic activity of the complement, reduced total protein in blood, subpleural emphysema and atelectasis foci, bronchitis, bronchopneumonia	Kosmider et al. 1973
guinea pig, English, a total of 34♂, no other details	12, 15, 18 months, 1, 5, 25 ml/m^3 (**6 h/day, 5 days/week**)	**≥ 1 ml/m^3**: slight thickening of the alveolar septa as a result of inflammation, some alveolar dilatation	Wagner et al. 1965
guinea pig, New England, groups of 4, no other details	5.5 months, 5.0, 15 ml/m^3 (**4 or 7.5 h/day, 5 days/week**)	**5 ml/m^3**: increased serum antibodies against lung tissue with increasing concentration and exposure duration; some dilation of the terminal bronchioli; tracheal inflammation; pneumonitis	Balchum et al. 1965
dog, beagle, 6♀, no other details	68 months, 0.64 ml NO$_2$/m^3 + 0.25 ml NO/m^3 (**16 h/day**)	**0.64 ml NO$_2$/m^3 + 0.25 ml NO/m^3**: total lung capacity increased, surface density of the alveoli and volumetric density of the parenchymal tissue reduced, loss of cilia, hyperplasia bronchiolar (non-cilia-bearing) cells	Hyde et al. 1978
dog, mongrel, groups of 10♂, no other details	up to 18 months, 1.0, 5.0 ml/m^3 (**6 h/day, 5 days/week**)	**≥ 1 ml/m^3**: dilated alveoli and alveolar ducts, oedema, thickening of the alveolar septa, inflammatory cells; but no clear difference to the controls	Wagner et al. 1965

Table 3. continued

Species, strain, number of animals per sex and group	Exposure conditions	Results	References
monkey, squirrel monkey, 4♂ exposed, 3♂ controls, 600–900 g in weight	about 16 months, 1.0 ml/m^3 (**continuous**)	**1 ml/m^3**: before and during exposure the animals were given 5 injections with the virus (adapted for monkeys); evidence of slight emphysema and thickened bronchial and bronchiolar epithelium; no changes detected with the electron microscope, increased antibodies (serum neutralizing)	Ehrlich and Fenters 1973, Fenters *et al.* 1973
monkey, baboon, 4♂,♀, 3–4 years old	6 months, 2.0 ml/m^3 (**8 h/day, 5 days/week**)	**2 ml/m^3**: impairment of the reaction of AM to the antigen or mitogen-induced cytokine MIF (macrophage migration inhibitory factor)	Greene and Schneider 1978
monkey, rhesus, 4♂/♀, maturing	14 months, 2.0 ml/m^3 (**continuous**)	**2 ml/m^3**: hypertrophy of the bronchiolar epithelium, erythrocyte count increased	Furiosi *et al.* 1973
monkey, squirrel monkey, 600–800 g in weight	1.2 months, 5.0, 10 ml/m^3 (**continuous**)	**5 ml/m^3**: after infection with *Klebsiella pneumoniae* or influenza virus increased mortality (virus-induced 1/3, *Klebsiella*-induced 2/7); in control animals no mortality; bacteria detected in the lungs of surviving animals	Henry *et al.* 1970
monkey, squirrel monkey, 7♂, 600–800 g in weight	3 months, 5, 10 ml/m^3 (**continuous**)	infection with viruses adapted for the mouse or monkey: **controls**: 1/14 (7 %) died; **5 ml/m^3**: 1/7 (14 %) died; in the surviving animals after a further 37 days exposure initially a reduced tidal volume and increased respiration rate **10 ml/m^3**: 9/10 (90 %) died; in the surviving animal a reduced tidal volume and increased respiration rate	Ehrlich and Fenters 1973
monkey, squirrel monkey, 7♂, 600–800 g in weight	6 months, 5.0 ml/m^3 (**continuous**)	**5 ml/m^3**: during exposure the animals were infected 5 times with viruses adapted for mice; initial decrease in antibodies (serum neutralizing), which normalized on day 133	Fenters *et al.* 1971
monkey, cynomolgus monkey, groups of 10♂, no other details	90 days, 5.0, 10 ml/m^3 (**continuous**)	**≥ 5 ml/m^3**: hyperplasia of the bronchiolar epithelium; some focal pulmonary oedema **10 ml/m^3**: impaired distribution of ventilation in the lung, increased respiration rate, reduced tidal volume	Busey *et al.* 1974 (abstract), Coate and Badger 1974 (abstract)

AM: alveolar macrophages; $\Delta FEF_{25\%}$: forced expiratory flow at 25 % of the vital capacity; G6PD: glucose-6-phosphate dehydrogenase; GSH: glutathione; LDH: lactate dehydrogenase; MetHb: methaemoglobin; NPSH: non-protein sulfhydrils; SD: Sprague-Dawley; SOD: superoxide dismutase

Rats

The continuous exposure of groups of 3 to 4 male Wistar rats (2 months old at the beginning of exposure) to NO_2 concentrations in the air of 0.04 ml/m^3 for 9, 18 or 27 months did not produce any significant changes in the lungs. The median thickness of the air–blood barrier in the lungs, however, had a tendency to increase, and the interstitial components and the pattern of the variations of the type II cells were, although slight in form, similar to those of the next highest exposure group (0.4 ml/m^3). The increased exhalation of ethane indicates lipid peroxidation. After NO_2 concentrations of 0.4 ml/m^3 and more, the median thickness of the air–blood barrier was slightly increased after 18 months and significantly increased after 27 months. Also some interstitial oedema, slight changes in the bronchiolar and alveolar epithelium, and evidence of oxidative stress were found. After 4 ml/m^3 hypertrophy and hyperplasia of the bronchiolar epithelium and hyperplasia of Clara cells, interstitial fibrosis and hypertrophy of type I and type II cells were observed. The authors detected 2 phases in the effects of this exposure group. In the first phase, there was a reduced number of type I cells and an increased cell volume after exposure for 9 months. After 18 months the number of type II cells and their volume was increased, the interstice was enlarged and changes in the epithelium were observed. In phase 2, evidence of recovery was detected with the electron microscope after exposure for 18 to 27 months. It was, however, evident that the total volume of interstitial tissue had decreased and interstitial oedema had been replaced by collagen fibres, which indicates possible functional impairment and a progression of the effects. The activity of antioxidative, protective enzymes was reduced. The authors conclude that an extensive study is necessary to clarify whether the histopathological changes in the low exposure groups are the result of the longer exposure or increased sensitivity of the ageing lungs because of the action of delayed repair mechanisms (Kubota et al. 1987, Sagai and Ichinose 1987, Sagai et al. 1984). The limiting factor of this study is the relatively low number of animals (a maximum of 12 animals per exposure group with 3 examination times).

In other investigations histopathological changes were detected in the lungs at NO_2 concentrations as low as 0.5 ml/m^3 (Hayashi et al. 1987, Yamamoto and Takahashi 1984) and 0.8 ml/m^3 (Freeman et al. 1966). Many other studies describe effects at higher exposure concentrations (see Table 3). The age of the animals at the beginning of exposure seems to have an influence on the severity of the effects. Young rats, whose lungs are still developing, react particularly sensitively to NO_2-induced changes (Stevens et al. 1988). A study with rats of different ages (1 to 21 months old) showed that type I and type II pneumocytes, interstitial cells, the interstitial matrix and the capillary endothelium react to different extents depending on the age of the animals and the level of exposure to NO_2, so that different impairments and repair processes are observed. Also, individual compartments do not produce a simple dose-dependent increase or decrease in effects, but there seems to be a multiphasic reaction pattern (Kyono and Kawai 1982). Old rats, as a result of delayed repair mechanisms, were found to be more sensitive than middle-aged rats (Evans et al. 1977).

Hamsters

In Syrian hamsters slight alveolar changes were observed after exposure to NO_2 concentrations of 2 ml/m^3 (8 hours a day, 5 days a week) for 8 weeks (Lafuma et al. 1987). In a 15-month study golden hamsters were exposed to NO_2 concentrations of 1 ml/m^3. Only slight changes (congestion, slight interstitial pneumonia) were observed, whose origin, according to the authors, was unclear (Wagner et al. 1965); the findings cannot, therefore, be evaluated.

Guinea pigs

Immunological changes were observed in guinea pigs exposed to NO_2 concentrations of 1 ml/m^3 for 6 months (Kosmider et al. 1973). After up to 18 months exposure, slight histopathological changes were observed in the lungs at 1 ml/m^3 (Wagner et al. 1965). Marked changes in the lungs and pneumonia were described (Balchum et al. 1965) after 5.5 months exposure to 5 ml/m^3.

Dogs

In dogs exposed for 68 months to NO_2 concentrations of 0.64 ml/m^3 in combination with NO in concentrations of 0.25 ml/m^3, a loss of cilia and hyperplasia of bronchiolar (non-cilia-bearing) cells were detected (Hyde et al. 1978). Unclear findings in the lungs were obtained in another study after exposure for 18 months to NO_2 concentrations of 1 and 5 ml/m^3 (Wagner et al. 1965).

Monkeys

Immunological changes (increased serum neutralizing antibodies) were detected in monkeys after 5 virus injections and exposure to NO_2 concentrations of 1 ml/m^3 for 1 year (Ehrlich and Fenters 1973, Fenters et al. 1973). After exposure of the animals to NO_2 concentrations of 2 ml/m^3 for 6 months, the reaction of alveolar macrophages to the antigen or mitogen-induced cytokine MIF (macrophage migration inhibitory factor) was changed (Greene and Schneider 1978) and after exposure to 5 ml/m^3 for more than one month the mortality of the animals after infection with *Klebsiella pneumoniae* or influenza virus was increased (Ehrlich and Fenters 1973, Fenters et al. 1971, Henry et al. 1970). Histopathological changes were observed in the lungs of monkeys after exposure to NO_2 concentrations of 1 ml/m^3 for about 16 months (evidence of slight emphysema and thickened bronchial and bronchiolar epithelium; Fenters et al. 1973), after 14 months exposure to 2 ml/m^3 (hypertrophy of the bronchiolar epithelium; Furiosi et al. 1973) and after 90 days exposure to 5 ml/m^3 and more (hyperplasia of the bronchiolar epithelium; some focal pulmonary oedema; Busey et al. 1974, Coate and Badger 1974; both abstracts).

5.3 Local effects on skin and mucous membranes

In the respiratory tract NO_2 is hydrolysed to form nitrous acid and nitric acid, both of which are irritative.

5.4 Reproductive and developmental toxicity

There are a few studies of the toxic effects of NO_2 on fertility and development. However, they do not fulfil present-day requirements and are therefore of only limited use.

5.4.1 Fertility

Six male rats were exposed to NO_2 concentrations of 1 ml/m^3 for 21 days (7 hours per day, 5 days a week). Light microscopic examination of the testicular tissue did not reveal any differences between the exposed animals and the 5 controls. No changes were detected in the exposed animals with regard to spermatogenesis, germ cell atrophy or Leydig cell abnormalities. The level of vitamin B12 in serum was comparable in the two groups (Kripke and Sherwin 1984).

The influence of NO_2 on the oestrus cycle was investigated in groups of 10 female rats. The duration of the oestrus cycle was determined over 24 days before the exposure to NO_2. The animals were then exposed to NO_2 concentrations of 0.13 or 2.4 mg/m^3 (0.07 and 1.2 ml/m^3) for 12 hours a day over 3 months. At 0.07 ml/m^3 there were no effects on the oestrus cycle compared to that of the controls. After 1.2 ml/m^3, however, the cycle was longer: with a cycle in the controls of 5.3 days, it was extended to up to 6.4 days after exposure for one month, 9.0 days after exposure for 2 months and 9.1 days after exposure for 3 months. In the third month of exposure the increased cycle duration was attributed to a 20 % increase in the oestrus and an 85 % increase in the oestrus interval. After a three-month recovery period the average duration of the oestrus cycle no longer differed from the control values. Histopathological examinations at the end of exposure revealed changes in the adrenal gland, the thyroid gland and the pituitary gland in the group exposed to 1.2 ml/m^3. A reduction in the number of functional primordial follicles was detected in the ovaries and dystrophy of the glandular epithelium in the uterus. At the end of the 3-month exposure to 1.2 ml/m^3, 7 of the exposed rats were mated with untreated males. It was reported that the exposure had no influence on the number of females which became pregnant. The litter size at birth was reduced to 5.1 (after 1.2 ml/m^3) compared to the control value of 8.0 pups. The weight of the newborn pups and the body weight gains in the first 12 days of life were also significantly reduced (20–30 %) (Barlow and Sullivan 1982, Shalamberidze and Tsereteli 1971).

5.4.2 Developmental toxicity

The studies of the developmental toxicity of NO_2 in mice and rats are reviewed in Table 4. None of these studies fulfil present-day requirements and most have grave limitations (in particular the lack of methodological data and the inadequate representation of data). These studies are therefore of only very limited use.

One author carried out studies with mice of the effects on prenatal development (Singh 1984) and postnatal development of the offspring (Singh 1988) after exposure of the dams to NO_2. Exposure to NO_2 on days 7 to 18 and 8 to 18 of gestation was very high (22 and 45 ml/m^3 air). The concentration of 22 ml/m^3 caused haematomas, reduced body weights and increased mortality in the foetuses, while 45 ml/m^3 led to delays in postnatal development of the pups (Singh 1984, 1988). Lower concentrations were not investigated.

In rats exposed over the whole gestation period to NO_2 concentrations of 0.045 or 0.43 ml/m^3, increased "intrauterine mortality" was reported even at the low concentration. At the higher concentration reduced maternal and foetal body and liver weights, and stillbirths were observed (Gofmekler et al. 1977). Comparable effects were described by the working group of Tabacova et al. (see below) only after 10-fold higher concentrations of 0.53 and 5.3 ml/m^3.

The toxic effects of NO_2 on prenatal development in rats were investigated by another working group; the results, however, are only available as an abstract. This study reported that daily exposure to NO_2 concentrations of 1 or 10 mg/m^3 (0.53 or 5.3 ml/m^3) for 6 hours during the gestation period resulted in lipid peroxidation in the placenta, a small, dark placenta and postimplantation losses, reduced foetal weights, delayed ossification and hydrocephalus (Tabacova and Balabaeva 1986, 1988).

There is only one published study of the toxic effects of NO_2 on postnatal development (Tabacova et al. 1985). Wistar rats were exposed to concentrations of 0.05, 0.1, 1.0 or 10 mg/m^3 (0.027, 0.053, 0.53, 5.3 ml/m^3) for 6 hours a day during the whole gestation period. After NO_2 concentrations of 0.053 ml/m^3 air and more, changes were observed in posture and gait in 9-day-old and 14-day-old pups, which were no longer visible after the animals were weaned. Evidence of biochemical changes (e.g. lipid peroxidation) was not observed in this exposure group. After NO_2 concentrations of 0.53 ml/m^3, lipid peroxidation in the liver and delays in postnatal development were observed in the pups. At the high concentration of 5.3 ml/m^3 the viability and body weight gains of the pups, determined on day 21 after birth, were reduced; after 2 and 3 months these differences were no longer found (Tabacova et al. 1985).

Table 4. Studies of the developmental toxicity of NO_2 in mice and rats

Species, strain, number of animals per sex and group	Exposure	Results	References
mouse, CD-1, groups of 18–22♀	GD 8–18, 0, 22, 45 ml/m³, investigation GD 18	≥ **22 ml/m³** dams: no externally visible signs of toxicity; foetuses: haematomas, reduced body weights, increased mortality	Singh 1984
mouse, CD-1, groups of 18–22♀	GD 7–18, 0, 22, 45 ml/m³, postnatal investigation	≥ **22 ml/m³** newborn pups: reduced body weights, increased mortality **45 ml/m³** dams: no externally visible signs of toxicity; pups: slightly decreased number of live offspring; longer time needed for righting reflex (PND 1), longer time needed for negative geotaxis (PND 10), reduced aerial righting score (PND 12), reduced activity (PND 28)	Singh 1988
rat, no other details	GD 1–21, 0.018, 0.045, 0.43 ml/m³ investigation GD 21	≥ **0.045 ml/m³** dams, foetuses: increased intrauterine mortality **0.43 ml/m³** dams: reduced body and liver weights; foetuses: reduced foetal weights, increased mortality	Gofmekler et al. 1977 (see also Barlow and Sullivan 1982)
rat, no other details	GD 1–21, 0.53, 5.3 ml/m³, (6 h/day), investigation GD 21	≥ **0.53 ml/m³** dams: lipid peroxidation in the lungs and placenta, placenta smaller and darker, evidence of degeneration in the placenta; increased postimplantation losses; foetuses: reduced foetal weights, delayed ossification, hydrocephalus	Tabacova and Balabaeva 1986, 1988 (both abstracts)
rat, Wistar, groups of 20♀	GD 1–21, 0.03, 0.05, 0.53, 5.3 ml/m³, (6 h/day, 7 d/week); postnatal investigation of the offspring	≥ **0.05 ml/m³** pups: changes in the open field (posture and gait PND 9, 14) (1), which disappeared after the animals were weaned; no biochemical changes were observed (MetHb, GSH, lipid peroxidation, liver enzymes) (2) ≥ **0.53 ml/m³** pups: cutting of the incisor teeth and opening of the eyes delayed, neuromotor development impaired (righting, negative geotaxis), behavioural changes in the open field (PND 9 but no longer PND 14), hexobarbital sleeping period increased (PND 7, 14, 21), lipid peroxides in liver increased (PND 30) (1) **5.3 ml/m³** pups: viability and weight gains of the pups slightly but significantly reduced (PND 21); after 2 and 3 months no longer any differences; neuromotor development impaired (startle reflex to acoustic stimulus, air righting, hindlimb support), CYP450 and aminopyrine-N-demethylase activity reduced, O_2 consumption reduced, no effects on MetHb and GSH (1)	(1) Tabacova et al. 1985, (2) Tabacova et al. 1987 (abstract)

CYP450: cytochrome P450; GD: gestation day; GSH: glutathione; MetHb: methaemoglobin; PND: postnatal day

In a follow-up study, no morphological or biochemical changes were observed apart from early transient behavioural changes in newborn rats whose mothers were exposed to NO_2 concentrations of 0.1 mg/m^3 (0.053 ml/m^3) for 6 hours a day during the whole gestation period. Increased MetHb, lipid peroxidation, the inhibition of liver enzymes and further biochemical changes were reported after the offspring were exposed at the age of 2 years to NO_2 concentrations of 0.05 ml/m^3 air for 6 hours a day for 30 days. The authors of this study, available only as an abstract, attributed this to the increased sensitivity of the ageing organism (Tabacova et al. 1987). The studies do not exclude the possibility that toxic effects on development can occur after maternal exposure to NO_2 concentrations in the range of 0.5 ml/m^3. These studies, as a result of their limitations, do not allow a decisive evaluation to be made; prenatal and postnatal studies carried out according to present-day standards are urgently needed.

5.5 Genotoxicity

5.5.1 *In vitro*

Studies of *in vitro* genotoxicity of NO_2 are reviewed in Table 5. NO_2 yielded positive results in bacterial indicator tests and mutagenicity tests. Tests for the induction of the SOS response with both *Escherichia coli* and *Salmonella typhimurium* produced positive results (Kosaka et al. 1985, 1986, 1987, Nakamura et al. 1987). NO_2 also produced gene mutations in different strains of *E. coli* and *S. typhimurium* (Biggart and Rinehart 1987, Isomura et al. 1984, Kosaka et al. 1985, 1986, 1987, Victorin and Stahlberg 1988).

In V79 cells, the induction of DNA strand breaks was detected using the alkaline elution technique after exposure to NO_2 concentrations of 10 ml/m^3 and more for 20 minutes (Bittrich et al. 1993, Görsdorf et al. 1990). NO_2 also induced increased sister chromatid exchange (SCE) in V79 cells after exposure to concentrations of 1 ml/m^3 and more for 2 hours (Shiraishi and Bandow 1985). In another study the gassing of V79 cells with NO_2 for 10 minutes led to the induction of both SCE (≥ 5 ml/m^3) and also of chromosomal aberrations (≥ 10 ml/m^3) (Tsuda et al. 1981). The authors hold the NO_2 responsible for the observed clastogenic effects, not the nitrite ion formed in water. Chromosomal aberrations were also only observed if the cells were washed with physiological saline before exposing them so that the NO_2, according to the authors, could thus not react with components of the culture medium (Tsuda et al. 1981).

Table 5. The genotoxicity of NO_2 in vitro

Test system		Exposure conditions	Results		References
			without MA	with MA	
SOS	S. typhimurium TA1535/pSK1002	10, 30, 90 ml/m³, 30 min, cell suspension[1]	+ ($\geq$ 30 ml/m³)	not investigated	Kosaka et al. 1985
	S. typhimurium TA1535/pSK1002	50 ml/m³, 30 min, cell suspension[1]	+	not investigated	Nakamura et al. 1987
	E. coli/pSK1002	90 ml/m³, 30 min, cell suspension[1]	+	not investigated	Kosaka et al. 1986, 1987
BMT	S. typhimurium TA1535	37.7 ml/m³, 10–60 min, cell suspension[1] or 5 min, culture dish[2]	–	not investigated	Biggart and Rinehart 1987
		377 ml/m³, 30–90 min, cell suspension[1] or 10–30 min, culture dish[2]	+	not investigated	
	S. typhimurium TA100	0.5–20 ml/m³, 6–7 h, culture dish[2]	not investigated	+ ($\geq$ 10 ml/m³)	Victorin and Stahlberg 1988
	S. typhimurium TA102, TA104		not investigated	–	
	S. typhimurium TA100, TA102	45, 90, 180 ml/m³, 30 min, cell suspension[1]	–	not investigated	Kosaka et al. 1985
		10, 45, 90 ml/m³, culture dish[2]	–	not investigated	
	S. typhimurium TA100	2–10 ml/m³, 40 min, culture dish[2]	+ ($\geq$ 4 ml/m³)	not investigated	Isomura et al. 1984
	S. typhimurium TA1535		+ ($\geq$ 8 ml/m³)	not investigated	
	S. typhimurium TA1535	30 min, culture dish[2]	+	not investigated	Arroyo et al. 1992
	E. coli WP2 trp⁺	60–240 ml/m³, 30 min, cell suspension[1]	+ ($\geq$ 60 ml/m³)	not investigated	Kosaka et al. 1986, 1987
SSB	V79 cells	5–200 ml/m³, 20 min, culture dish[2]	+ ($\geq$ 10 ml/m³)	not investigated	Görsdorf et al. 1990
	V79 cells	200 ml/m³, 10 min, culture dish[2]	+	not investigated	Bittrich et al. 1993
	alveolar macrophages	20 ml/m³, 2 h, culture dish[2]	+	not investigated	Walles et al. 1995
SCE	V79 cells	0.5, 1, 2, 4, 6, 8 ml/m³, 2 h, culture bottle[3]	+ ($\geq$ 1 ml/m³)	not investigated	Shiraishi and Bandow 1985
	V79 cells	5–100 ml/m³, 10 min, culture dish[2]	+ ($\geq$ 5 ml/m³)	not investigated	Tsuda et al. 1981

Table 5. continued

Test system		Exposure conditions	Results		References
			without MA	with MA	
CA	V79 cells	5–100 ml/m^3, 10 min, culture dish^1	+ ($\geq$ 10 ml/m^3)	not investigated	Tsuda et al. 1981
gene mutation	Don cells	2–20 ml/m^3, 10 min, culture dish^2	–	not investigated	Isomura et al. 1976

BMT: bacterial mutagenicity test; CA: chromosomal aberrations; SSB: DNA single strand breaks; MA: metabolic activation; SCE: sister chromatid exchange; SOS: SOS chromotest
^{1} infusion of NO$_2$ in cell suspension before plating
^{2} exposure of cells in culture dishes to NO$_2$
^{3} infusion of NO$_2$ in rotating culture bottles with a single layer of cells

In Don cells of the Chinese hamster the incidence of 8-azaguanine-resistant colonies was marginally increased (a maximum of 5 resistent colonies/10^5 survivors) after exposure to NO$_2$ concentrations of 2 to 8 ml/m^3 for 10 minutes, while the mutation frequency was no longer increased at higher, strongly cytotoxic concentrations (up to 20 ml/m^3) (Isomura et al. 1976). In a later publication the authors reported that they may have underestimated the number of mutants as a result of the short expression period (48 hours) (Isomura et al. 1984).

5.5.2 *In vivo*

In *Drosophila melanogaster* NO$_2$ did not induce any X-chromosomal recessive lethal mutations either after exposure to concentrations of 500 to 700 ml/m^3 air for 1 hour or after exposure to 50 to 280 ml/m^3 for 2 days or to 50 to 560 ml/m^3 for around 10 to 12 days until the next generation had hatched (Inoue et al. 1981). No somatic mutations were induced after exposure to 50 ml/m^3 for 5 and 19 hours (Victorin et al. 1990). If, however, the flies were fed 0.1 M methylurea or 0.1 M ethylurea 2 days before, exposure to NO$_2$ concentrations of 150 to 280 ml/m^3 for 3 hours had mutagenic effects (Inoue et al. 1981).

The induction of DNA strand breaks was investigated in lung cells of mice exposed to NO$_2$ concentrations of 30 ml/m^3 for 16 hours or 50 ml/m^3 for 5 hours. With the alkaline extraction method, but not the alkaline elution method, a significant increase in DNA single strand breaks was found (Walles et al. 1995). In alveolar macrophages of rats exposed to 1.2 ml/m^3 for three days, no significant increase in the incidence of DNA strand breaks was found (Bermudez et al. 1999).

Primary lung cells of rats exposed to NO$_2$ were investigated *ex vivo* for the induction of gene mutations (resistance against ouabain) and chromosomal aberrations. 5 rats were exposed for 3 hours to NO$_2$ concentrations of 8, 15, 21 or 27 ml/m^3. 40 animals served as controls. 18 hours after the end of exposure, primary cultures of the lung cells were prepared and cultured for 5 days. There was a dose-dependent increase in mutation frequency relative to in the controls: 1.7-fold at 8 ml/m^3, up to 21-fold at 27 ml/m^3. Also a dose-dependent increase in chromosomal aberrations (chromatid breaks and exchange, chromosomal breaks, multiple aberrations) was observed after culturing for 3 days. Aberrations of chromatid type were most frequent; the frequency of breaks at 8 ml/m^3 was increased 2.5-fold and at 27 ml/m^3 12-fold (Isomura *et al.* 1984). The experimental design of the study is, however, unusual and this makes it difficult to give a decisive interpretation of the results. The frequency of chromosomal aberrations was determined at a very late point: the animals were not killed until 18 hours after the exposure and the primary lung cells were cultivated for 5 days before they were seeded for the chromosomal aberration test. The preparation and evaluation of the metaphases was carried out three days later. It seems unusual that aberrations were observed at such a late point in time, especially as most of the chromatid breaks and other instable aberrations normally occur during the first mitosis after exposure. According to the authors, the cells divided at least 4 to 5 times after exposure. They suggest that persisting damage in the cells could be the reason for the long-term genomic instability observed. A positive control was not examined. The induction of ouabain-resistant cells in the gene mutation test is regarded as very specific for base-pair substitution and has proved to be comparatively insensitive in standard tests. The test conditions used here (expression period, selection period) were not validated as a result of the test being carried out as an *ex vivo* variant, and there was no positive control. The number of cells used for the mutant selection was, as a result of the low plating efficiency of the lung cells used (10 to 15 %), very small at around 2×10^5. Despite these limitations, the results of the study can be regarded as evidence of local mutagenic effects of NO$_2$ *in vivo*.

An *in vivo* micronucleus test exposed mice to NO$_2$ concentrations of 20 ml/m^3 air for 23 hours. The bone marrow was analysed 30 hours after starting exposure. No increase in the number of micronuclei was detected (Victorin *et al.* 1990).

C3H mice were exposed for 6 hours to NO$_2$ concentrations of 0.1, 1, 5 or 10 ml/m^3. Blood samples were taken by cardiocentesis immediately after the exposure, and one and two weeks later. The blood samples of several animals were mixed for chromosomal analysis. Eight weeks after the exposure, spermatocytes were prepared for chromosomal analysis. Neither in leukocytes nor in spermatocytes was the induction of chromosomal aberrations by NO$_2$ observed (Gooch *et al.* 1977).

5.7 Carcinogenicity

The possibility that inhaled nitrous gases could react with amines of the mucous membranes of the respiratory tract to form carcinogenic nitrosamines was noted early on (Druckrey and Preussmann 1962). Valid long-term studies of the carcinogenicity of NO$_2$ are, however, not available. Long-term studies with limited validity did not yield evidence of carcinogenic effects of NO$_2$. Initiation promotion studies yielded evidence of tumour-promoting effects of NO$_2$ in the rat lung.

Initiation promotion studies

In an initiation promotion experiment the effects of NO$_2$ in the rat lung after initiation with *N*-bis(2-hydroxypropyl)nitrosamine (BHPN) were investigated. Male Wistar rats were given single intraperitoneal doses of BHPN of 500 mg/kg body weight in physiological saline for initiation or were injected with only physiological saline as controls. The animals were then exposed continuously to NO_2 concentrations of 0, 0.04, 0.4 or 4 ml/m^3 air for 17 months. Initiation with BHPN alone led to increased mortality, hyperplasia of the alveolar cells in particular and in one animal to a pulmonary adenoma. NO_2 concentrations of 0.04 and 0.4 ml/m^3 without previous initiation with BHPN did not cause effects in the lungs. Even after previous initiation with BHPN, amplification of the BHPN-induced effects was not observed at these NO_2 concentrations (relative to in the BHPN controls). NO_2 concentrations of 4 ml/m^3 led to slight hyperplasia of the alveolar cells in only one animal (1/10). After previous initiation with BHPN (relative to in the BHPN controls) a statistically significant increase in hyperplasia of the bronchiolar mucosa (17/40) and an increase in pulmonary adenomas (4/40) were observed, and a carcinoma was detected in one animal (1/40) (Table 6). The authors conclude that an NO_2 concentration of 4 ml/m^3 can promote BHPN-initiated cells to produce lung tumours. In the nose, initiation with BHPN already caused tumours in 97 % to 100 % of the animals, so that the effects of the NO_2 treatment could not be documented (Ichinose et al. 1991).

Table 6. The incidence (and percentage) of lung changes in Wistar rats after initiation with single doses of BHPN and promotion with NO_2 for 17 months (Ichinose et al. 1991)

Exposure		n	Mortality	Hyperplasia alveolar cells		Lung tumours			
BHPN (mg/kg)	NO_2 (ml/m^3)			slight/ moderate	severe	bronchiolar mucosa	adenomas	carcinomas	total
–	0.0	10	1 (10)	0	0	0	0	0	0
–	0.04	10	1 (10)	0	0	0	0	0	0
–	0.4	10	1 (10)	0	0	0	0	0	0
–	4.0	10	1 (10)	1 (10)	0	0	0	0	0
500	0.0	40	12 (30)	10 (25)	7 (18)	1 (2.5)	1 (2.5)	0	1 (2.5)
500	0.04	40	10 (25)	21 (53)	10 (25)	1 (2.5)	1 (2.5)	0	1 (2.5)
500	0.4	40	9 (23)	30 (75)	5 (13)	1 (2.5)	0	0	0
500	4.0	40	13 (33)	19 (48)	11 (28)	17 (43)*	4 (10)	1 (2.5)	5 (13)

* significant difference compared to the group with BHPN initiation without NO_2 promotion according to Fisher's exact test

Table 7. The incidence (and percentage) of lung changes in Wistar rats after initiation with single doses of BHPN and promotion with ozone and NO_2 for 13 months and an 11-month follow-up period (Ichinose and Sagai 1992)

Treatment			n	Morta-lity	Hyperplasia alveolar cells		Lung tumours				Metastasis of other tumours
BHPN (mg/kg)	exposure (ml/m³)						bron-chiolar mucosa	ade-nomas	carci-nomas	total	
	ozone	NO_2			slight	severe					
–	–	–	35	11 (31)	5 (14)	0	0	0	0	0	0
–	0.05	–	36	11 (31)	8 (22)	0	0	0	0	0	0
–	0.05	0.4	36	5 (17)	8 (22)	1 (3)	0	0	0	0	0
500	–	–	36	23 (64)	6 (17)	3 (8)	0	0	0	0	0
500	0.05	–	36	20 (56)	11 (31)	4 (11)	1 (3)	1 (3)	2 (6)	3 (8)	0
500	0.05	0.4	36	21 (59)	7 (19)	6 (17)	2 (6)	5 (14)	1 (3)	5 (14)*	2 (6)

* significant difference compared to the group with BHPN initiation without NO_2 promotion according to the χ^2 test

In a further study by the same working group, after single injections of BHPN in doses of 500 mg/kg body weight male Wistar rats were exposed for 13 months (10 hours a day) either to ozone concentrations of 0.05 ml/m³ alone or in combination with NO_2 in concentrations of 0.4 ml/m³ and then kept for 11 months under ambient air conditions. Initiation with BHPN led to increased mortality. The increase in the incidence of lung tumours (3/36) observed after initiation with BHPN and promotion with ozone was further increased after the combined exposure to ozone and NO_2 (5/36) and was found to be significant compared to the incidence in the control group with BHPN initiation (0/36). After the combined exposure to ozone and NO_2 metastases of other tumours were observed in two animals (Table 7). The authors conclude that ozone and NO_2 have synergistic tumour-promoting effects (Ichinose and Sagai 1992); this, however, does not seem justified in view of the small difference in the number of animals with tumours of 3/36 (BHPN initiation and promotion with ozone) to 5/36 (BHPN initiation and promotion with ozone and NO_2). The high mortality in this study by the end of the experiment is, compared to the low mortality in the study of Ichinose et al. (1991), the result of the much longer duration of the study.

Intratracheal doses of carbon black particles coated with diesel exhaust particles were administered to 24 male F344 rats once a week for four weeks. The animals were then exposed to NO_2 concentrations of 6 ml/m³ air for 16 hours a day for 10 months. 18 months after the beginning of the experiment the lungs were histopathologically examined and DNA adducts of the diesel motor emissions were analysed by means of P^{32} postlabelling. A significant increase in alveolar adenomas (6/24, controls 0/29) and DNA adducts of the diesel motor emissions were detected in the animals treated with the particles from diesel motor emissions and exposed to NO_2. The authors conclude that NO_2 promotes the DNA damage and lung tumours induced by diesel motor emissions, e.g. by producing DNA-damaging hydroxyl radicals or promoting the permeation of substances into the cells (Ohyama et al. 1999). This study does not allow any statement to be made, however, about whether the effects may also have been caused by polycyclic nitroaromatics, which could have been formed during the instillation of the particles from diesel motor emissions and subsequent exposure to NO_2.

5 male F344 rats were given 4 intratracheal doses of a suspension of a particle fraction from the urban environment of Tokyo. After exposure to NO_2 concentrations of 6 ml/m^3 or clean air for 11 months and a follow-up period of 7 months, hyperplasia and papillomas of the pulmonary epithelial cells were observed independent of the NO_2 exposure (Ito et al. 1997).

Groups of 20 Syrian hamsters were exposed to NO_2 concentrations of 15 ml/m^3 for 23 hours a day for 24 weeks, with or without previous initiation with diethyl nitrosamine (DEN) (2 × 20 mg/kg body weight). The animals were examined 8 weeks after the end of exposure. The incidence of lung tumours was not increased relative to that in the corresponding controls after either exposure to NO_2 alone or previous initiation (Witschi et al. 1993).

In another study with Syrian hamsters the effects on the lungs of a mixture of NO_2 and SO_2 in concentrations of 5 ml/m^3 and 10 ml/m^3, respectively, were investigated in combination with diesel soot extract with or without initiation with DEN (Heinrich et al. 1989). The incidence of tumours was not increased, either with or without initiation with DEN.

Other short-term carcinogenicity studies

The following short-term studies are not suitable for evaluating the carcinogenic effects of NO_2 for various reasons. The exposure of 24 CAF1/Jax mice (a strain sensitive to lung tumours) to NO_2 concentrations of 5 ml/m^3 for 6 hours a day for 16 weeks did not cause an increase in lung tumours relative to the number found in 24 control animals (Wagner et al. 1965). The meaningfulness of this study is limited by the fact that only one (relatively low) concentration of NO_2 was used and the exposure duration of 4 months was relatively short, but sufficient for detecting urethane-induced lung tumours.

Groups of 30 female A/J mice, a strain also sensitive to lung tumours, were exposed to NO_2 concentrations of 0, 1, 5 or 10 ml/m^3 for 6 hours a day on 5 days a week for 6 months. The percentage of animals with tumours in the control group and the high concentration group was the same (30 %). In the high concentration group, however, the number of tumours per affected lung was slightly, but significantly increased compared to that in the controls (1.55 compared to 1.39) (Adkins et al. 1986). The validity of this study is questionable e.g. as a result of methodological shortcomings (such as wrong numbers given, results inadequately represented). Also the informative value of this experimental model must be critically evaluated. While some undeniably carcinogenic substances (1,2-dibromoethane, ethylene oxide, vinyl chloride) yielded positive results, carbon disulfide, which to date is not regarded as carcinogenic, was also positive.

There are several studies with mice available from one working group which show that exposure to NO_2 produces conditions in the lungs of the animals which favour the metastasis of intravenously administered melanoma cells (Richters and Kuraitis 1981, 1983, Richters and Richters 1983, 1989, Richters et al. 1985). The authors exposed SWR and C57Bl/6J mice first of all to NO_2 concentrations of 0.4 and 0.8 ml/m^3 for 10 to 12 weeks. Then they injected melanoma cells of the line B16-F10 into the caudal vein of the animals and determined the incidence of melanoma cell foci in the lungs after 2 and 3 weeks. The animals pretreated with NO_2 developed more melanoma cell foci than the animals not pretreated. Also, a second working group reported in a summary that the metastatic burden (i.e. the product of the number and size of the metastatic nodules) was increased by a factor of 8 in the C57Bl/6 mice pretreated with NO_2 compared to in the control animals (not pretreated) (Weinbaum et al. 1987). Similar results were obtained using ozone for the pretreatment; even with the low ozone concentration of 0.1 ml/m³ the metastasis of fibrosarcoma cells (line NR-FS) increased (Kobayashi et al. 1987). It was concluded from the results that both NO_2 and ozone promote the metastasis of cancer cells in the lungs by damaging the endothelial cells of the lung capillaries and thus increasing capillary permeability.

Long-term studies

No valid long-term studies of the carcinogenic effects of the substance are available.

After continuous exposure of groups of 4 male Wistar rats to NO_2 concentrations of 0.04, 0.4 or 4 ml/m³ air for 27 months (Kubota et al. 1987; see Section 5.2) the occurrence of adenomas or carcinomas was not reported; this was not, however, a carcinogenicity study with an adequate number of animals.

After the continuous exposure of Sprague-Dawley rats first of all to NO_2 concentrations of 2 ml/m³ (for 69 days) and then to 0.8 ml/m³ for life (Freeman et al. 1968a) numerous tumours were reported, which were distributed randomly between the control group and exposed group (Freeman et al. 1968b); more details, however, were not given. The study cannot, therefore, be evaluated.

In female NMRI and A/Heston-J mice, two strains with a high incidence of pulmonary adenomas, the carcinogenic effects on the lungs of long-term, intermittent inhalation exposure to NO_2 concentrations of 40 ml/m³ were investigated (Henschler and Ross 1966). The first experiment, in which NMRI mice were exposed to NO_2 concentrations of 40 ml/m³ twice a week (A), every 10 days (B), or every 30 days (C) for 48 hours, cannot be included in the final evaluation as a result of the lack of control animals. The authors discuss in detail, however, the observed histopathological NO_2-typical changes in the lungs, which were evidently unusual in the untreated mice of this strain. They described nodular cellular proliferation distributed over the whole lung, which emanated from the joining sections between the bronchiolus and alveolar duct (the place where the highest NO_2 concentrations were found; see Section 3) and extended into the parenchyma as "ovoid nodules". All bronchioli were surrounded by such proliferations. After staining with a silver salt, alveolar structures were visible. For this reason the authors regarded most of the underlying cells as "swollen alveolar epithelia". A few were described as "septal cell elements". There were no inflammatory cellular components. In the light of more recent results it is evident that the cell clusters were not (or not exclusively) made up of alveolar epithelia, but of neuroendocrine cells, as these can also be stained with silver salts (see Section 2). That they occur in the lungs was, however, unknown at the time of the publication. This is also in accordance with the described distribution of the proliferations in the bronchioli ("no bronchiolus without proliferation"), which was found to be characteristic for neuroendocrine cells in other studies (Kleinerman et al. 1981). The proliferation in the animals exposed according to mode A was moderately atypical. In animals from mode B the cell appearance was much more irregular; cellular and nuclear polymorphia were increased and atypical mitosis was found. Characteristic of the proliferation in the animals exposed according to mode C were extensive mitosis, including much that was atypical, and broad cellular and nuclear atypia. From a morphological point of view, the cell appearance corresponded to that of a malignant process, with the difference that the growths regressed to a large extent after the exposure to NO_2 ceased. In no case was a transition to infiltrating growth observed. Independent of this proliferation, the number of spontaneously occurring adenomas in NMRI mice was increased after exposure to NO_2 (A: 35 %, B: 40 %, C: 54 %); the highest incidence of adenomas was observed, however, in the animals with the lowest exposure frequency (C). The authors interpreted the increased incidence of adenomas in group C with caution, as survival was markedly increased in this group compared to in the animals of the other two groups (14.4 months compared to 10.4 months). In some cases the adenomas occurred at the site of the proliferation, but a causal relationship could not be deduced (Henschler and Ross 1966).

In a second experiment groups of 50 female NMRI mice and 25 female A/Heston-J mice were exposed to NO_2 concentrations of 0 or 40 ml/m^3 every 10 days for 48 hours. The NMRI mice exposed to NO_2 died on average much earlier than the NMRI control animals (18.5 compared to 21.3 months) and compared to the control animals were found to have a lower incidence of pulmomary adenomas (57 % compared to 72 %) and the skin (29 % compared to 42 %), but increased leukosis (6 cases compared to 2). In A/Heston-J mice survival and the number of animals with tumours was comparable in the controls and animals exposed to NO_2, but the number of tumours per animal was higher in the control animals. The authors concluded that NO_2 does not have carcinogenic effects (Henschler and Ross 1966). Although this investigation has a high concentration–time product, it is of only limited value as only one very high concentration (40 ml/m^3) was tested and both the exposure interval of 10 days and the exposure duration of 48 hours are not relevant to the workplace.

There are no data available for the allergenic effects of nitrogen dioxide.

6 Manifesto (MAK value/classification)

As a result of the relatively low water solubility of NO_2 the terminal airways are the target organ. The mechanism of action of NO_2 corresponds to that of ozone. NO_2 is a reactive radical, which reacts with proteins, DNA and nucleosides, can initiate radical reactions and produces reactive oxygen species. The concentration dependence of the oxidative cell damage is determined by the antioxidative capacity of the alveolar fluid and the target cells. *In vitro* NO_2 was found to be clearly genotoxic. From an *in vivo* study with mice, which, however, is only of limited validity, there is evidence of local genotoxic effects in the lungs (Isomura et al. 1984). Valid long-term studies of the carcinogenic effects of the substance are not available. In two initiation promotion experiments with the rat, evidence was found of tumour-promoting effects in the lungs after NO_2 concentrations of 4 ml/m^3 (Ichinose *et al.* 1991) and 6 ml/m^3 (Ohyama *et al.* 1999), but not after 0.4 or 0.04 ml/m^3 (Ichinose *et al.* 1991).

From the histopathological examinations it is evident that the effects continue as far as fibrosis and (in special investigations) preneoplastic and neoplastic changes. It is unclear, however, whether the radical reactions lead primarily to DNA damage or to proliferation. This must be clarified before the substance can be classified in Carcinogenicity Category 4 or 5.

In view of the genotoxic effects *in vitro* and the suspected genotoxic effects *in vivo*, the tumour-promoting effects of NO_2 and the similarity with the mechanism of action of ozone, NO_2 is classified in Category 3B for carcinogenic substances.

For further clarification of the mechanism of action and the dose–response relationships, information is needed in particular about the bronchoalveolar antioxidative capacities in experimental animals and man, local genotoxicity and the possibility of differentiating between the primary DNA damage and the cell proliferation induced by NO_2.

As a result of the genotoxic effects of NO_2 *in vitro* and the uncertainty about their importance for the situation *in vivo*, it is not possible at present to derive a MAK value for NO_2. A possible threshold value is, however, suggested. The experience in man with long-term exposure to NO_2 is not suitable for evaluating a threshold value as a result of the exposure to a mixture of substances. In studies with animals, continuous long-term exposure to NO_2 concentrations as low as 0.4 ml/m^3 still caused relevant histopathological effects in the lungs of ageing rats (Kubota *et al.* 1987). In a relatively sensitive strain of mouse histopathological effects were observed in the lungs even after exposure to NO_2 concentrations of 0.34 ml/m^3 air for 6 hours a day for 6 weeks (Sherwin and Richters 1982). In view of the only slight effects observed at these concentrations, the continuous exposure used in the rat experiment and the relatively sensitive mouse strain, a workplace-relevant threshold value for NO_2 in the range of 0.1 to 0.5 ml/m^3 is suggested. The different sensitivity of man as a result of the different respiratory physiology and the composition of the alveolar fluid are not taken into account here.

Studies with volunteers are suitable for limiting short-term exposure. Increased bronchial reactivity was observed after exposure to NO_2 concentrations of 1.5 ml/m^3 air and more for 3 hours (Frampton *et al.* 1989b), increased airway resistance after exposure to 2 ml/m^3 for 4 hours for 4 days (Blomberg *et al.* 1999). Marked changes in the bronchoalveolar lavage, the relevant method for determining the toxic effects of NO_2 in the alveolar fluid, were evident after NO_2 concentrations of 1.5 ml/m^3 and more (Sandström *et al.* 1992) and 2 ml/m^3 and more (Blomberg *et al.* 1999, Devlin *et al.* 1999, Solomon *et al.* 2000). After NO_2 concentrations of 0.6 ml/m^3 there was occasional, but inconsistent evidence of inflammation (Frampton *et al.* 1989a, 1989b, 1991, 2002), which needs further clarification. Lower concentrations were not investigated with regard to inflammation. As the first signs of inflammation are observed at 0.6 ml/m^3 air, it is recommended that short-term exposure not be permitted above 0.5 ml/m^3.

In view of the mechanism of action, the absorption through the skin of gaseous NO_2 should not cause systemic effects, even if 100 % of the inhaled dose were additionally absorbed through the skin. Designation with an "H" is not, therefore, necessary.

There is no evidence of sensitization of the skin or airways by NO_2. Designation with an "Sh" or "Sa" is therefore not necessary.

As no MAK value has been established, the substance is also not classified in a Pregnancy risk group. From the available animal studies of the developmental toxicity of NO_2 there is evidence that toxic effects on development can occur after maternal exposure to NO_2 concentrations in the range of 0.5 ml/m^3 air and below. As a result of their limitations, these studies are not sufficient for a decisive evaluation. Valid studies of prenatal and postnatal toxicity after exposure to NO_2 of the dams during pregnancy are therefore necessary.

As a result of the information available for the reactivity and genotoxicity of nitrogen dioxide it can be assumed that the genotoxic effects of NO_2 *in vivo* are limited to the airways. Classification in a category for germ cell mutagens is therefore not necessary.

7 References

Abbey DE, Nishino N, McDonnell WF, Burchette RJ, Knutsen SF, Lawrence Beeson W, Yang JX (1999) Long-term inhalable particles and other air pollutants related to mortality in nonsmokers. *Am J Respir Crit Care Med 159*: 373–382
Adams WC, Brookes KA, Schelegle ES (1987) Effects of NO_2 alone and in combination with O_3 on young men and women. *J Appl Physiol 62*: 1698–1704
Adkins B Jr, Van Stee EW, Simmons JE, Eustis SL (1986) Oncogenic response of strain A/J mice to inhaled chemicals. *J Toxicol Environ Health 17*: 331–322
Ahmed T, Marchette B, Danta I, Birch S, Dougherty RL, Schreck R, Sackner MA (1982) Effect of 0.1 ppm NO_2 on bronchial reactivity in normals and subjects with bronchial asthma. *Am Rev Resp Dis 125*, Suppl: 152
Ahmed T, Danta I, Dougherty RL, Schreck R, Sackner MA (1983) Effect of NO_2 (0.1 ppm) on specific bronchial reactivity to ragweed antigen in subjects with allergic asthma. *Am Rev Resp Dis 127*, Suppl: 160
Aranyi C, Fenters J, Ehrlich R, Gardner D (1976) Scanning electron microscopy of alveolar macrophages after exposure to oxygen, nitrogen dioxide, and ozone. *Environ Health Perspect 16*: 180
Arner EC, Rhoades RA (1973) Long-term nitrogen dioxide exposure. Effects on lung lipids and mechanical properties. *Arch Environ Health 26*: 156–160
Arroyo PL, Hatch-Pigott V, Mower HF, Cooney RV (1992) Mutagenicity of nitric oxide and its inhibition by antioxidants. *Mutat Res 281*: 193–202
Avol EL, Linn WS, Peng RC, Valencia G, Little D, Hackney JD (1988) Laboratory study of asthmatic volunteers exposed to nitrogen dioxide and to ambient air pollution. *Am Ind Hyg Assoc J 49*: 143–149
Avol EL, Linn WS, Peng RC, Whynot JD, Shamoo DA, Little DE, Smith MN, Hackney JD (1989) Experimental exposures of young asthmatic volunteers to 0.3 ppm nitrogen dioxide and to ambient air pollution. *Toxicol Ind Health 5*: 1025–1034
Ayaz KL, Csallany AS (1978) Long-term NO_2 exposure of mice in the presence and absence of vitamin E. II. Effect of glutathione peroxidase. *Arch Environ Health 33*: 292–296
Azoulay E, Soler P, Blayo MC (1978) The absence of lung damage in rats after chronic exposure to 2 ppm nitrogen dioxide. *Bull Eur Physiopathol Respir 14*: 311–325
Balchum OJ, Buckley RD, Sherwin R, Gardner M (1965) Nitrogen dioxide inhalation and lung antibodies. *Arch Environ Health 10*: 274–277

Barlow SM, Sullivan FM (1982) Nitrogen dioxide. in: *Reproductive Hazards of Industrial Chemicals*, Academic Press, London, New York, 417–421

Barth PJ, Müller B, Wagner U, Bittinger A (1994) Assessment of proliferative activity in type II pneumocytes after inhalation of NO_2 by AgNOR-analysis. *Exp Toxicol Pathol 46*: 335–342

Barth PJ, Müller B (1999) Effects of nitrogen dioxide exposure on Clara cell proliferation and morphology. *Pathol Res Pract 195*: 487–493

Bauer MA, Utell MJ, Morrow PE, Speers DM, Gibb FR (1986) Inhalation of 0.30 ppm nitrogen dioxide potentiates exercise-induced bronchospasm in asthmatics. *Am Rev Respir Dis 134*: 1203–1208

Becklake MR, Goldman HI, Bosman AR, Frend CC (1957) The long-term effects of exposure to nitrous fumes. *Am Rev Tuberculosis 76*: 398–409

Beil M, Ulmer WT (1976) Wirkung von NO_2 im MAK-Bereich auf Atemmechanik und bronchiale Acetylcholinempfindlichkeit bei Normalpersonen (Effect of NO_2 in workroom concentrations on respiratory mechanics and bronchial susceptibility to acetylcholine in normal persons) (German). *Int Arch Occup Environ Health 38*: 31–44

Bermudez E, Ferng S-F, Castro CE, Mustafa MG (1999) DNA strand breaks caused by exposure to ozone and nitrogen dioxide. *Environ Res Section A 81*: 72–80

Biggart NW, Rinehart RR (1987) Comparison between aqueous-phase and gas-phase exposure protocols for determining the mutagenic potential of nitrogen dioxide and the gas fraction of welding fumes. *Mutat Res 188*: 175–184

Bittrich H, Matzig AK, Kraker I, Appel KE (1993) NO_2-induced DNA single strand breaks are inhibited by antioxidative vitamins in V79 cells. *Chem Biol Interact 86*: 199–211

Blair WH, Henry MC, Ehrlich R (1969) Chronic toxicity of nitrogen dioxide. *Arch Environ Health 18*: 186–192

Blomberg A (2000) Airway inflammatory and antioxidant response to oxidative and particulate air pollutants – experimental exposure studies in humans. *Clin Exp Allergy 30*: 310–317

Blomberg A, Krishna MT, Helleday R, Soderberg M, Ledin MC, Kelly FJ, Frew AJ, Holgate ST, Sandstrom T (1999) Persistent airway inflammation but accommodated antioxidant and lung function responses after repeated daily exposure to nitrogen dioxide. *Am J Respir Crit Care Med 159*: 536–543

Busey WM, Coate WB, Badger DW (1974) Histopathologic effects of nitrogen dioxide exposure and heat stress in Cynomolgus monkeys. *Toxicol Appl Pharmacol 29*: 130

Bylin G, Lindvall T, Rehn T, Sundin B (1985) Effects of short-term exposure to ambient nitrogen dioxide concentrations on human bronchial reactivity and lung function. *Eur J Respir Dis 66*: 205–217

Bylin G, Hedenstierna G, Lindvall T, Sundin B (1988) Ambient nitrogen dioxide concentrations increase bronchial responsiveness in subjects with mild asthma. *Eur Respir J 1*: 606–612

Carson TR, Rosenholtz MS, Wilinski FT, Weeks MH (1962) The response of animals inhaling nitrogen dioxide for single, short-term exposures. *Ind Hyg J 23*: 257–462

Carson JL, Collier AM, Hu SC, Delvin RB (1993) Effect of nitrogen dioxide on human nasal epithelium. *Am J Respir Cell Mol Biol 9*: 264–270

Castleman WL, Dungworth DL, Schwartz LW, Tyler WS (1980) Acute respiratory bronchiolitis. An ultrastructural and autoradiographic study of epithelial cell injury and renewal in Rhesus monkeys exposed to ozone. *Am J Pathol 98*: 811–840

Challis BC, Shuker DEG (1982) Formation of N-nitrosamines and N-nitramines by gaseous NO_2. *Acta Cient Compostelana 19*: 153–166

Chaney S, Blomquist W, DeWitt P, Muller K (1981) Biochemical changes in humans upon exposure to nitrogen dioxide while at rest. *Arch Environ Health 36*: 53–58

Chang L-Y, Graham JA, Miller FJ, Ospital JJ, Crapo JD (1986) Effects of subchronic inhalation of low concentrations of nitrogen dioxide. *Toxicol Appl Pharmacol 83*: 46–61

Chang L-Y, Mercer RR, Stockstill BL, Miller FJ, Graham JA, Ospital JJ, Crapo JD (1988) Effects of low levels of NO_2 on terminal bronchiolar cells and its relative toxicity compared with O_3. *Toxicol Appl Pharmacol 96*: 451–464

Coate WB, Badger DW (1974) Physiological effects of nitrogen dioxide exposure and heat stress in Cynomolgus monkeys. *Toxicol Appl Pharmacol 29*: 130–131

Coffin DL, Gardner DE, Blommer EJ (1976) Time-dose-response for nitrogen dioxide exposure in an infectivity model system. *Environ Health Perspect 13*: 11–15

Coffin DL, Gardner DE, Sodorenko GI, Pinigin MA (1977) Role of time as a factor in the toxicity of chemical compounds in intermittent and continuous exposures. Part II. Effects of intermittent exposure. *J Toxicol Environ Health 3*: 821–828

Crapo JD, Barry BE, Chang L-Y, Mercer RR (1984) Alterations in lung structure caused by inhalation of oxidants. *J Toxicol Environ Health 13*: 301–321

Csallany AS, Ayaz KL (1978) Long-term NO_2 exposure of mice in the presence and absence of vitamin E. I. Effect on body weights and lipofuscin in pigments. *Arch Environ Health 33*: 285–291

Devlin RB, Horstman DP, Gerrity TR, Becker S, Madden MC, Biscardi F, Hatch GE, Koren HS (1999) Inflammatory response in humans exposed to 2.0 ppm nitrogen dioxide. *Inhalat Toxicol 11*: 89–109

Drechsler-Parks DM, Bedi JF, Horvath SM (1987) Pulmonary function responses of older men and women to NO_2. *Environ Res 44*: 206–212

Druckrey H, Preussmann R (1962) Zur Entstehung carcinogener Nitrosamine am Beispiel des Tabakrauches (The formation of carcinogenic nitrosamines using the example of tobacco smoke) (German). *Naturwissenschaften 49*: 498–499

Ehrlich R, Fenters JD (1973) Influence of nitrogen dioxide on experimental influenza in squirrel monkeys. in: *Proceedings of the 3rd international clean air congress*, Düsseldorf, A11–A13

Ehrlich R, Henry MC (1968) Chronic toxicity of nitrogen dioxide. *Arch Environ Health 17*: 860–865

Ehrlich R, Silverstein E, Maigetter R, Fenters JD (1975) Immunologic response in vaccinated mice during long-term exposure to nitrogen dioxide. *Environ Res 10*: 217–223

Ehrlich R, Findlay LC, Gardner DE (1979) Effects of repeated exposures to peak concentrations of nitrogen dioxide and ozone on resistance to streptococcal pneumonia. *J Toxicol Environ Health 5*: 631–642

Elsayed NM (1994) Toxicity of nitrogen dioxide: an introduction. *Toxicology 89*: 161–174

Evans MJ, Stephens RJ, Cabral LJ, Freeman G (1972) Cell renewal in the lungs of rats exposed to low levels of NO_2. *Arch Environ Health 24*: 180–188

Evans MJ, Cabral-Anderson LJ, Freeman G (1977) Effects of NO_2 on the lungs of aging rats. II. Cell proliferation. *Exp Mol Pathol 27*: 366–376

Feldman JG (1974) The combined action on a human body of a mixture of the main components of motor traffic exhaust gases (carbon monoxide, nitrogen dioxide, formaldehyde and hexane). *Gig Sanit 10*: 7–10

Fenters JD, Ehrlich R, Findlay JC, Spangler J, Tolkacz V (1971) Serologic response in squirrel monkeys exposed to nitrogen dioxide and influenza virus. *Am Rev Respir Dis 104*: 448–451

Fenters JD, Findlay JC, Port CD, Ehrlich R, Coffin DL (1973) Chronic exposure to nitrogen dioxide. *Arch Environ Health 27*: 85–89

Folinsbee LJ, Horvath SM, Bedi JF, Delehunt JC (1978) Effect of 0.62 ppm NO_2 on cardiopulmonary function in young male nonsmokers. *Environ Res 15*: 199–205

Frampton MW, Smeglin AM, Roberts NJ Jr, Finkelstein JN, Morrow PE, Utell MJ (1989a) Nitrogen dioxide exposure *in vivo* and human alveolar macrophage inactivation of influenza virus *in vitro*. *Environ Res 48*: 179–192

Frampton MW, Finkelstein JN, Roberts NJ, Smeglin AM, Morrow PE, Utell MJ (1989b) Effects of nitrogen dioxide exposure on bronchoalveolar lavage proteins in humans. *Am J Respir Cell Mol Biol 1*: 499–505

Frampton MW, Morrow PE, Cox C, Gibb FR, Speers DM, Utell MJ (1991) Effects of nitrogen dioxide exposure on pulmonary function and airway reactivity in normal humans. *Am Rev Respir Dis 143*: 522–527

Frampton MW, Boscia J, Roberts NJ Jr, Azadniv M, Torres A, Cox C, Morrow PE, Nichols J, Chalupa D, Frasier LM, Gibb FR, Speers DM, Tsai Y, Utell MJ (2002) Nitrogen dioxide exposure: effects on airway and blood cells. *Am J Physiol Lung Cell Mol Physiol 282*: L 155–165

Freeman G, Furiosi NJ, Haydon GB (1966) Effects of continuous exposure of 0.8 ppm NO_2 on respiration of rats. *Arch Environ Health 13*: 454–456

Freeman G, Stephens RJ, Crane SC, Furiosi NJ (1968a) Lesion of the lung in rats continuously exposed to two parts per million of nitrogen dioxide. *Arch Environ Health 17*: 181–192

Freeman G, Crane SC, Stephens RJ, Furiosi NJ (1968b) Pathogenesis of the nitrogen dioxide-induced lesion in the rat lung: a review and presentation of new observations. *Am Rev Respir Dis 98*: 429–443

Fujimaki H, Shimizu F, Kubota K (1982) Effect of subacute exposure to NO_2 on lymphocytes required for antibody responses. *Environ Res 29*: 280–286

Furiosi NJ, Crane SC, Freeman G (1973) Mixed sodium chloride aerosol and nitrogen dioxide in air. *Arch Environ Health 27*: 405–408

Gamble J, Jones W, Hudak J (1983) An epidemiological study of salt miners in diesel and nondiesel mines. *Am J Ind Med 4*: 435–458

Gamble J, Jones W, Minshall S (1987) Epidemiological-environmental study of diesel bus garage workers: acute effects of NO_2 and respirable particulate on the respiratory system. *Environ Res 42*: 201–214

Gardner DE (1982) Use of experimental airborne infections for monitoring altered host defenses. *Environ Health Perspect 43*: 99–107

Gardner DE, Coffin DL, Pinigin MA, Sidorenko GI (1977) Role of time as a factor in the toxicity of chemical compounds in intermittent and continuous exposures. Part I. Effects of continuous exposure. *J Toxicol Environ Health 3*: 811–820

Gardner DE, Miller FJ, Blommer EJ, Coffin DL (1979) Influence of exposure mode on the toxicity of NO_2. *Environ Health Perspect 30*: 23–29

Gofmekler VA, Brekhman II, Golotin BG, Sheparev AA, Krivelevich EB, Kamynina PN, Dobrjakoba AI, Gonenko VA (1977) The embryotropic action of nitrogen dioxide and a complex of atmospheric pollutants (Russian). *Gig Sanit 12*: 22–27

Goings SA, Kulle TJ, Bascom R, Sauder LR, Green DJ, Hebel JR, Clements ML (1989) Effect of nitrogen dioxide exposure on susceptibility to influenza A virus infection in healthy adults. *Am Rev Respir Dis 139*: 1075–1081

Gooch PC, Luippold HE, Creasia DA, Brewen HG (1977) Observations on mouse chromosomes following nitrogen dioxide inhalation. *Mutat Res 48*: 117–120

Görsdorf S, Appel KE, Engeholm C, Obe G (1990) Nitrogen dioxide induces DNA single-strand breaks in cultured Chinese hamster cells. *Carcinogenesis 11*: 37–41

Grayson RR (1956) Silage gas poisoning: nitrogen dioxide pneumonia, a new disease in agricultural workers. *Ann Intern Med 45*: 393–408

Greene ND, Schneider SL (1978) Effects of NO_2 on the response of baboon alveolar macrophages to migration inhibitory factors. *J Toxicol Environ Health 4*: 869–880

Gregory RE, Pickrell JA, Hahn FF, Hobbs CH (1983) Pulmonary effects of intermittent subacute exposure to low-level nitrogen dioxide. *J Toxicol Environ Health 11*: 405–414

Hackney JD, Thiede FC, Linn WS, Pedersen EE, Spier CE, Law DC, Fischer DA (1978) Experimental studies on human health effects of air pollutants. IV. Short-term physiological and clinical effects of nitrogen dioxide exposure. *Arch Environ Health 33*: 176–181

Halliwell B (1999) Oxygen and nitrogen are pro-carcinogens. Damage to DNA by reactive oxygen, chlorine and nitrogen species: measurement, mechanism and the effects of nutrition. *Mutat Res 443*: 37–52

Hayashi Y, Kohno T, Ohwada H (1987) Morphological effects of nitrogen dioxide in the rat lung. *Environ Health Perspect 73*: 135–145

Haydon GB, Freeman G, Furiosi NJ (1965) Covert pathogenesis of NO_2 induced emphysema in the rat. *Arch Environ Health 11*: 776–783

Hazucha MJ, Ginsberg JF, McDonnell WF, Haak ED, Pimmel RL, House DE, Bromberg PA (1982) Changes in bronchial reactivity of asthmatics and normals following exposure to 0.1 ppm NO_2. Stud Environ Sci 21: 387–400

Hazucha MJ, Ginsberg JF, McDonnell WF, Haak ED, Pimmel RL, Salaam SA, House DE, Bromberg PA (1983) Effects of 0.1 ppm nitrogen dioxide on airways of normal and asthmatic subjects. J Appl Physiol 54: 730–739

Hazucha MJ, Folinsbee LJ, Seal E, Bromberg PA (1994) Lung function response of healthy women after sequential exposures to NO_2 and O_3. Am J Respir Crit Care Med 150: 642–647

Heinrich U, Mohr U, Fuhst R, Brockmeyer C (1989) Investigation of a potential cotumorigenic effect of the dioxides of nitrogen and sulfur, and of diesel-engine exhaust, on the respiratory tract of Syrian golden hamsters. Health Effects Institute, HEI Research Report Number 26, Boston, MA, USA

Helleday R, Huberman D, Blomberg A, Stjernberg N, Sandstrom T (1995) Nitrogen dioxide exposure impairs the frequency of the mucociliary activity in healthy subjects. Eur Respir J 8: 1664–1668

Henry MC, Findlay J, Spangler J, Ehrlich R (1970) Chronic toxicity of NO_2 in squirrel monkeys. III. Effect on resistance to bacterial and viral infection. Arch Environ Health 20: 566–570

Henschler D, Lüdtke W (1963) Methämoglobinbildung durch Einatmung niederer Konzentrationen nitroser Gase (Methemoglobin formation by inspiration of low concentrations of nitrous gas) (German). Int Arch Gewerbepathol Gewerbehyg 20: 362–370

Henschler D, Ross W (1966) Zur Frage einer cancerogenen Wirkung inhalierter Stickstoffoxyde (On the problem of cancerogenic effect of inhaled nitrogen oxides) (German). Naunyn-Schmiedebergs Arch Exp Pathol Pharmakol 253: 495–507

Henschler D, Stier A, Beck H, Neumann W (1960) Geruchsschwellen einiger wichtiger Reizgase (Schwefeldioxid, Ozon, Stickstoffdioxid) und Erscheinungen bei der Einwirkung geringer Konzentrationen auf den Menschen (The odour threshold of some important irritant gases (sulfur dioxide, ozone, nitrogen dioxide) and the manifestations of the effect of small concentrations on man) (German). Arch Gewerbepath Gewerbehyg 17: 547–570

Hyde DM, Orthoefer J, Dungworth D, Tyler W, Carter R, Lum H (1978) Morphometric and morphologic evaluation of pulmonary lesions in beagle dogs chronically exposed to high ambient levels of air pollutants. Lab Invest 38: 455–469

Ichinose T, Sagai M (1982) Studies on biochemical effects of nitrogen dioxide. III. Changes of the antioxidative protective systems in rat lungs and of lipid peroxidation by chronic exposure. Toxicol Appl Pharmacol 66: 1–8

Ichinose T, Sagai M (1992) Combined exposure to NO_2, O_3 and H_2SO_4-aerosol and lung tumor formation in rats. Toxicology 74: 173–184

Ichinose T, Fujii K, Sagai M (1991) Experimental studies on tumor promotion by nitrogen dioxide. Toxicology 67: 211–225

Inoue H, Fukunaga A, Okubo S (1981) Mutagenic effects of nitrogen dioxide combined with methylurea and ethylurea in Drosophila melanogaster. Mutat Res 88: 281–290

Iqbal ZM, Dahl K, Epstein SS (1981) Biosynthesis of dimethylnitrosamine in dimethylamine-treated mice after exposure to nitrogen dioxide. J Natl Cancer Inst 67: 137–141

Isomura K, Fukase O, Watanabe H (1976) Cytotoxicities and mutagenicities of gaseous air pollutants on cultured cells (Japanese). J Jpn Soc Air Pollut 11: 59–64

Isomura K, Chikahira M, Teranishi K, Hamada K (1984) Induction of mutations and chromosome aberrations in lung cells following in vivo exposure of rats to nitrogen oxides. Mutat Res 136: 119–125

Ito T, Ohyama K-I, Kusano T, Usuda Y, Nozawa A, Hayashi H, Ohji H, Kitamura H, Kanisawa M (1997) Pulmonary endocrine cell hyperplasia and papilloma in rats induced by intratracheal injections of extract from particulate air pollutants. Exp Toxicol Pathol 49: 65–70

Jacobsen M, Smith TA, Hurley JP, Robertson A, Roscow R (1988) Respiratory infections in coal miners exposed to nitrogen oxides. Institute of Health Effects, Research Report No. 18, Cambridge, MA, USA. in: WHO (1997)

Jenkins HS, Devalia JL, Mister RL, Bevan AM, Rusznak C, Davies RJ (1999) The effect of exposure to ozone and nitrogen dioxide on the airway response of atopic asthmatics to inhaled allergens. *Am J Respir Crit Care Med 160*: 33–39

Johnson DA, Winters RS, Lee KR, Smith CE (1990a) *Oxidant effects on rat and human lung proteinase inhibitors*. Health Effect Institute, HEI Research Report Number 37, Boston, MA, USA

Johnson DA, Frampton MW, Winters RS, Morrow PE, Utell MJ (1990b) Inhalation of nitrogen dioxide fails to reduce the activity of human lung alpha-1-proteinase inhibitor. *Am Rev Respir Dis 142*: 758–162

Jörres R, Magnussen H (1991) Effect of 0.25 ppm nitrogen dioxide on the airway response to methacholine in asymptomatic asthmatic patients. *Lung 169*: 77–85

Jörres R, Magnussen H (1990) Airways response of asthmatics after a 30 min exposure, at resting ventilation, to 0.25 ppm NO_2 or 0.5 ppm SO_2. *Eur Respir J 3*: 132–137

Jörres R, Nowak D, Grimminger F, Seeger W, Oldigs M, Magnussen H (1995) The effect of 1 ppm nitrogen dioxide on bronchoalveolar lavage cells and inflammatory mediators in normal and asthmatic subjects. *Eur Respir J 8*: 416–424

Kagawa J (1983) Respiratory effects of two-hour exposure with intermittent exercise to ozone, sulfur dioxide and nitrogen dioxide alone and in combination in normal subjects. *Am Ind Hyg Assoc J 44*: 14–20

Kanoh T, Fukuda M, Mizoguchi I, Kinouchi T, Nishifuji K, Ohnishi Y (1987) Detection of mutagenic compounds in the urine of mice administered pyrene during exposure to NO_2. *Jpn J Cancer Res 78*: 1057–1062

Kanoh T, Fukuda M, Hayami E, Kinouchi T, Nishifuji K, Ohnishi Y (1990) Nitro reaction in mice injected with pyrene during exposure to nitrogen dioxide. *Mutat Res 245*: 1–4

Kerr HD, Kulle TJ, McIlhany ML, Swidersky P (1979) Effects of nitrogen dioxide on pulmonary function in human subjects: an environmental chamber study. *Environ Res 19*: 392–404

Kim SU, Koenig JQ, Pierson WE, Hanley QS (1991) Acute pulmonary effects of nitrogen dioxide exposure during exercise in competitive athletes. *Chest 99*: 815–819

Kim PK, Zamora R, Petrosko P, Billiar TR (2001) The regulatory role of nitric oxide in apoptosis. *Int Immunopharmacol 1*: 1421–1441

Kirsch M, Korth H-G, Sustmann R, de Groot H (2002) The pathobiochemistry of nitrogen dioxide. *Biol Chem 383*: 389–399

Kleeberger SR, Zhang L-Y, Jakab GJ (1997) Differential susceptibility to oxidant exposure in inbred strains of mice: nitrogen dioxide versus ozone. *Inhalat Toxicol 9*: 601–621

Kleinerman J, Marchevsky AM, Thornton J (1981) Quantitative studies of APUD cells in airways of rats. *Amer Rev Respir Dis 124*: 458–462

Kleinman MT, Bailey RM, Linn WS, Anderson KR, Whynot JD, Shamoo DA, Hackney JD (1983) Effects of 0.2 ppm nitrogen dioxide on pulmonary function and response to broncho-provocation in asthmatics. *J Toxicol Environ Health 12*: 815–826

Kobayashi T, Todoaki T, Sato H (1987) Enhancement of pulmonary metastasis of murine fibrosarcoma NR-FS by ozone exposure. *J Toxicol Environ Health 20*: 135–145

Koenig JQ, Covert DS, Morgan MS, Horike M, Horike N, Marshall SG, Pierson WE (1985) Acute effects of 0.12 ppm ozone and 0.12 ppm nitrogen dioxide on pulmonary function in healthy and asthmatic adolescents. *Am Rev Respir Dis 132*: 648–651

Koenig JQ, Covert DS, Marshall SG, van Belle G, Pierson WE (1987) The effects of ozone and nitrogen dioxide on pulmonary function in healthy and in asthmatic adolescents. *Am Rev Respir Dis 136*: 1152–1157

Koenig JQ, Covert DS, Smith MS, van Belle G, Pierson WE (1988) The pulmonary effects of ozone and nitrogen dioxide alone and combined in healthy and asthmatic adolescent subjects. *Toxicol Ind Health 4*: 521–532

Kosaka H, Oda Y, Uozumi M (1985) Induction of umuC gene expression by nitrogen dioxide in *Salmonella typhimurium*. *Mutat Res 142*: 99–102

Kosaka H, Yamamoto K, Oda Y, Uozumi M (1986) Induction of SOS functions by nitrogen dioxide in *Escherichia coli* with different DNA-repair capacities. *Mutat Res 162*: 1–5

Kosaka H, Uozumi M, Nakajima T (1987) Induction of SOS functions in *Escherichia coli* and biosynthesis of nitrosamine in rabbits by nitrogen dioxide. *Environ Health Perspect 73*: 153–156

Kosmider S, Misiewicz A, Felus E, Drozdz M, Ludyga K (1973) Experimentelle und klinische Untersuchungen über den Einfluß der Stickstoffoxyde auf die Immunität (Experimental and clinical studies on the effects of nitrogen oxides on immunity) (German). *Int Arch Arbeitsmed 31*: 9–23

Kripke BJ, Sherwin RP (1984) Nitrogen dioxide exposure – influence in rat testes. *Anesth Analg 63*: 526–528

Kröncke KD, Fehsel K, Suschek C, Kolb-Bachofen V (2001) Inducible nitric oxide synthase-derived nitric oxide in gene regulation, cell death and cell survival. *Int Immunopharmacol 1*: 1407–1420

Kubota K, Murakami M, Takenaka S, Kawai K, Kyono H (1987) Effects of long-term nitrogen dioxide exposure on rat lung: morphological observations. *Environ Health Perspect 73*: 157–169

Kulle TJ (1982) Effects of nitrogen dioxide on pulmonary function in normal healthy humans and subjects with asthma and chronic disease. *Stud Environ Sci 21*: 477–486

Kuraitis KV, Richters A (1989) Spleen cellularity shifts from the inhalation of 0.25–0.35 ppm nitrogen dioxide. *J Environ Pathol Toxicol 9*: 1–11

Kyono H, Kawai K (1982) Morphometric study on age-dependent pulmonary lesions in rats exposed to nitrogen dioxide. *Ind Health 20*: 73–99

Lafuma C, Harf A, Lange F, Bozzi L, Poncy JL, Bignon J (1987) Effects of low-level NO_2 chronic exposure on elastase-induced emphysema. *Environ Res 43*: 75–84

Larcan A, Calamai M, Lambert H, Mentrer B (1970) Pneumopathie par inhalation de vapeurs nitroseuses (combustion de poupées de celluloide) (Lung disease caused by inhalation of nitrous fumes (combustion of celluloid manikins)) (French). *Poumon Coeur 26*: 957–960

Linn WS, Solomon JC, Trim SC, Spier CE, Shamoo DA, Venet TG, Avol EL, Hackney JD (1985a) Effects of exposure to 4 ppm nitrogen dioxide in healthy and asthmatic volunteers. *Arch Environ Health 40*: 234–239

Linn WS, Shamoo DA, Spier CE, Valencia LM, Anzar UT, Venet TG, Avol EL, Hackney JD (1985b) Controlled exposure of volunteers with chronic obstructive pulmonary disease to nitrogen dioxide. *Arch Environ Health 40*: 313–317

Linn WS, Shamoo DA, Avol EL, Whynot JD, Anderson KR, Venet TG, Hackney JD (1986) Dose–response study of asthmatic volunteers exposed to nitrogen dioxide during intermittent exercise. *Arch Environ Health 41*: 292–296

Lynch PL, Bruns DE, Boyd JC, Savory J (1998) Chiron 800 system CO-oximeter module overestimates methemoglobin concentrations in neonatal samples containing fetal hemoglobin. *Clin Chem 44*: 1569–1570

Maigetter RZ, Fenters JD, Findlay JC, Ehrlich R, Gardner DE (1978) Effect of exposure to nitrogen dioxide on T and B cells in mouse spleen. *Toxicol Lett 2*: 157–161

Menzel DB, Abou-Donia MB, Roe CR, Ehrlich R, Gardner DE, Coffin DL (1977) Biochemical indices of nitrogen dioxide intoxication of guinea pigs following low level-long term exposure. in: Dimitriades B (Ed.) *International conference on photochemical oxidant pollution and its control, proceedings, Vol. II*, 577–587, US Environmental Protection Agency, EPA 600/3-77-001b, Research Triangle Park, NC, USA

Mercer RR, Costa DL, Crapo JD (1995) Effects of prolonged exposure to low doses of nitric oxide or nitrogen dioxide on the alveolar sept of adult rat lung. *Lab Invest 73*: 20–28

Miller FJ, Graham JA, Raub JA, Illing JW (1987) Evaluating the toxicity of urban patterns of oxidant gases. II. Effects in mice from chronic exposure to nitrogen dioxide. *J Toxicol Environ Health 21*: 99–112

Milne JEH (1969) Nitrogen dioxide inhalation and *bronchiolitis obliterans*. A review of the literature and report of cases. *J Occup Med 11*: 538–547

Mirvish SS, Ramm MD, Sams JP, Babcook DM (1988) Nitrosamine formation from amines applied to the skin of mice after and before exposure to nitrogen dioxide. *Cancer Res 48*: 1095–1099

Mohsenin V (1987) Effect of vitamin C on NO_2-induced airway hyperresponsiveness in normal subjects. *Am Rev Respir Dis 136*: 1408–1411

Mohsenin V (1988) Airway responses to 2.0 ppm nitrogen dioxide in normal subjects. *Arch Environ Health 43*: 242–246

Mohsenin V, Gee JB (1987) Acute effect of nitrogen dioxide exposure on the functional activity of alpha-1-protease inhibitor in bronchoalveolar lavage fluid of normal subjects. *Am Rev Respir Dis 136*: 646–650

Mücke H-G, Wagner HM (1998) VI-1. Anorganische Gase/Stickstoffdioxid (Inorganic gases/nitrogen dioxide) (German). in: Wichmann HE, Schlipköter HW, Fülgraff G (Eds) *Handbuch Umweltmedizin*, 14th Supplement 10/98, ecomed-Verlag, Landsberg/Lech

Müller B (1969) Nitrogen dioxide intoxication after a mining accident. *Respiration 26*: 249–261

Nakajima T, Oda H, Kusumoto S, Nogami H (1980) Biological effects of nitrogen dioxide and nitric oxide. in: Lee SD (Ed.) Nitrogen dioxides and their effects on health. *Ann Arbor Science Publisher Inc*, 121–141

Nakamura S, Oda Y, Shimada T, Oki I, Sugimoto A (1987) SOS-inducing activity of chemical carcinogens and mutagens in *Salmonella typhimurium* TA1535/PSK1002. Examination with 151 chemicals. *Mutat Res 192*: 239–246

von Nieding G, Krekeler H, Smidt U, Muysers K (1970) Akute Wirkung von 5 ppm NO_2 auf die Lungen- und Kreislauffunktion des gesunden Menschen (Acute effects of 5 ppm NO_2 on the function of the lung and circulation of healthy subjects) (German). *Int Arch Arbeitsmed 27*: 234–243

von Nieding, G, Wagner HM, Krekeler H, Smidt U, Muysers K (1971) Grenzwertbestimmung der akuten NO_2-Wirkung auf den respiratorischen Gasaustausch und die Atemwegswiderstände des chronisch lungenkranken Menschen(Minimum concentrations of NO_2 causing acute effects on the respiratory gas exchange and airway resistance in patients with chronic bronchitis) (German). *Int Arch Arbeitsmed 27*: 338–348

von Nieding G, Wagner HM (1975) Vergleich der Wirkung von Stickstoffdioxid und Stickstoffmonoxid auf die Lungenfunktion des Menschen (Comparison of the effects of nitrogen dioxide and nitrogen monoxide on pulmonary function in humans) (German). *Staub-Reinhalt Luft 35*: 175–178

von Nieding G, Wagner HM (1979) Effects of NO_2 on chronic bronchitis. *Environ Health Perspect 29*: 137–242

Norkus EP, Boyle S, Kuenzig W, Mergens WJ (1984) Formation of N-nitrosomorpholine in mice treated with morpholine and exposed to nitrogen dioxide. *Carcinogenesis 5*: 549–554

Norwood WD, Wisehart DE, Earl CA, Adeley FE, Anderson DE (1966) Nitrogen dioxide poisoning due to metal-cutting with oxyacetylene torch. *J Occup Med 8*: 301–306

Nyberg F, Gustavsson P, Jarup L, Bellander T, Berglind N, Jakobsson R, Pershagen G (2000) Urban air pollution and lung cancer in Stockholm. *Epidemiology 11*: 487–495

Ohyama K, Ito T, Kanisawa M (1999) The roles of diesel exhaust particle extracts and the promotive effects of NO_2 and/or SO_2 exposure on rat lung tumorigenesis. *Cancer Lett 139*: 189–197; Erratum: *Cancer Lett 147*: 229

Orehek J, Massari JP, Gayrard P, Grimaud C, Charpin J (1976) Effect of short-term, low-level nitrogen dioxide exposure on bronchial sensitivity of asthmatic patients. *J Clin Invest 57*: 301–307

Palisano JR, Kleinerman J (1980) APUD cells and neuroepithelial bodies in hamster lung: methods, quantitation, and response to injury. *Thorax 35*: 363–370

Pan QS, Fang ZP, Huang FJ (2000) Identification, localization and morphology of APUD cells in gastroenteropancreatic system of stomach-containing teleosts. *World J Gastroenterol 6*: 842–847

Park HS, Huh SH, Kim MS, Lee SH, Choi EJ (2000) Nitric oxide negatively regulates c-Jun N-terminal kinase/stress-activated protein kinase by means of S-nitrosylation. *Proc Natl Acad Sci USA 97*: 14382–14387

Pavelchak N, Church L, Roerig S, London M, Welles W, Casey G (1999) Silo gas exposure in New York state following the dry growing season of 1995. *Appl Occup Environ Hyg 14*: 34–38

Pitts JN, van Cauwenberghe KA, Grosjean D, Schmid JP, Fitz DR, Belser WL, Knudson GB, Hynds PM (1978) Atmospheric reactions of polycyclic aromatic hydrocarbons: facile formation of mutagenic nitro derivatives. *Science 202*: 515–519

Port CD, Coffin DL, Kane P (1977) A comparative study of experimental and spontaneous emphysema. *J Toxicol Environ Health 2*: 589–604

Posin C, Clarks K, Jones MP, Patterson JY, Buckley RD, Hackney JD (1978) Nitrogen dioxide inhalation and human blood biochemistry. *Arch Environ Health 33*: 318–324

Ramirez J, Dowell AR (1971) Silo-filler's disease: nitrogen dioxide-induced lung injury. Long-term follow-up and review of the literature. *Ann Intern Med 74*: 569–576

Rasmussen TR, Kjaergaard SK, Tarp U, Pedersen OF (1992) Delayed effects of NO_2 exposure on alveolar permeability and glutathione peroxidase in healthy humans. *Am Rev Respir Dis 146*: 654–659

Rejthar L, Rejthar A (1975) Histological changes of terminal bronchioles in rats during the exposure to nitrogen dioxide. *Exp Pathol 10*: 245–250

Reznick-Schuller (1976) Proliferation of endocrine (APUD-type) cells during early N-diethyl-nitrosamine-induced lung carcinogenesis in hamsters. *Cancer Lett 1*: 255–258

Reznick-Schuller (1977a) Sequential morphologic alterations in the bronchial epithelium of Syrian golden hamsters during N-nitrosomorpholine-induced pulmonary tumorigenesis. *Am J Pathol 89*: 59–66

Reznick-Schuller (1977b) Ultrastructural alterations of APUD cells during nitrosamine-induced lung carcinogenesis. *J Pathol 121*: 79–82

Richters A, Damji KS (1988) Changes in T-lymphocyte subpopulations and natural killer cells following exposure to ambient levels of nitrogen dioxide. *J Toxicol Environ Health 25*: 247–256

Richters A, Damji KS (1990) The relationship between inhalation of nitrogen dioxide, the immune system, and progression of a spontaneously occurring lymphoma in AKR mice. *J Environ Pathol Toxicol Oncol 10*: 225–230

Richters A, Kuraitis K (1981) Inhalation of NO_2 and blood borne cancer cell spread to the lungs. *Arch Environ Health 36*: 36–39

Richters A, Kuraitis K (1983) Air pollutants and the facilitation of cancer metastasis. *Environ Health Perspect 52*: 165–168

Richters A, Richters V (1983) A new relationship between air pollutant inhalation and cancer. *Arch Environ Health 38*: 69–75

Richters A, Richters V (1989) Nitrogen dioxide (NO_2) inhalation, formation of microthrombi in lungs and cancer metastasis. *J Environ Pathol Toxicol Oncol 9*: 45–51

Richters A, Richters V, Alley WP (1985) The mortality rate from lung metastasis in animals inhaling nitrogen dioxide (NO_2). *J Surg Oncol 28*: 63–66

Ritz B, Yu F, Fruin S, Chapa G, Shaw GM, Harris JA (2002) Ambient air pollution and risk of birth defects in Southern California. *Am J Epidemiol 155*: 17–25

Robertson A, Dodgson J, Collings P, Seaton A (1984) Exposure to oxides of nitrogen: respiratory symptoms and lung function in British coal miners. *Br J Ind Med 41*: 214–219

Roger LJ, Horstman DH, McDonnell W, Kehrl H, Ives PJ, Seal E, Chapman R, Massaro E (1990) Pulmonary function, airway responsiveness, and respiratory symptoms in asthmatics following exercise in NO_2. *Toxicol Ind Health 6*: 155–171

Rojas A (1992) No increase in chromosome aberrations in lymphocytes from workers exposed to nitrogen fertilisers. *Mutat Res 281*: 133–135

Rombout PJA, Dormans JAMA, Marra M, van Esc GJ (1986) Influence of exposure regimen on nitrogen dioxide-induced morphological changes in the rat lung. *Environ Res 41*: 466–480

Rubinstein I, Bigby BG, Reiss TF, Boushey HA Jr (1990a) Short-term exposure to 0.3 ppm nitrogen dioxide does not potentiate airway responsiveness to sulfur dioxide in asthmatic subjects. *Am Rev Respir Dis 141*: 381–385

Rubinstein I, Reiss TF, Bigby BG, Stites DP, Boushey HA (1990b) Effect of repeated exposure to 0.6 ppm nitrogen dioxide on lymphocyte phenotypes in the blood and bronchoalveolar lavage fluid of healthy subjects. *Clin Res 38*: 139A

Sagai M, Ichinose T (1987) Lipid peroxidation and antioxidative protection mechanism in rat lungs upon acute and chronic exposure to nitrogen dioxide. *Environ Health Perspect* 73: 179–189

Sagai M, Ichinose T, Kubota K (1984) Studies on the biochemical effects of nitrogen dioxide. *Toxicol Appl Pharmacol 73*: 444–456

Sandström T, Andersson MC, Kolmodin-Hedman B, Stjernberg N, Angstroem T (1990) Bronchoalveolar mastocytosis and lymphocytosis after nitrogen dioxide exposure in man: a time–kinetic study. *Eur Respir J 3*: 138–143

Sandström T, Stjernberg N, Eklund A, Ledin M-C, Bjermer L, Kolmodin-Hedman B, Lindstrom K, Rosenhall L, Angstrom T (1991) Inflammatory cell response in bronchoalveolar lavage fluid after nitrogen dioxide exposure of healthy subjects: A dose–response study. *Eur Respir J 4*: 332–339

Sandström T, Ledin MC, Thomasson L, Helleday R, Stjernberg N (1992) Reductions in lymphocyte subpopulations after repeated exposure to 1.5 ppm nitrogen dioxide. *Br J Ind Med 49*: 850–854

Sattler M, Salgia R (2003) Molecular and cellular biology of small cell lung cancer. *Sem Oncol 30*: 57–71

Schindler H, Bogdan C (2001) NO as a signaling molecule: effects on kinases. *Internat Immunopharmacol 1*: 1443–1455

Schuller HM, Witschi HP, Nylen E, Joshi PA, Correa E, Becker KL (1990) Pathobiology of lung tumors induced in hamsters by 4-(methylnitrosamino)-1-(3-pyridyl)-1-butanone and the modulating effect of hyperoxia. *Cancer Res 50*: 1960–1965

Schwela D (2000) Air pollution and health in urban areas. *Rev Environ Health 15*: 13–42

Selgrade MK, Daniels MJ, Grose EC (1991) Evaluation of immunotoxicity of an urban profile of nitrogen dioxide: acute, subchronic and chronic studies. *Inhalat Toxicol 3*: 389–403

Shalamberidze OP (1967) Reflex effects of mixtures of sulfur and nitrogen dioxide (Russian). *Gig Sanit 32*: 10–15

Shalamberidze OP, Tsereteli NT (1971) Effect of small concentrations of sulfurous gas and nitrogen dioxide on the estrual cycle and the genital function of animals in experiments (Russian). *Gig Sanit 8*: 13–17

Sherwin RP, Richters V (1982) Hyperplasia of type 2 pneumocytes following 0.34 ppm nitrogen dioxide exposure: quantitation by image analysis. *Arch Environ Health 37*: 306–315

Sherwin RP, Richters V (1995a) Effects of 0.25 ppm nitrogen dioxide on the developing mouse lung. Part 1: Quantitation of type 2 cells and measurements of alveoar walls. *Inhalat Toxicol 7*: 1173–1182

Sherwin RP, Richters V (1995b) Effects of 0.25 ppm nitrogen dioxide on the developing mouse lung. Part 2: Quantitation of elastic tissue and alveolar walls. *Inhalat Toxicol 7*: 1183–1194

Sherwin RP, Shih JC, Lee JD, Ransom R (1986) Serotonin content of the lungs, brains, and blood of mice exposed to 0.45 ppm nitrogen dioxide. *J Am Coll Toxicol 5*: 583–588

Shiraishi F, Bandow H (1985) The genetic effects of the photochemical reaction products of propylene plus NO_2 on cultured Chinese hamster cells exposed *in vitro*. *J Toxicol Environ Health 15*: 531–538

Singh J (1984) Teratological evaluation of nitrogen dioxide. *Proc Inst Environ Sci 30*: 229–231

Singh J (1988) Nitrogen dioxide exposure alters neonatal development. *Neurotoxicology 9*: 545–549

Solomon C, Christian DL, Chen LL, Welch BS, Kleinman MT, Dunham E, Erle DJ, Balmes JR (2000) Effect of serial-day exposure to nitrogen dioxide on airway and blood leukocytes and lymphocyte subsets. *Eur Respir J 15*: 922–928

Stacy RW, Seal E, House DE, Green J, Roger LJ, Raggio L (1983) A survey of effects of gaseous and aerosol pollutants on pulmonary function of normal males. *Arch Environ Health 38*: 104–115

Stephens RJ, Freeman G, Crane SC, Furiosi NJ (1971a) Ultrasturctural changes in the terminal bronchiole of the rat during continuous, low-level exposure to nitrogen dioxide. *Exp Mol Pathol 14*: 1–19

Stephens RJ, Freeman G, Evans MJ (1971b) Ultrastructural changes in connective tissue in lungs of rats exposed to NO_2. *Arch Intern Med 127*: 873–883

Stephens RJ, Freeman G, Evans MJ (1972) Early response of lungs to low levels of nitrogen dioxide. *Arch Environ Health 24*: 160–179

Stevens MA, Menache MG, Crapo JD, Miller FJ, Graham JA (1988) Pulmonary function in juvenile and young adult rats exposed to low-level NO_2 with diurnal spikes. *J Toxicol Environ Health 23*: 229–240

Stresemann E, von Nieding G (1970) Akute Wirkung von 5 ppm NO_2 auf den Atemwegswiderstand des Menschen (Acute effect of 5 ppm NO_2 on airway resistance in humans) (German). *Staub Reinh Luft 30*: 259–260

Tabacova S, Balabaeva L (1986) Nitrogen dioxide embryotoxicity and lipid peroxidation. *Teratology 33*: 58A

Tabacova S, Balabaeva L (1988) Nitrogen dioxide embryotoxicity and lipid peroxidation. *Teratology 38*: 29A

Tabacova S, Nikiforov B, Balabaeva L (1985) Postnatal effects of maternal exposure to nitrogen dioxide. *Neurobehav Toxicol Teratol 7*: 785–789

Tabacova S, Nikiforov B, Balabaeva L (1987) Nitrogen dioxide prenatal exposure versus exposure in old age. *Teratology 36*: 32A

Tabacova S, Baird DD, Bablabaeva L (1998) Exposure to oxidized nitrogen: lipid peroxidation and neonatal health risk. *Arch Environ Health 53*: 214–221

Tepper JS, Costa DL, Winsett DW, Stevens MA, Doerfler DL, Watkinson WP (1993) Near-lifetime exposure of the rat to a simulated urban profile of nitrogen dioxide: pulmonary function evaluation. *Fundam Appl Toxicol 20*: 88–96

Tokiwa H, Kitamori S, Nakagawa R, Ohnishi Y (1983) Mutagens in airborne particulate pollutants and nitro derivatives produced by exposure of aromatic compounds to gaseous pollutants. *Environ Sci Res 27*: 555–567

Touze MD, Desjars P, Baron D, Tasseau F, Delajartre AY, Nicolas F (1983) Collective acute poisoning by nitrous gases (French). *Toxicol Eur Res 5*: 220–224

Tse RL, Bockman AA (1970) Nitrogen dioxide toxicity. Report of four cases in firemen. *J Am Med Assoc 25*: 1341–1344

Tsuda H, Kushi A, Yoshida D, Goto F (1981) Chromosomal aberrations and sister-chromatid exchanges induced by gaseous nitrogen dioxide in cultured Chinese hamster cells. *Mutat Res 89*: 303–309

Uozumi M, Kusumoto T, Kimura T, Nakamura A, Nakajima T (1982) Formation of N-nitrosodimethylamine by NOx exposure in rats and rabbit pretreated with aminopyrine. *IARC Sci Publ 41*: 425–432

Vagaggini B, Paggiaro PL, Giannini D, Franco AD, Cianchetti S, Carnevali S, Taccola M, Bacci E, Bancalari L, Dente FL, Giuntini C (1994) Effects of short-term exposure to NO_2 (0.3–1 ppm) on the hypertonic saline induced sputum in normal, asthmatic and rhinitic subjects (Italian). *Arch Sci Lav 10*: 655–660

Victorin K, Stahlberg M (1988) A method for studying the mutagenicity of some gaseous compounds in *Salmonella typhimurium*. *Environ Mol Mutagen 11*: 65–77

Victorin K, Busk L, Cederberg H, Magnusson J (1990) Genotoxic activity of 1,3-butadiene and nitrogen dioxide and their photochemical reaction products in *Drosophila* and in the mouse bone marrow micronucleus assay. *Mutat Res 228*: 203–209

Wagner JH (1917) *Bronchiolitis obliterans* following the inhalation of acid fumes. *Am J Med Sci 154*: 511–522

Wagner WD, Duncan BR, Wright PG, Stokinger HE (1965) Experimental study of threshold limit of NO_2. *Arch Environ Health 10*: 455–466

Walles SA, Victorin K, Lundborg M (1995) DNA damage in lung cells *in vivo* and *in vitro* by 1,3-butadiene and nitrogen dioxide and their photochemical reaction products. *Mutat Res 328*: 11–19

Wantz JM, Dechoux J (1982) Œdème pulmonaire lèsionnel par vapeurs nitreuses chez les mineurs de carbon (Lesional pulmonary oedema produced in coal miners by nitrous gases) (French). *Arch Mal Prof 43*: 367–373

Weinbaum G, Oppenheim D, Arai K (1987) Stimulation of pulmonary metastasis by exposure to nitrogen dioxide. *Amer Rev Resp Dis 135*: A141

WHO (World Health Organisation) (1997) Nitrogen oxides (second edition). *Environmental Health Criteria 188*, WHO, Geneva

Wistuba II, Gazdar AF, Minna JD (2001) Molecular genetics of small cell lung carcinoma. *Sem Oncol 28(2), Suppl 4*: 3–13

Witschi H (1988) Ozone, nitrogen dioxide and lung cancer: a review of some recent issues and problems. *Toxicology 48*: 1–20

Witschi H, Breider MA, Schuller HM (1993) *Failure of ozone and nitrogen dioxide to enhance lung tumor development in hamsters.* Health Effect Institute, HEI Research Report Number 60, Boston, MA, USA

Yamamoto I, Takahashi M (1984) Ultrastructural observations of rat lung exposed to nitrogen dioxide for 7 months. *Kitasato Arch Exp Med 57*: 57–65

completed 19.02.2003

Propargyl alcohol

MAK value (1969)	2 ml/m^3 (ppm) ≙ 4.7 mg/m^3
Peak limitation (2000)	Category I, excursion factor 2
Absorption through the skin (1969)	H
Sensitization	–
Carcinogenicity	–
Prenatal toxicity (1991)	see Section IIc of the *List of MAK and BAT Values*
Germ cell mutagenicity	–
BAT value	–
Synonyms	–
Chemical name (CAS)	2-propyn-1-ol
CAS number	107-19-7
Structural formula	HC≡C–CH$_2$OH
Molecular formula	C$_3$H$_4$O
Molecular weight	56.06
Melting point	−48 °C (GAF 1988) −51.8 °C (Lide and Frederikse 1996) −53 °C (BASF AG 1996)
Boiling point	113.6–115 °C
Density at 20°C	0.947–0.9484 mg/cm^3
Vapour pressure at 20°C	11.6 hPa
pH	7 (at 330 g/l, 20°C)
log P$_{ow}$*	−0.35
1 ml/m^3 (ppm) ≙ 2.33 mg/m^3	**1 mg/m^3 ≙ 0.430 ml/m^3 (ppm)**

The following review is based on a toxicological evaluation for propargyl alcohol from the BG Chemie (2000).

---

* *n*-octanol/water distribution coefficient

1 Toxic Effects and Mode of Action

Propargyl alcohol is toxic after single oral doses, dermal applications and inhalation exposures. The most frequent symptoms are apathy, shortness of breath, irritation of the mucous membranes and bleeding in the internal organs and brain. Propargyl alcohol has irritative to caustic effects on the skin, depending on the concentration of the substance. In the eye, the undiluted substance is caustic.

After repeated inhalation exposure or oral administration of the substance, the target organs were found to be the liver and kidneys and also the blood. The toxic effects were increased relative kidney weights without a histopathological correlate, increased leukocyte counts, increased alanine aminotransferase activity, increased relative liver weights and histopathological liver changes. Inhalation exposure of the rat for 90 days yielded a NOAEL (no observed adverse effect level) of 11.8 mg/m^3, oral administration of the substance a NOAEL of 5 mg/kg body weight.

In vitro, propargyl alcohol was not found to be mutagenic but clastogenic. In the mouse, the micronucleus test yielded negative results after oral administration of the substance.

There are no data available for the mechanism of action of propargyl alcohol.

2 Toxicokinetics and Metabolism

Propargyl alcohol is probably metabolized to propargyl aldehyde. This oxidation has been demonstrated *in vitro* in experiments with catalase obtained from bovine liver.

In vitro experiments with phenobarbital-induced rat liver microsomes yielded evidence of low levels of binding of propargyl alcohol to cytochrome P450 (BG Chemie 1999).

3 Effects in Man

There are no data for the effects in man of exposure to propargyl alcohol.

4 Animal Experiments and *in vitro* Studies

4.1 Acute toxicity

Propargyl alcohol was found to be toxic after oral administration, dermal application and inhalation exposure in experiments with various animal species. The oral LD_{50} for the rat is 35–110 mg/kg body weight and for the mouse 50 mg/kg body weight. The dermal LD_{50} for the rabbit is 16–190 mg/kg body weight. The 2-hour LC_{50} for the rat is 2000 mg/m^3.

The most frequent symptoms in the rat and mouse after inhalation exposure, oral administration and dermal application were apathy, shortness of breath, prostration, irritation of the mucous membranes, hyperaemia and bleeding in the internal organs and brain.

4.2 Subacute, subchronic and chronic toxicity

The results of toxicity studies with repeated exposure are summarized in Table 1.

4.2.1 Inhalation

In a medium-term inhalation study with rats exposed to concentrations of 0, 1, 5 or 25 ml/m^3 (corresponding to analytical concentrations of 0, 2.35, 11.8 or 58.75 mg/m^3) for 90 days, at the highest concentration of 25 ml/m^3 increased relative liver and kidney weights were observed, but there was no histopathological correlate (BASF AG 1992). In this study the NOAEL was 5 ml/m^3 (11.8 mg/m^3).

Another 90-day inhalation study with rats revealed increased leukocyte counts, increased alanine aminotransferase activity, increased relative liver weights and histopathological liver changes after exposure of the animals to concentrations of 80 ml/m^3 (corresponding to about 188 mg/m^3) (DOW 1964).

Groups of 10 Swiss mice (20–25 g) were exposed to concentrations of 88 ml/m^3 (RD_{50}, i.e. the concentration which reduces the respiration rate by 50 %; corresponding to about 206.8 mg/m^3) or 25.3 ml/m^3 (1/3 RD_{50}; corresponding to about 59.5 mg/m^3) for 6 hours a day, on 4, 9 or 14 days. The respiratory and olfactory epithelia of the nasal cavity, and the trachea and lungs were subsequently examined histopathologically. Changes were found in both the respiratory and olfactory epithelium of the nose, but not in the trachea or lungs (no further details). The histopathological results showed that the severity of the effects in the nasal region varied with the exposure duration, but not with the concentration (Zissu 1995).

Table 1. Repeated exposure to propargyl alcohol

Species/number	Concentration/dose	Symptoms	References
rat, Sprague-Dawley 10 ♂, 10 ♀ per group	oral (gavage) 14 days 0.1, 1.0, 10 mg/kg body weight and day	no substance-related changes	Komsta et al. 1989
rat, Wistar 10 ♂, 10 ♀ per group	oral (gavage) 4 weeks 5, 15, 45 mg/kg body weight and day	≥ 5 mg/kg body weight: liver weights increased, kidney weights increased ≥ 15 mg/kg body weight: anaemia 45 mg/kg body weight: ALT increased, AP increased, GLDH increased, cholinesterase inhibition in ♀ rats, histology: liver cell damage	Bayer AG 1984
rat, Wistar 10 ♂, 10 ♀ per group	oral (gavage) 4 weeks 60 mg/kg body weight 1–3 days, 50 mg/kg body weight 4–21 days, 60 mg/kg body weight 22–28 days	apathy, atonia, bloody salivation, GLDH increased, N-demethylase decreased, cytochrome P450 decreased	Bayer AG 1984
rat, CD 10 ♂, 10 ♀ per group	oral (gavage) 28 days 5, 15, 50 mg/kg body weight	≥ 15 mg/kg body weight: megalocytosis of the liver 50 mg/kg body weight: one animal died, Hb decreased, leukocyte count increased, number of neutrophils increased, ALT increased, AST increased, LDH increased	TRL 1988
rat, CD, 20 ♂, 20 ♀ per group	oral (gavage) 90 days 5, 15, 50 mg/kg body weight and day	50 mg/kg body weight: 4/20 ♂ died, body weights decreased, relative liver weights decreased, relative kidney weights decreased in ♂ animals, karyomegaly in the tubule epithelia of the kidneys ≥ 15 mg/kg body weight: vacuolization of the liver, NOAEL 5 mg/kg body weight	TRL 1988
rabbit (no other details)	oral (gavage) 20 days 9.5 mg/kg body weight	no deaths (no other details)	BASF AG 1987
rabbit (no other details)	oral (gavage) 20 days 18.9 mg/kg body weight	100 % mortality (no other details), bleeding in the gastrointestinal tract and heart muscle, signs of lung congestion in the dead animals	BASF AG 1987
cat, 2 (no other details)	oral (gavage) 16 or 17 days 9.5 mg/kg body weight	2/2 died, symptoms: vomiting, lack of appetite, weight loss, liver function disorders (no other details)	BASF AG 1987

Table 1. continued

Species/number	Concentration/dose	Symptoms	References
rat, strain not stated 5 ♂, 5 ♀ per group	inhalation 5 days (1 h/day) about 2977 mg/m^3	4/10 died, mucosal irritation, no effects in surviving animals	BASF AG 1965
mouse, strain not stated 5 ♂, 5 ♀ per group	inhalation 5 days (1 h/day) about 2977 mg/m^3	7/10 died, mucosal irritation, liver damage, no effects in surviving animals	BASF AG 1965
guinea pig, 4	inhalation 5 days (1 h/day) about 2977 mg/m^3	no deaths, mucosal irritation	BASF AG 1965
rabbit, 2 (no other details)	inhalation 5 days (1 h/day) about 2977 mg/m^3	1/2 died, mucosal irritation, fatty degeneration of the liver, no effects in surviving animal	BASF AG 1965
cat, 1	inhalation 5 days (1 h/day) about 2977 mg/m^3	1/1 died, histology: liver dystrophy	BASF AG 1965
rat, Wistar 5 ♂, 5 ♀ per group	inhalation 14 days (6 h/day) 22, 115, 456 mg/m^3	456 mg/m^3: apathy, closed eyelids, red nasal secretion, irregular breathing, 1 ♀ animal died, body weights decreased, ALT increased, AST increased, relative liver weights increased, relative kidney weights increased, damage to the nasal mucosa and the liver	BASF AG 1990
mouse, Swiss 10 ♂, 10 ♀ per group	inhalation 4, 9, 14 days (6 h/day) 59.5, 206.8 mg/m^3	changes in the respiratory and olfactory epithelia of the nose, but not in the trachea or lungs	Zissu 1995
rat, Wistar, 10 ♂, 10 ♀ per group	inhalation 90 days (6 h/day, 5 days/week) 2.35, 11.8, 58.75 mg/m^3	58.75 mg/m^3: body weights decreased, relative liver weights increased, relative kidney weights increased, in ♀ serum cholinesterase decreased NOAEL 11.8 mg/m^3	BASF AG 1992
rat, Wistar 12 ♂, 12 ♀ per group	inhalation 89 days (7 h/day, 5 days/week) 188 mg/m^3	leukocyte count increased, ALT increased, relative liver weights increased, only in ♀ relative kidney weights increased, degenerative liver changes	DOW 1964
rabbit albino 2–3 ♂, 2–3 ♀ per group	dermal 91 days (8 h/day, 5 days/week) 1, 3, 10 mg/kg body weight	no substance-related changes	IBL 1965

ALT	alanine aminotransferase	AST	aspartate aminotransferase
AP	alkaline phosphatase	LDH	lactate dehydrogenase
GLDH	γ-glutamate dehydrogenase	Hb	haemoglobin

4.2.2 Ingestion

In rats given oral propargyl alcohol doses of 0, 5, 15 or 50 mg/kg body weight daily for 90 days, doses of 15 mg/kg body weight or more caused changes in haematological parameters (in the form of anaemia), increased relative liver weights, and histopathological changes in the liver (megalocytosis) and kidneys (karyomegaly in the tubule epithelia); at 50 mg/kg body weight, in addition increased relative kidney weights (male animals) and changes in clinico-chemical parameters were observed. In this study the LOAEL (lowest observed adverse effect level) was 5 mg/kg body weight (TRL 1988). Similar effects were found after rats were given oral doses of the substance for 28 days (see Table 1).

4.2.3 Dermal absorption

Epicutaneous application of propargyl alcohol to the intact or scarified skin of rabbits for 91 days did not produce treatment-related effects up to the highest tested dose of 20 mg/kg body weight and day (IBL 1965).

4.3 Local effects on skin and mucous membranes

Undiluted propargyl alcohol instilled into the conjunctival sac of the rabbit eye caused pain and erythema, oedema and clouding of the cornea. The effects were still marked 8 days after the application. A 10 % aqueous solution produced mild pain reactions and only transient irritation. A 1 % solution did not have any effects (BG Chemie 1999).

Undiluted propargyl alcohol applied to rabbit skin (dorsal) caused oedema, hyperaemia and necrosis. 8 days after the application, marked necrosis and anaemia were evident. A 10 % solution caused slight hyperaemia and oedema, a 1 % solution did not have any effects (BASF AG 1963, DOW 1952, 1964).

4.4 Genotoxicity

4.4.1 *In vitro*

In the *Salmonella* mutagenicity test, propargyl alcohol (4–2500 µg/plate) was not found to have mutagenic potential in the strains TA97, TA98, TA100, TA102, TA1535, TA1537 or TA1538 in the presence or absence of a metabolic activation system. In the strain hisD3052, however, without activation the substance had a weak mutagenic effect (propargyl aldehyde 1370 revertants/µmol, propargyl alcohol 15 revertants/µmol) (Basu and Marnett 1984). In the test for chromosomal aberrations in CHO cells (a cell line derived from Chinese hamster ovary), propargyl alcohol (168, 336 and 561 µg/ml without metabolic activation and 22, 34, 45 and 56 µg/ml with metabolic activation) yielded positive results (Blakey *et al.* 1994). It was not investigated to what extent this was the result of the formation of the aldehyde.

4.4.2 *In vivo*

In a micronucleus test, 15 male and 15 female NMRI mice (mean initial weights 29.1 and 23.3 g, respectively) were given propargyl alcohol (purity 99.4 %) in aqueous solution in single gavage doses of 70 mg/kg body weight. This dose was found in pilot studies to be the maximum tolerable dose. After 24, 48 and 72 hours, 5 male and 5 female animals were killed and the femur bone marrow was examined for micronuclei (1000 polychromatic erythrocytes/animal). Groups of 5 male and 5 female animals served as negative and positive controls (59 mg Endoxan®/kg body weight). After 24 and 72 hours a very slight, but statistically significant increase in the incidence of polychromatic erythrocytes containing micronuclei was detected in the female mice; the values were, however, still within the range of the historical control values from this laboratory. The number of normochromatic erythrocytes with micronuclei was not increased. Also the ratio of polychromatic to normochromatic erythrocytes was unchanged in both sexes. These results do not demonstrate that propargyl alcohol is clastogenic *in vivo* (Hoechst 1990).

Also in another micronucleus test, propargyl alcohol was not clastogenic. Groups of 5 male and 5 female 17-week-old C57BL mice were given gavage doses of propargyl alcohol (97 %) of 0 mg/kg (controls, olive oil), 24, 48 or 72 mg/kg body weight (corresponding to 0 %, 25 %, 50 % and 75 % of the LD_{50}) twice 24 hours apart. 36 hours after the second dose the femur bone marrow was prepared and 500 polychromatic erythrocytes/animal were examined for micronuclei. The highest dose was lethal for all male mice, so that these animals could not be evaluated. In the other dose group the ratio of polychromatic to normochromatic erythrocytes was unchanged, and there was no increase in the numbers of polychromatic erythrocytes containing micronuclei (Blakey *et al.* 1994).

There are no data available for the allergenic effects, the toxic effects on reproduction or the carcinogenicity of propargyl alcohol.

5 Manifesto (MAK value/classification)

Results of studies of exposed persons which would be suitable for the derivation of a MAK value are not available. *In vitro*, progargyl alcohol induced chromosomal aberrations in CHO cells. The results of the micronucleus test *in vivo* are to be considered as negative. There is thus no evidence of genotoxic potential for the substance *in vivo*.

After repeated inhalation exposure or oral administration of propargyl alcohol, the target organs were found to be the liver and kidneys as well as the blood. In a 90-day inhalation study with rats, no effects were observed at a concentration of 11.8 mg/m^3 (5 ml/m^3). The MAK value for propargyl alcohol of 4.7 mg/m^3 (2 ml/m^3) has therefore been retained.

Depending on the concentration, propargyl alcohol is irritating to caustic for the rabbit skin and eye. The LOEL established for histopathological changes in the nasal epithelium of the mouse was 25 ml/m^3. The animals were exposed to 25 or 88 ml/m^3 for 4 to 14 days and the histopathological changes were dependent more on the duration of exposure than on the concentration. On the basis of this finding and the well documented NOEL of 5 ml/m^3 for the rat propargyl alcohol is classified in peak limitation category 1 with an excursion factor of 2.

In experiments with animals propargyl alcohol is absorbed percutaneously in relevant amounts. Designation with an "H" has therefore also been retained.

There are no data available for sensitizing effects of the substance, and so no designation can be given.

As there are no data for reproductive toxicity, the substance is listed in Section IIc of the *List of MAK and BAT Values*.

6 References

BASF AG (Gewerbehygienisch-Pharmakologisches Institut) (1963) *Gewerbetoxikologische Vorprüfung – Propargylalkohol* (Preliminary toxicological tests – Propargyl alcohol) (German). unpublished report, Experiment No. XIII 62

BASF AG (Gewerbehygienisch-Pharmakologisches Institut) (1965) *Bericht über die Prüfung der Inhalationstoxizität von Propargylalkohol im Vergleich zu Allylalkohol* (Comparative inhalation toxicity of propargyl and allyl alcohols) (German). unpublished report No. XIII 62-63

BASF AG (Gewerbehygiene und Toxikologie) (1987) Communication to the Berufsgenossenschaft der chemischen Industrie, 18.02.1987 (German)

BASF AG (Abteilung Toxikologie) (1990) *Range-finding Studie zur Inhalationstoxizität von Propargylalkohol als Dampf an Ratten – 14-Tage Versuch.* (Range-finding study on the inhalation toxicity of propargyl alcohol vapour in the rat – 14-day study) (German) unpublished report, project No. 3610969/88060 carried out for the Berufsgenossenschaft der chemischen Industrie

BASF AG (Abteilung Toxikologie) (1992) *Study on the inhalation toxicity of propargylalkohol as a vapor in rats – 90-day-test.* unpublished report, project No. 5010969/88100 carried out for the Berufsgenossenschaft der chemischen Industrie

BASF AG (1996) *Sicherheitsdatenblatt Propargylalkohol hochkonz* (Safety data sheet for highly concentrated propargyl alcohol) (German).

Basu AK, Marnett JL (1984) Molecular requirements of the mutagenicity of malondialdehyde and related amines. *Cancer Res 44*: 2848–2854

Bayer AG (Institut für Toxikologie) (1984) *Propargylalkohol – Subakute orale Toxizitätsversuche an Ratten* (Propargyl alcohol – subacute oral toxicity studies with rats) (German). unpublished report No. 12653

BG Chemie (Berufsgenossenschaft der chemischen Industrie) (2000) Propargyl alcohol in: *Toxicological evaluations Vol. 2 Potential health hazards of existing chemicals, 121–134* (English translation of *Toxikologische Bewertung Nr. 116)*, BG Chemie, Heidelberg

Blakey DH, Maus KL, Bell R, Bayley J, Douglas GR, Nestmann ER (1994) Mutagenic activity of 3 industrial chemicals in a battery of *in vitro* and *in vivo* tests. *Mutat Res 320*: 273–283

DOW (The Dow Chemical Company) Biochemical Research Department (1952) *Results of range finding toxicological tests on propargyl alcohol.* NTIS/OTS 0510184, NTIS, Springfield, VA, USA

DOW (The Dow Chemical Company) Biochemical Research Laboratory (1964) *Results of repeated exposure of male and female rats to 80 ppm of propargyl alcohol in air.* NTIS/OTS 0510182, NTIS, Springfield, VA, USA

GAF (General Anilin and Film Corporation GmbH, Deutschland) (1988) *DIN-Sicherheitsdatenblatt Propargyl Alkohol* (DIN safety data sheet for propargyl alcohol) (German)

Hoechst AG (Pharma Research Toxicology and Pathology) (1990) *Propargylalkohol – Micronucleus test in male and female NMRI mice after oral administration.* unpublished report No. 90.0017, carried out for the Berufsgenossenschaft der chemischen Industrie

IBL (Industrial Biology Laboratories, Inc. USA) (1965) *90-day chronic skin absorption study with propargyl alcohol 11-72063 – B. 155*, report No. 2695 carried out for the GAF. NTIS/OTS 0510181, NTIS, Springfield, VA, USA

Komsta E, Secours VE, Chu I, Valli VE, Morris R, Harrison J, Baranowski E, Villeneuve DC (1989) Short-term toxicity of nine industrial chemicals. *Bull Environ Contam Toxicol 43*: 87–94

Lide DR, Frederikse HPR (Eds) (1996) *CRC handbook of chemistry and physics*, CRC Press, Boca Raton, New York, 3–294

TRL (Toxicity Research Laboratories, Ltd., USA) (1988) *Rat oral subchronic toxicity study.* unpublished report, TRL Study No. 042–004 carried out for the Dynamac Corporation

Zissu D (1995) Histopathological changes in the respiratory tract of mice exposed to ten families of airborne chemicals. *J Appl Toxicol 15*: 207–203

completed 01.03.1999

Vinyl acetate

MAK value	–
Peak limitation	–
Absorption through the skin	–
Sensitization	–
Carcinogenicity (2002)	Category 3A
Prenatal toxicity	–
Germ cell mutagenicity	–
BAT value	–
Synonyms	acetic acid vinyl ester
Chemical name (CAS)	acetic acid ethenyl ester
CAS number	108-05-4

As a result of recent studies of the mechanisms of action and carcinogenicity of vinyl acetate, the substance, and particularly its carcinogenicity, has been re-evaluated. As acetaldehyde is formed during the metabolism of vinyl acetate, important data for this substance are also included. The toxicokinetics and metabolism, genotoxicity, reproductive toxicity and carcinogenicity of vinyl acetate were dealt with in the documentation from 1983 and the supplement from 1991 (see the chapter in Volume 5 of the present series). Reference is made to the previous documentation; this is a supplementary chapter.

1 Toxic Effects and Mode of Action

Summary of the data from the documentation from 1983, the supplement from 1991 and recent data: vinyl acetate is absorbed very well by the epithelia of the nasal cavity and is metabolized there. In the organism, vinyl acetate is hydrolysed to acetaldehyde and acetic acid. Acetaldehyde is, in turn, oxidized to acetic acid, which then enters the C2 intermediary metabolism.

Vinyl acetate is of only low acute toxicity after ingestion and dermal absorption, while inhalation of higher concentrations is acutely toxic. Longer exposure to the substance can have irritative to caustic effects on the skin and irritative effects in the eye.

The NOAEL (no observed adverse effect level) for sensory irritation of the respiratory tract in man is 10 ml/m^3.

Cell proliferation of the oral mucosa and the respiratory and olfactory epithelia is observed after repeated exposure. With repeated inhalation exposure the NOAEL for local and systemic effects is 50 ml/m^3 in the rat and mouse. Inhalation of vinyl acetate concentrations of 1000 ml/m^3 was maternally toxic and foetotoxic in Sprague-Dawley rats. Oral administration of 5000 mg/l drinking water caused maternal toxicity but not prenatal toxicity in Sprague-Dawley rats. In a 2-generation study, the fertility of F_0 Sprague-Dawley rats was not impaired at this concentration, while that of the offspring was.

Vinyl acetate causes DNA–protein cross-links *in vitro*. This effect is attributed to the metabolite (hydrolysis product) acetaldehyde. In the *Salmonella* mutagenicity test, the substance is not mutagenic. Vinyl acetate produces positive results in the micronucleus test, the chromosomal aberration test and SCE (sister chromatid exchange) test *in vitro*. *In vivo*, after very high single intraperitoneal doses, micronuclei and SCE were observed in the bone marrow cells of mice, but not after inhalation exposure or the administration of vinyl acetate in drinking water. A micronucleus test *in vivo* in germ cells (spermatids) yielded negative results.

In a long-term study, vinyl acetate caused local tumours in the rat after inhalation of a markedly irritative concentration and in the rat and mouse after oral administration of concentrations which were not cytotoxic, but already systemic-toxic.

2 Mechanism of Action

Vinyl acetate is rapidly cleaved in the organism by ubiquitous esterases to form acetaldehyde and acetic acid. The local irritation and cytotoxicity of vinyl acetate are explained by the intracellular acidity resulting from the metabolically formed acetic acid (Kuykendall and Bogdanffy 1992, Kuykendall *et al.* 1993). The threshold level for these effects is accordingly determined by the physiological buffer capacity.

The clastogenic effects of vinyl acetate may be caused by acetaldehyde (DNA–protein cross-links, DNA cross-links) (Norppa *et al.* 1985) or as a result of the lowering of the intracellular pH value by the metabolite acetic acid (Morita *et al.* 1990). In experiments vinyl acetate was shown to lower the intracellular pH value in hepatocytes (Bogdanffy 2002).

As neither acetaldehyde nor vinyl acetate cause point mutation, but both have clastogenic effects, and the half-life of acetaldehyde-induced DNA–protein cross-links is short (6.5 hours; Kuykendall *et al.* 1993), it has been suggested that the lowering of intracellular pH by the metabolite acetic acid is more important for the extent of clastogenicity produced by high doses of vinyl acetate than the effects of acetaldehyde itself (Bogdanffy *et al.* 1999, 2001). This has, however, not yet been verified and how the reduction in pH contributes to this is still unclear. The available information indicates, however, that together with the local irritation and the induction of cell proliferation (Bogdanffy et al. 1997, Valentine et al. 2002) the clastogenicity is probably responsible for the formation of local tumours. Whether the genotoxic or the epigenetic mechanism is decisive, is at present unclear. It is also unclear what the threshold levels for the named effects in man are. Examination of the tumour incidences of the relevant carcinogenicity studies (Bogdanffy et al. 1994a, 1994b, Maltoni et al. 1997, Ministry of Labour of Japan 1998) reveals, however, that a non-linear dose–response relationship is to be expected (Valentine et al. 2002).

The following findings do not seem compatible with the postulated major importance of acetic acid and an epigenetic mechanism of tumour formation based on this (Bogdanffy et al. 1999): 1) the pH value in the respiratory epithelium was predicted to drop to a greater extent than that in the olfactory epithelium (Plowchalk et al. 1997). Fewer tumours developed in the respiratory epithelium, however, than in the olfactory epithelium; 2) in the respiratory epithelium a small number of tumours was observed, but no cytotoxicity; 3) models showed that not only the level of acetic acid, but also the level of vinyl acetate in the olfactory epithelium increase non-linearly with the concentration of vinyl acetate (Bogdanffy et al. 1999). Without further information, it cannot, therefore, be assumed that the concentration dependency of tumour incidences in the nasal epithelium, which are likewise non-linear, is to be attributed alone to cytotoxicity during tumour development.

3 Toxicokinetics and Metabolism

The amount of inhaled vinyl acetate absorbed was determined in the upper respiratory tract of anaesthetized rats. The exposure concentrations were 73 to 2190 ml/m^3 with inhalation for one hour. After exposure for about 8 minutes, an equilibrium was reached for vinyl acetate in nasal tissue. The concentration-dependent absorption of vinyl acetate by the epithelia was non-linear and between 36 % and 94 %, with the highest percentage absorbed at the lowest vinyl acetate concentrations (at 76 ml/m^3 or less, more than 93 % vinyl acetate was absorbed). At all vinyl acetate concentrations acetaldehyde was detected as a metabolite in the exhaled air (Plowchalk et al. 1997). Rapid hydrolytic cleavage of vinyl acetate was carried out by ubiquitous esterases (Fedtke and Wiegand 1990). The species differences between rats and mice were small for the kinetic parameters of the esterases in the nasal epithelium (Bogdanffy and Taylor 1993).

In mucosa obtained from the oral cavity of rats and mice, carboxylesterase was ubiquitously present, but with local differences in the activity of the enzyme. The species differences in the activity of the enzyme were only small (Morris et al. 2002).

According to a PBPK model for the dosimetry of vinyl acetate and its metabolites acetaldehyde and acetic acid in the nasal cavity, a significant drop in the intracellular pH value in the olfactory epithelium of the rat was stated to occur only above 50 ml/m^3 (Bogdanffy et al. 1999). However, this model has the following weaknesses, which make its validity seem questionable: there are no data for distribution coefficients. The mass transfer coefficients listed are not given in the source quoted. The calculations cannot therefore be checked. Although V_{max} and K_m are based on data from experiments, adjustment for the values determined is inadequate, and the predictions of the model are therefore doubtful. For the model of vinyl acetate metabolism in the human nose, two V_{max} values were allometrically extrapolated from animals to man. In addition it was assumed that the corresponding K_m values in man are identical to those in the rat; there are, however, no experimental data to back this up.

There are no experimental data available for dermal absorption. The models of Fiserova-Bergerova *et al.* (1990) and Guy and Potts (1993) predict absorption of 541 and 68 mg, respectively, after exposure to a saturated aqueous solution of 2000 cm^2 for one hour.

4 Effects in Man

In a cohort study 4806 persons employed in a factory producing chemicals between 1942 and 1973 and exposed to 19 chemicals (including vinyl acetate) were investigated. No statistically verifiable evidence of carcinogenic effects of vinyl acetate on the respiratory system was detected (Waxweiler *et al.* 1981).

In another study with embedded case-control studies, in which 29139 men were investigated who were employed in the production of chemicals, no decisive results were obtained. Only cancer of the lymphatic and haematopoietic tissue was investigated. Also here the difficulty was that most individuals were exposed to several chemicals. 14 persons were exposed to vinyl acetate (Ott *et al.* 1989).

In workers who had frequent and intensive skin contact with vinyl acetate, no allergic skin reactions were detected (BAuA 2002). There is no information available about the sensitizing potential of vinyl acetate on the respiratory tract in man.

5 Animal experiments and *in vitro* Studies

5.1 Acute toxicity

5.1.1 Inhalation

In an inhalation study with rats, a 4-hour LC$_{50}$ value of 4490 ml/m^3 (15800 mg/m^3) was determined. Inhalation of 8000 ml/m^3 led to the death of all animals during exposure.

The clinical symptoms were gasping for air after 10 minutes, debilitation and convulsions after 25 minutes, death after 50–90 minutes. During the inhalation of 4000 ml/m^3 2 of 6 male rats and 2 of 6 female rats died. The clinical symptoms were laboured breathing after 20 minutes, convulsions after 2.5 hours, death after 3 hours. No rats died after inhalation of 2000 ml/m^3 (clinical symptoms: reddened extremities; Union Carbide 1969).

After inhalation of vinyl acetate concentrations of 4000 ml/m^3 (14100 mg/m^3), 2 to 4 (no other details) of 6 rats died during the 4-hour exposure period (Carpenter et al. 1949). Exposure to an atmosphere saturated with vinyl acetate (temperature 20°C) caused the death of 0/12 rats after 1 minute, 10/12 rats after 3 minutes and 6/6 rats after 10 minutes (BAuA 2002).

5.1.2 Ingestion

An oral LD$_{50}$ value for vinyl acetate of 3.73 ml/kg body weight (3470 mg/kg body weight) was determined in rats. After doses of 8 ml/kg body weight, all rats died within 24 hours. After doses of 4 ml/kg body weight, 3 of the 5 rats died within 24 hours. After 2 ml/kg body weight, mortality was not observed. The symptoms included motoric inactivity, accumulation of fluid in the lungs and abdominal intestines, and a spotted liver (Union Carbide 1969).

An oral LD$_{50}$ value of 3.76 ml/kg body weight (about 3500 mg/kg body weight) was determined in a study for rats given a 2 % or 20 % vinyl acetate emulsion (BAuA 2002).

5.1.3 Dermal absorption

A dermal LD$_{50}$ value of 8.0 ml/kg body weight (7440 mg/kg body weight) was determined for vinyl acetate in rabbits after application of the substance for 24 hours. After application of 16 ml/kg body weight all animals died, after 8 ml/kg body weight 2 of 4 animals died within 2 days; neither deaths nor clinical symptoms were observed after 4 ml/kg body weight (Union Carbide 1969).

5.2 Subacute, subchronic and chronic toxicity

5.2.1 Inhalation

To investigate the effects on cell proliferation in the olfactory and respiratory epithelia of the nasal cavity, rats were exposed to vinyl acetate once or repeatedly for up to 4 weeks (50, 200, 600 or 1000 ml/m^3; 6 hours/day). The NOAEL for all effects was 200 ml/m^3. With higher concentrations and single exposures cell proliferation was markedly increased. With exposure for 5 days, the NOAEL was 1000 ml/m^3, with exposure for 20 days 200 ml/m^3 (olfactory epithelium only). After exposure to concentrations of

600 ml/m^3 or more for 4 weeks, regenerative hyperplasia was observed in the olfactory epithelium, and necrosis at 1000 ml/m^3 (Bogdanffy et al. 1997).

The 2-year study described below was presented earlier as an unpublished report (Owen et al. 1988, see Volume 5 of the present series) and is described here in more detail. Sprague-Dawley rats and CD-1 mice were exposed to vinyl acetate in concentrations of 50, 200 or 600 ml/m^3 over a period of 2 years (6 hours/day, 5 days/week) (see Section 5.7.2). At concentrations of 600 ml/m^3 reduced urine volumes were determined in the rats. This was attributed to reduced food and water consumption, but no data were collected for this. Reduced body weights were determined in mice after concentrations of 200 ml/m^3 or more and in rats after 600 ml/m^3. Vinyl acetate affected only the upper and lower respiratory tract (Tables 1 and 2). After concentrations of 200 ml/m^3 or more, atrophy, regenerative processes, inflammation and metaplasia in the olfactory epithelium, and basal cell hyperplasia were found in rats and mice. Hyperplastic and metaplastic changes in the trachea, together with epithelial desquamation and fibrotic reactions of the tracheal epithelium were observed only in mice at 600 ml/m^3. Similar changes were found in the bronchial and bronchiolar airways of rats and mice at 600 ml/m^3. In addition, histiocytic cell accumulation in the alveoli and the pulmonary interstitium was observed, which is possibly connected with the increased lung weights (Bogdanffy et al. 1994b). The following NOAELs were obtained: local effects: 50 ml/m^3 in rats and mice; systemic effects (end point reduced body weights): 200 ml/m^3 in rats, 50 ml/m^3 in mice.

Table 1. The number of rats with non-neoplastic histological lesions in the nose and lungs after exposure to vinyl acetate for 2 years (Bogdanffy et al. 1994b)

	male animals				female animals			
	0	50 ml/m^3	200 ml/m^3	600 ml/m^3	0	50 ml/m^3	200 ml/m^3	600 ml/m^3
Lungs	(58)	(59)	(60)	(60)	(60)	(60)	(60)	(59)
bronchial exfoliation								
very slight	0	0	0	8[b]	0	0	0	0
slight	0	0	0	26[b]	0	0	0	4
moderate	0	0	0	2	0	0	0	0
intraluminal fibrotic deposits								
very slight	0	0	0	16[c]	0	0	0	3
slight	0	0	0	14[c]	0	0	0	28[c]
moderate	0	0	0	1	0	0	0	8[b]
severe	0	0	0	0	0	0	0	1
pigmented macrophages								
very slight	1	0	0	0	0	0	0	1
slight	1	3	3	35[c]	6	4	1	10
moderate	0	0	1	2	0	0	0	4
peribronchiolar/perivascular lymphoid aggregation								
very slight	5	1	0	0	0	1	2	0
slight	15	18	21	14	11	14	14	23[b]
moderate	1	4	1	2	2	1	2	5

Table 1. continued

	male animals				female animals			
	0	50 ml/m³	200 ml/m³	600 ml/m³	0	50 ml/m³	200 ml/m³	600 ml/m³
Nose	(59)	(60)	(59)	(59)	(60)	(60)	(60)	(59)
olfactory epithelium: atrophy								
very slight	0	1	4	0	0	1	4	0
slight	0	2	47^c	7^a	0	0	23^c	18^c
moderate	0	0	2	33^c	0	0	0	30^c
severe	0	0	0	10^b	0	0	0	3
olfactory epithelium: squamous cell metaplasia								
very slight	0	0	0	2	0	0	5	4
slight	0	0	0	12^b	0	0	0	26^c
moderate	0	0	0	9^b	0	0	0	7^b
severe	0	0	0	1	0	0	0	0
olfactory epithelium: regeneration								
very slight	0	0	3	0	0	0	3	2
slight	0	0	30^c	1	0	0	16^c	7^b
moderate	0	0	2	0	0	0	3	0
olfactory epithelium: inflammatory cell infiltrates								
very slight	0	0	0	1	0	0	0	0
slight	0	0	0	7^a	0	0	0	5^a
moderate	0	0	0	1	0	0	0	2
epithelial nest-like infolds								
very slight	0	0	0	0	0	0	1	0
slight	0	0	1	0	0	0	0	0
moderate	0	0	15^c	5	0	0	5	5^a
severe	0	0	1	5	0	0	0	2
olfactory epithelium: leukocyte exudate								
very slight	0	0	0	0	0	0	1	0
slight	0	0	0	11^b	0	0	0	5^a
moderate	0	0	0	2	0	0	1	3
severe	0	0	0	1	0	0	0	0
basal cell hyperplasia								
very slight	2	5	3	1	0	0	7^a	0
slight	0	0	40^c	21^c	0	0	24^c	35^c
moderate	0	0	11^c	22^c	0	0	3	16^c
severe	0	0	0	2	0	0	0	0

^{a} $p < 0.05$, ^{b} $p < 0.01$, ^{c} $p < 0.001$ (Fisher's exact test)
number of animals investigated in brackets

5.2.2 Ingestion

In a 13-week study with F344 rats and BDF_1 mice, no treatment-related histopathological changes were reported. The vinyl acetate concentrations in drinking water were 600, 1500, 3800, 10000 or 24000 mg/l (assuming water consumption to be 10 % in rats and 15 % in mice, this corresponds with a daily consumption of 2400 mg/kg body weight in rats and 3600 mg/kg body weight in mice; no other details; Ministry of Labour of Japan 1998).

Table 2. The number of mice with non-neoplastic histological lesions in the nose and lungs after exposure to vinyl acetate for 2 years (Bogdanffy et al. 1994b)

	male animals				female animals			
	0	50 ml/m³	200 ml/m³	600 ml/m³	0	50 ml/m³	200 ml/m³	600 ml/m³
Lungs	(51)	(51)	(56)	(53)	(56)	(55)	(55)	(51)
accumulation of alveolar macrophages								
very slight	5	1	4	3	5	2	6	1
slight	10	2[a]	4	7	3	8	4	10
moderate	0	4	8[b]	4	2	1	1	12[b]
severe	1	1	4	0	1	3	1	1
intraalveolar eosinophilic material								
very slight	0	0	3	1	0	0	2	1
slight	3	1	1	19[c]	0	0	0	7[b]
moderate	0	0	0	10[b]	0	0	1	15[c]
severe	0	0	0	2	0	0	0	1
accumulation of brown pigmented macrophages								
very slight	2	2	1	11[a]	3	5	1	2
slight	0	0	5	12[c]	1	1	4	21[c]
moderate	0	0	1	1	0	0	0	2
intraluminal fibroepithelial deposits								
very slight	0	1	2	3	1	0	0	6
slight	0	0	0	17[c]	0	2	1	19[c]
moderate	0	0	0	3	0	0	0	7[b]
bronchial/bronchiolar epithelial flattening or exfoliation								
very slight	0	0	0	4	0	0	0	4[a]
slight	1	0	0	25[c]	0	0	0	28[c]
moderate	0	0	0	7[a]	0	0	0	4[a]
severe	0	0	0	0	0	0	0	1
bronchial/bronchiolar epithelial disorganization								
very slight	0	0	0	0	0	0	0	5[a]
slight	0	0	0	11[b]	0	1	0	18[c]
moderate	0	0	0	4	0	0	0	0

Table 2. continued

	male animals				female animals			
	0	50 ml/m³	200 ml/m³	600 ml/m³	0	50 ml/m³	200 ml/m³	600 ml/m³
Nose	(52)	(48)	(53)	(50)	(56)	(57)	(55)	(51)
inflammatory exudate	0	0	2	15^c	0	0	1	5^b
inflammatory infiltration in the mucosa	1	0	0	12^b	1	2	0	5
hyperplasia of the submucosal glands								
slight	3	3	28^c	25^c	2	5	42^c	35^c
moderate	0	0	8^b	15^c	0	0	7^b	13^c
olfactory epithelium (mainly dorsal meatus): atrophy								
very slight	0	0	2	0	0	0	0	0
slight	0	0	5	0	2	4	8	0
moderate	0	0	28^c	2	0	0	26^c	0
severe	0	0	4	3	0	0	4	1
olfactory epithelium: atrophy (delocalized)								
very slight	0	0	1	0	0	0	3	0
slight	1	0	8^a	5	0	0	12^c	5^a
moderate	0	0	4	39^c	0	0	2	45^c
naso/maxilloturbinal region: squamous cell metaplasia								
very slight	0	0	0	0	0	0	0	1
slight	1	1	2	13^b	4	2	0	13^a
moderate	0	1	0	11^c	0	0	0	6^b
severe	0	0	0	0	0	0	0	1
substitution of the olfactory epithelia by respiratory epithelia								
slight	0	0	5	11^c	0	0	15^c	10^c
moderate	0	0	1	0	0	1	5^a	10^c
severe	0	0	0	0	1	0	0	0
Trachea/bronchi	(49)	(46)	(51)	(48)	(55)	(56)	(52)	(48)
epithelial hyperplasia	0	0	2	19^c	1	1	0	11^c

^{a} $p < 0.05$, ^{b} $p < 0.01$, ^{c} $p < 0.001$ (Fisher's exact test)
number of animals investigated in brackets

Groups of 20 male F344 rats (CDF) and BDF$_1$ mice (BGD2F1/CrlBr) were given drinking water containing vinyl acetate in concentrations of 1000, 5000, 10000 or 24000 mg/l for 90 days. The highest concentration was limited by the solubility. This study was designed in particular to determine threshold concentrations for local cell proliferation processes. Based on the drinking water consumption, the daily doses for rats were calculated to be 81, 350, 660 and 1400 mg/kg body weight and for mice 250,

1200, 2300 and 5300 mg/kg body weight. Statistically significant reductions in food consumption were observed in the rats exposed to 10000 and 24000 mg/l. There was a statistically significant reduction in the body weights of the rats of all groups. In mice, no effects on water and food consumption, or body weight gains were detected. No treatment-related histopathological changes were found in the oral cavity, oesophagus or forestomach of the animals. In rats, increased proliferation was found at concentrations of 24000 mg/l in the mucosa (BrdU incorporation; but below a doubling of the labelling index) in the lower jaw on days 1 and 29 (but not on day 8) and in the upper jaw on days 29 and 92. In the mice, increases in cell proliferation of 2.4 and 3.4 times those found in the controls occurred only after 92 days in the mucosa of the lower jaw following concentrations of 10000 and 24000 mg/l (Valentine et al. 2002).

A 78-week study with Swiss mice (see Section 5.7.2) with vinyl acetate concentrations of 1000 or 5000 mg/l drinking water yielded a NOAEL for systemic effects of 5000 mg/l drinking water (750 mg/kg body weight and day; Maltoni et al. 1997).

Groups of 60 Sprague-Dawley rats per sex and dose were given vinyl acetate with the drinking water in concentrations of 200, 1000 or 5000 mg/l (♂: 10, 47 and 202 mg/kg body weight and day; ♀: 16, 76 and 302 mg/kg body weight and day), beginning *in utero*, for a period of 2 years. The solutions were freshly prepared each day. The clinical symptoms, body weights, food and water consumption, haematology, clinical chemistry, urine composition, and gross pathological and microscopic changes were determined. Rats of the 1000 and 5000 mg/l groups consumed less water. In the 5000 mg/l groups, food consumption and body weights were reduced (♂: –19 %; ♀: –11 %). No treatment-related changes in mortality, in the haematological and clinico-chemical parameters or the urine status were observed. There was also no evidence of systemic organ toxicity. Some organ weights were changed in the 1000 and 5000 mg/l groups; no histopathological changes were found, however, and the changes were attributed to the lower body weights. The NOAEL was 1000 mg/l (Bogdanffy et al. 1994a).

In the 2-year drinking water study of the Ministry of Labour of Japan, mice and rats (see Section 5.7.2) were given vinyl acetate with the drinking water in concentrations of 400, 2000 or 10000 mg/l. In the highest concentration group, a decrease in drinking water and food consumption, and in body weight gains was determined in the male and female rats. Drinking water consumption and body weight gains were also reduced in the mice of this concentration group. Survival of the treated animals did not differ from that of the control animals. There were no reports of histopathological non-neoplastic or non-preneoplastic changes (Ministry of Labour of Japan 1998).

5.3 Local effects on skin and mucous membranes

Vinyl acetate had irritative and caustic effects on the skin of rabbits and in the eye of guinea pigs, depending on the duration of contact. The following description of unpublished studies was taken from the EU Risk Assessment Report (BAuA 2002).

5.3.1 Skin

In a Draize skin test, the application of 0.5 ml vinyl acetate to the intact skin of 6 rabbits caused irritation. Oedema (grade 4) and subdermal haemorrhage were observed 4, 24 and 72 hours after the application. Yellow discoloration of the skin prevented a valid determination of the severity of the erythema.

In another study of skin irritation in rabbits (number not specified), slight erythema was observed after exposure of the skin to undiluted vinyl acetate (amount unknown) for 5 to 15 minutes. Slight erythema and slight oedema were observed on day 1 after exposure for 20 hours. These effects increased and 8 days later necrosis was detected.

Two of 4 rabbits died within 2 days after dermal application of 8 ml/kg body weight; these animals were found to have necrosis at the site of application (see Section 5.1.3; Union Carbide 1969).

5.3.2 Eyes

It appears that the irritative potential of vinyl acetate on the eyes is similar to that on the skin: instillation of 1 to 2 drops of the undiluted substance into the eyes of rabbits led to corneal clouding, reddening and severe oedema of the conjunctiva 24 hours later. These effects disappeared after 8 days.

5.4 Allergenic effects

Positive results were obtained in a Bühler test with groups of 10 female and 10 male Hartley guinea pigs. Induction was carried out according to the original test protocol, and not the test guidelines, by means of occlusive application of undiluted vinyl acetate nine times within 3 weeks for 6 hours, instead of one application a week for three weeks. On provocation 2 weeks after the last induction treatment, 6 of 20 animals reacted to 25 % vinyl acetate with slight, confluent erythema or moderate, patchy erythema. The other animals had slight, patchy erythema. 7 of 10 vehicle controls and 9 of 10 untreated control animals also reacted with slight, patchy erythema and 7 of 10 untreated animals also on repeated provocation. In pilot studies to determine the irritation threshold, 3 of 4 animals reacted to 25 % and 10 % vinyl acetate with a similar level of erythema. During the induction phase, severe irritative reactions were observed at some application sites (Vinyl Acetate Toxicology Group 1995). It is therefore probable that the reactions were also or above all irritative reactions.

There are no studies available of the sensitizing effects of vinyl acetate on the airways.

5.5 Reproductive and developmental toxicity

There are no more recent studies available. Vinyl acetate was classified in Pregnancy risk group D in 1991.

5.6 Genotoxicity

Vinyl acetate

The available studies of the mutagenicity and genotoxicity of vinyl acetate were interpreted by the Commission as summarized below (see also Volume 5 of the present series).

In procaryotic cell systems *in vitro* (*Salmonella* mutagenicity test, SOS chromotest) no mutagenic potential was found. On incubation with eukaryotic cells, however, dose-dependent effects such as chromosomal aberrations, sister chromatid exchange and micronuclei are observed. Evidence of the occurrence of DNA cross-links was obtained after the incubation of very high concentrations of vinyl acetate with human lymphocytes. Under *in vivo* conditions no stable DNA adducts were found. On the other hand, after mice were given single intraperitoneal injections of 1000 or 2000 mg/kg body weight, an increase in the micronucleus count in bone marrow cells was found. In the same experiment no micronuclei were found in the male germ cells (spermatids).

After rats and mice inhaled the substance or were given oral doses in the drinking water for 3 months, no micronuclei were found in the bone marrow (Gale *et al.* 1980a, 1980b, Owen *et al.* 1980a, 1980b, see Volume 5 of the present series).

The following studies were not described in Volume 5. After the incubation of plasmid DNA, calf thymus histone proteins and rat liver microsomes with 1 to 100 mM vinyl acetate, DNA–protein cross-links were formed as a result of the formation of acetaldehyde. After incubation with acetaldehyde, a reduction in the pH value promoted cross-links (Kuykendall and Bogdanffy 1992). Also in isolated cells of the respiratory and olfactory epithelium of rats, vinyl acetate induced DNA–protein cross-links; their formation could be reduced by the inhibition of the carboxyl esterases *in vivo* and *in vitro*. When equimolar amounts of vinyl acetate and acetaldehyde were used, more cross-links were formed by vinyl acetate. As more acetic acid is formed with vinyl acetate than with acetaldehyde, the authors suspect that the greater drop in the pH value of vinyl acetate is responsible. Vinyl acetate was cytotoxic for both epithelia. The aldehyde scavenger semicarbacide could not reduce the cytotoxicity, 50 mM acetaldehyde alone was not cytotoxic, but 50 mM acetic acid was (Kuykendall *et al.* 1993).

In an inadequately documented study it was reported that after single intraperitoneal doses of vinyl acetate of 160 mg/kg body weight, chromosomal aberrations were found in the bone marrow of rats 26 hours after administration of the substance (BAuA 2002).

Likewise, the incidence of SCE in bone marrow was increased in rats 21 hours after intraperitoneal administration of vinyl acetate doses of 370–560 mg/kg body weight (Takeshita *et al.* 1986).

Acetaldehyde

As acetaldehyde, a metabolite of vinyl acetate, is thought to be responsible for the genotoxic effects (clastogenicity) after exposure to vinyl acetate, studies with acetaldehyde are mentioned below (see also the chapter "Acetaldehyde" in Volume 3 of the present series).

For the standard strains of *Salmonella typhimurium* investigated, acetaldehyde was not found to be mutagenic. In mammalian cells of various origin, acetaldehyde induced *in vitro* sister chromatid exchange, chromosomal aberration and gene mutation (WHO 1995). *In vitro* studies showed that the genotoxic effects (sister chromatid exchange, chromosomal aberration) of acetaldehyde occurred at concentrations equimolar to those found in studies with vinyl acetate (He and Lambert 1985, Norppa *et al.* 1985).

Acetaldehyde caused DNA cross-links (Lambert *et al.* 1985, Ristow and Obe 1975).

In vivo, after intraperitoneal administration, acetaldehyde induced micronuclei in bone marrow cells of mice (LOEL (lowest observed effect level) 200 mg/kg body weight; Morita *et al.* 1997) and sister chromatid exchange in peripheral lymphocytes of the Chinese hamster (0.5 mg/kg body weight; Korte *et al.* 1981).

Evaluation of genotoxicity

Comparative *in vitro* studies with acetaldehyde indicate that the genotoxic effects of vinyl acetate are to be attributed to the metabolite acetaldehyde (He and Lambert 1985, Norppa *et al.* 1985). The clastogenic effects resulting from a drop in the pH value seem to be less important.

The overall picture obtained from studies of the genotoxicity of vinyl acetate *in vivo* is that systemic genotoxic effects after ingestion or inhalation were not detected. After high intraperitoneal doses resulting in death, however, an increase in micronuclei in bone marrow cells was observed; this is explained by the saturation of inactivation mechanisms. At high doses, mutagenic effects of vinyl acetate (induced by the metabolite acetaldehyde or a drop in the pH value) on tissues directly exposed locally cannot be excluded.

With regard to the metabolic formation of acetaldehyde, the detoxification capacity of the organism and the endogenous presence of ethanol and acetaldehyde must be borne in mind. In this context, reference is made to the MAK value documentation for ethanol from 1998 (in Volume 12 of the present series). The endogenous level of acetaldehyde in the blood is given in the literature as 0.3 mg/ml (Halvorson *et al.* 1993).

5.7 Carcinogenicity

5.7.1 Short-term tests

New-born Wistar rats (4–5 animals per sex) were given oral doses of vinyl acetate in condensed milk of 100 or 200 mg/kg body weight and day (99 % pure, impurities: 100 mg hydroquinone/l) twice a day over a period of 3 weeks. A subgroup of animals was additionally treated with phenobarbital in drinking water for 8 weeks to stimulate the growth of potential preneoplastic liver foci. 14 weeks after the beginning of the study the animals were killed and their livers examined. No ATPase-free and gamma-GT positive areas could be found in the liver as evidence of the development of preneoplastic liver foci (Laib and Bolt 1986).

5.7.2 Long-term studies

Inhalation

In a 10-month study with Wistar rats exposed to vinyl acetate concentrations of 2.8, 28 and 140 ml/m^3 (5 hours/day, 5 days/week), metaplasia of the bronchial epithelium was found in all treated animals. The medium and high concentration caused, in addition, liver changes in the form of fatty degeneration of the liver and proliferation of the endoplasmic reticulum, and histological changes in the biliary capillaries (Czajkowska *et al.* 1986, see Volume 5 of the present series). The results of this study are in direct contradiction to those found in the better documented study of Bogdanffy *et al.* (1994b) and are therefore not included in the evaluation of vinyl acetate.

The inhalation study with rats with negative results (Maltoni *et al.* 1974, see Volume 5) is only very inadequately documented, the exposure period of 52 weeks does not meet present-day standards and mortality was very high; therefore, no conclusions can be drawn from this study. The 2-year inhalation study with rats and mice was described previously (Owen 1988, see Volume 5) and in the meantime has been published. Table 3 shows the tumour incidences. This study revealed local tumorigenic effects after inhalation exposure of the rat to markedly irritative concentrations (see Table 1). In the olfactory epithelium 3 papillomas and 2 carcinomas were found, in the respiratory epithelium 2 inverted papillomas. 5 other carcinomas were found at other sites or were of unknown origin (no other details). In mice no tumours were found, but irritation of the nasal epithelia was observed (see Table 2) (Bogdanffy *et al.* 1994b).

Ingestion

There are several studies available of the effects of vinyl acetate after ingestion, but the quality of some is considerably lacking.

The half-life of vinyl acetate in aqueous solution (distilled water) is about one week (Bogdanffy *et al.* 1994a). This fact is important when evaluating some of the studies with oral administration.

The long-term study with oral administration already described previously (Shaw 1988, see Volume 5) has in the meantime been published (Bogdanffy *et al.* 1994a) and is described in more detail below. Groups of 60 Sprague-Dawley rats (per sex and dose) were given vinyl acetate with the drinking water in concentrations of 200, 1000 or 5000 mg/l (♂: 10, 47 and 202 mg/kg body weight and day; ♀: 16, 76 and 302 mg/kg body weight and day), beginning *in utero*, for a period of 2 years. In the male rats of the highest dose group, 2 squamous cell carcinomas of the oral cavity were found and a sarcoma in the stomach. The tumour incidence for oral cavity carcinomas was, however, within the range of the historical control data for the strain and age of the rats used (Bogdanffy et al. 1994a).

Table 3. Incidence of tumours of the respiratory tract in Crl:CD(SR)BR rats after exposure to vinyl acetate (Bogdanffy et al. 1994b)

		0	50 ml/m³	200 ml/m³	600 ml/m³
Nasal cavity	♂	(59)	(60)	(59)	(59)
	♀	(60)	(60)	(60)	(59)
inverted papilloma	♂	0	0	0	4
	♀	0	0	0	0
papilloma	♂	0	0	1	0
	♀	0	0	0	0
squamous cell carcinoma	♂	0	0	0	2
	♀	0	0	0	4
carcinoma *in situ*	♂	0	0	0	1
	♀	0	0	0	0
sum of all tumours	♂	0	0	1	7[a]
	♀	0	0	0	4
Larynx	♂	(59)	(60)	(60)	(60)
	♀	(60)	(60)	(60)	(59)
squamous cell carcinoma	♂	0	0	0	0
	♀	0	0	0	1

[a] $p < 0.01$, Fisher's exact test
number of animals investigated in brackets

The method used in the 2-year study with F344 rats given vinyl acetate in doses of 1000 and 2500 mg/l with the drinking water (Lijinski and Reuber 1983, see Volume 5) was already criticized earlier. The study revealed a clear increase (with a dose-dependent tendency) in tumours of the liver and uterus. The shortcomings in the planning and implementation of the experiment (a small number of animals, no analytical monitoring of the vinyl acetate concentration, different life-spans of the animals) make it impossible to interpret the results.

The following studies were not described in Volume 5. Groups of 50 male and 50 female F344 rats and BDF$_1$ mice were given vinyl acetate with the drinking water in doses of 400, 2000 or 10000 mg/l for a period of 104 weeks (rats, ♂: 16–48, 75–226 and 364–950 mg/kg body weight and day; ♀: 22–60, 109–266 and 478–1062 mg/kg body weight and day, mice, ♂: 32–85, 167–405 and 800–2081 mg/kg body weight and day, ♀: 45–125, 230–483 and 1024–2185 mg/kg body weight and day). The vinyl acetate solutions were prepared twice a week. After the end of exposure the animals were killed and the weights of the brain, lungs, liver, spleen, heart, kidneys, adrenal glands, testes and ovaries were determined. The histopathological examinations in all animals covered 37 organs and tissues including the nasal cavity. In the rats the increased tumour incidences in the oral cavity, the oesophagus and the stomach were regarded as treatment-related (see

Table 4). In the oral cavity a positive trend was found for the incidence of squamous cell carcinomas in female animals. The increase was statistically significant in the male animals of the 10000 mg/l group. At this concentration squamous cell papillomas were observed in 2/50 male animals, epithelial dysplasia in 2/50 female animals and squamous cell carcinomas of the oesophagus in 1/50 female animals. Preneoplastic lesions of the oesophagus and the stomach were evident only at this concentration. No stomach tumours were found. In the mice, exposure-related tumours were found in the oral cavity, oesophagus, stomach and larynx (see Table 5). Particularly after doses of 10000 mg/l, (depending on the organ) statistically significant squamous cell carcinomas, papillomas and hyperplasia, basal cell activation and epithelial dysplasia developed. The highest concentration was, however, above the maximum tolerated dose, as the body weight curves show a reduction in body weight gains of at least 10 %. The nominal vinyl acetate concentration was not analytically verified, however, nor was drinking water consumption determined (Ministry of Labour of Japan 1998).

Table 4. Percentage of F344 rats with tumours and preneoplastic lesions after doses of vinyl acetate with the drinking water for 2 years (Ministry of Labour of Japan 1998)

		0	400 mg/l	2000 mg/l	10000 mg/l
	♂	(50)	(50)	(50)	(50)
	♀	(50)	(50)	(50)	(49)
Oral cavity					
squamous cell carcinoma	♂	0	0	0	$10.0^{a,b}$
	♀	0	2.0^b	2.0^b	6.0^b
squamous cell papilloma	♂	0	0	0	0
	♀	0	0	0	4.0
basal cell activation	♂	0	0	0	4.0
	♀	0	0	0	2.0
epithelial dysplasia	♂	0	0	0	0
	♀	0	0	0	4.0
Oesophagus					
squamous cell carcinoma	♂	0	0	0	0
	♀	0	0	0	2.0
squamous cell hyperplasia	♂	0	0	0	2.0
	♀	0	0	0	2.0
basal cell activation	♂	0	0	0	0
	♀	0	0	0	8.0
Stomach					
basal cell activation	♂	0	0	0	4.0
	♀	0	0	0	10.0

[a] $p < 0.05$; Fisher's exact test. [b] $p < 0.05$; positive trend in the Peto test or Cochran-Armitage test number of animals investigated in brackets

Table 5. Percentage of BDF$_1$ mice with tumours and preneoplastic lesions after doses of vinyl acetate with the drinking water for 2 years (Ministry of Labour of Japan 1998)

		0	400 mg/l	2000 mg/l	10000 mg/l
	♂	(50)	(50)	(50)	(50)
	♀	(50)	(50)	(50)	(50)
Oral cavity					
squamous cell carcinoma	♂	0	0	0	26.0[b,c]
	♀	0	2.0	2.0	30.6[b,c]
squamous cell papilloma	♂	0	0	0	8.0[c]
	♀	0	0	0	6.0
squamous cell hyperplasia	♂	0	0	4.0	26.0
	♀	0	0	2.0	12.0
basal cell activation	♂	0	0	4.0	4.0
	♀	0	0	2.0	2.0
epithelial dysplasia	♂	0	0	0	48.0
	♀	0	0	0	34.6
Oesophagus					
squamous cell carcinoma	♂	0	0	0	14.0[a,b]
	♀	0	0	0	2.0
squamous cell papilloma	♂	0	0	0	0
	♀	0	0	2.0	0
squamous cell hyperplasia	♂	0	0	0	4.0
	♀	0	0	0	4.0
basal cell activation	♂	0	0	0	18.0
	♀	0	0	0	30.6
epithelial dysplasia	♂	0	0	0	4.0
	♀	0	0	0	14.2
Forestomach					
squamous cell carcinoma	♂	2.0	0	0	14.0[a,c]
	♀	0	0	0	6.1
squamous cell papilloma	♂	0	0	0	4.0
	♀	0	0	2.0	2.0
squamous cell hyperplasia	♂	0	0	0	6.0
	♀	0	4.0	0	8.1
basal cell activation	♂	0	0	0	2.0
	♀	0	0	0	2.0
epithelial dysplasia	♂	0	0	0	2.0
	♀	0	0	0	0

Table 5. continued

		0	400 mg/l	2000 mg/l	10000 mg/l
	♂	(50)	(50)	(50)	(50)
	♀	(50)	(50)	(50)	(50)
Larynx					
squamous cell carcinoma	♂	0	0	0	4.0
	♀	0	0	2.0	2.0
squamous cell hyperplasia	♂	0	0	0	2.0
	♀	0	0	0	0
basal cell activation	♂	0	0	0	6.0
	♀	0	0	0	12.2
epithelial dysplasia	♂	0	0	0	4.0
	♀	0	0	0	6.0

[a] $p < 0.05$; [b] $p < 0.01$ Fisher's exact test
[c] $p < 0.01$; positive trend in the Peto test or Cochran-Armitage test
number of animals investigated in brackets

Another long-term study consists of the summary of a study carried out very much earlier against the background of studies for vinyl chloride. In this study Swiss mice were given vinyl acetate with the drinking water in doses of 1000 or 5000 mg/l for a period of 78 weeks. The F_0 generation (13–14 male and 37 female animals/group) and their young (37–38 male and 44–48 female animals/group) were treated from day 12 of gestation. The animals were observed up to the natural end of their lives. The drinking water was freshly prepared each day. No treatment-related effects were found as regards survival, body weight and behaviour. An increased tumour incidence was found in the following organs (see Table 6): Zymbal gland, oral cavity, tongue, oesophagus, forestomach, lungs, liver and uterus (Maltoni *et al.* 1997). Interpretation of these results is difficult when the large number of spontaneous systemic tumours in the control animals of this study is taken into consideration. The Commission considers interpretation of the study impossible for the following reasons: mortality was also high in untreated animals in the second year of the experiment. The nominal vinyl acetate concentrations and the drinking water consumption were not monitored. The maximum tolerated dose was probably exceeded in the F_1 animals, as body weight gains were reduced by at least 10 % at the highest dose (at which the tumours were reported). However, as the oral cavity, tongue, oesophagus and forestomach are locations coinciding with those in the results obtained by the Ministry of Labour of Japan (1998), this study is nevertheless of supportive value.

Table 6. Number of Swiss mice with tumours and preneoplastic changes after doses of vinyl acetate in the drinking water for 78 weeks (Maltoni et al. 1997)

		0		1000 mg/l (150 mg/kg body weight and day)		5000 mg/l (750 mg/kg body weight and day)	
		F_0	F_1	F_0	F_1	F_0	F_1
	♂	(14)	(38)	(13)	(37)	(13)	(49)
	♀	(37)	(48)	(37)	(44)	(37)	(48)
Zymbal gland							
carcinoma	♂	0	0	0	0	0	2
	♀	0	0	0	2	1	4
epithelial dysplasia	♂	0	2	0	0	1	4
	♀	1	3	4	2	6	11^a
Oral cavity							
squamous cell carcinoma	♂	0	0	0	0	1	10^c
	♀	0	0	0	0	1	9^c
Tongue							
squamous cell carcinoma	♂	0	1	0	0	1	7
	♀	0	0	1	0	3	12^c
epithelial dysplasia	♂	0	0	0	0	0	4
	♀	0	0	0	1	3	7^b
Oesophagus							
squamous cell carcinoma	♂	0	0	0	0	0	12^b
	♀	0	0	0	0	6^a	18^b
acanthoma	♂	0	0	0	0	0	0
	♀	0	0	0	0	1	3
epithelial dysplasia	♂	0	0	0	0	4^a	4
	♀	0	0	0	0	6^a	7^b
Forestomach							
squamous cell carcinoma	♂	0	0	0	0	0	2
	♀	0	0	0	0	3	7^b
acanthoma	♂	0	0	0	1	1	8^a
	♀	0	0	0	0	5^a	11^c
epithelial dysplasia	♂	0	0	0	0	0	1
	♀	0	0	0	0	0	0

Table 6. continued

		0		1000 mg/l (150 mg/kg body weight and day)		5000 mg/l (750 mg/kg body weight and day)	
		F_0	F_1	F_0	F_1	F_0	F_1
	♂	(14)	(38)	(13)	(37)	(13)	(49)
	♀	(37)	(48)	(37)	(44)	(37)	(48)
Glandular stomach							
adenocarcinoma	♂	0	1	0	0	1	0
	♀	0	0	0	0	1	1
adenomatous polyp	♂	0	0	0	1	0	0
	♀	0	0	0	0	0	0
dysplasia of the glands	♂	0	0	0	0	0	3
	♀	0	0	0	1	0	0
Lungs							
adenocarcinoma	♂	2	1	2	0	2	2
	♀	0	0	1	1	1	3
adenoma	♂	2	6	2	6	3	11
	♀	3	6	6	3	6	11
animals with lung tumours	♂	4	7	4	6	4	13
	♀	3	6	6	4	7	11
Liver							
hepatocarcinoma	♂	3	10	4	8	2	17
	♀	0	1	0	0	0	2
Uterus							
adenocarcinoma	♀	1	5	2	6	6	8
leiomyosarcoma	♀	1	0	0	2	2	4
leiomyoma	♀	1	2	0	3	2	6
animals with uterus tumours	♀	1	5	2	8	8[a]	12[a]

[a] $p \leq 0.05$; [b] $p \leq 0.01$; [c] $p \leq 0.005$; Fisher's exact test
number of animals investigated in brackets

5.7.3 Carcinogenicity of acetaldehyde

In a carcinogenicity study with rats exposed to acetaldehyde vapour in concentrations of 750, 1500 or 3000 ml/m³ (gradually reduced to 1000 ml/m³) for 6 hours a day, on 5 days a week, over a period of 28 months, treatment-related effects such as increased mortality, delayed growth, nasal tumours and non-neoplastic changes were observed in each test group (see Table 7). At the end of the study the following concentration-dependent nasal changes were found: degeneration, metaplasia and adenocarcinomas of the olfactory epithelium at all concentrations. Metaplasia and squamous cell carcinomas were found in the respiratory epithelium at the two highest exposure concentrations, and slight to severe rhinitis and sinusitis at the highest concentration. In the larynx, in many animals of the middle and high concentration groups hyperplasia and metaplasia were found in the epithelia of the vocal chord region. One female rat of the 1500 ml/m³ group developed a carcinoma of the larynx *in situ*. A NOAEL for non-neoplastic effects was not determined (Woutersen *et al.* 1986).

Table 7. Number of Wistar rats with tumours of the respiratory tract after exposure to acetaldehyde vapour for 28 months (Woutersen *et al.* 1986)

		0	750 ml/m³	1500 ml/m³	3000/1000 ml/m³
Nose	♂	(49)	(52)	(53)	(49)
	♀	(50)	(48)	(53)	(53)
papilloma	♂	0	0	0	0
	♀	0	1	0	0
squamous cell carcinoma	♂	1	1	10ᵃ	15ᶜ
	♀	0	0	5	17ᶜ
carcinoma *in situ*	♂	0	0	0	1
	♀	0	1	3	5
adenocarcinoma	♂	0	16ᶜ	31ᶜ	21ᶜ
	♀	0	6ᵃ	26ᶜ	21ᶜ
Larynx	♂	(50)	(50)	(55)	(52)
	♀	(53)	(52)	(54)	(54)
carcinoma *in situ*	♂	0	0	0	0
	♀	0	1	0	0

ᵃ $p < 0.05$; ᵇ $p < 0.012$; ᶜ $p < 0.001$; Fisher's exact test
number of animals investigated in brackets

For an evaluation of the carcinogenicity of acetaldehyde, refer to the chapter "Acetaldehyde" in Volume 3 of the present series. As acetaldehyde is also a metabolite of ethanol, refer to the chapter "Ethanol" in Volume 12. As a result of the acetaldehyde formed, ethanol was classified in Carcinogenicity category 5.

The local irritative effects of vinyl acetate (on a molar basis) are markedly stronger than those of acetaldehyde. This is explained by the intracellular acidity resulting from metabolically formed acetic acid. Following the intracellular cleavage of vinyl acetate by esterase(s), one molecule of acetic acid is produced simultaneously with one of acetaldehyde, so that vinyl acetate causes markedly stronger intracellular acidification and more rapid exhaustion of the intracellular buffer reserves than acetaldehyde does.

6 Manifesto (MAK value/classification)

In two epidemiological studies with exposure to a mixture of substances, no statistically decisive evidence of carcinogenic effects of vinyl acetate in man was found. No differentiation was made between persons exposed to high levels and those to low levels, however, and the influence of smoking habits was not excluded. The data are therefore not meaningful.

In 2-year drinking water studies with F344 rats and BDF_1 mice, tumours of the oesophagus and oral cavity (rats, mice) and stomach and larynx (mice) were induced at the highest concentration of 10000 mg/l. This shows that local tumours can be induced with high oral exposure to vinyl acetate in the drinking water. An inhalation carcinogenicity study with rats and mice yielded local tumours of the nasal mucosa of the rat at 600 ml/m^3, while in the mouse no tumours were observed. Compared to that of acetaldehyde, the potential of vinyl acetate to produce tumours of the nasal mucosal epithelium of the rat is evidently smaller, as in the long-term inhalation study (28 months) with acetaldehyde at 750 ml/m^3 the incidence of adenocarcinomas of the nasal cavity (Woutersen *et al.* 1986) was much greater than in the long-term study (24 months) with vinyl acetate at 600 ml/m^3. On the other hand, the irritative effects of vinyl acetate are greater than those of acetaldehyde.

The findings in studies of the carcinogenicity of vinyl acetate are explained by the local cytotoxicity resulting from the local metabolism to acetaldehyde and acetic acid on the one hand, and by the genotoxic effects of its metabolite acetaldehyde on the other hand. The hypothesis that the epigenetic effects are more important than the genotoxic effects at present does not seem adequately demonstrated. There is much to indicate, however, that the carcinogenic effects of vinyl acetate are subject to a threshold, below which no notable contribution towards the cancer risk in man is to be expected. This is clear from the non-linear course of the dose–effect relationships in the carcinogenicity studies. There are, however, no studies available which validly predict the amounts of acetaldehyde and acetic acid formed at the site of action for vinyl acetate to be classified in Carcinogenicity category 5. In addition, the local irritation threshold in man must be better investigated. Until the necessary data become available, the substance has been provisionally classified in Carcinogenicity category 3A and the previous MAK value has been withdrawn.

The systemic NOAEL (reduced body weights in mice) from the inhalation carcinogenicity study is 50 ml/m^3 (175 mg/m^3). With 100 % retention this would be a daily dose of about 285 mg/kg body weight. Calculated for a person of 70 kg, this means about 12250 mg. The models of Fiserova-Bergerova *et al.* (1990) and Guy and Potts (1993) predict absorption of 541 and 68 mg, respectively, after exposure to a saturated aqueous solution of 2000 cm^2 for one hour. Thus, in the least favourable case, dermal exposure should lead only to absorption of less than one tenth of the critical amount. The volatility of the substance and the irritative effects are very high and dermal exposure to the undiluted substance for longer periods is therefore unlikely. Designation with an "H" is therefore not necessary.

There is no information available for possible sensitizing effects of vinyl acetate in man. The results of a Bühler test cannot be evaluated, as the possibility of false positive reactions cannot be excluded. There are no data available for the sensitizing effects on the respiratory tract; it is, therefore, not possible to give a definitive evaluation of the sensitizing effects of vinyl acetate. The substance is therefore not designated with an "S".

Vinyl acetate was found to have genotoxic effects *in vitro*, e.g. chromosomal aberrations, micronuclei and SCE, and gene mutations were observed in mammalian cell cultures. *In vivo*, after very high single intraperitoneal doses, micronuclei were observed in the bone marrow cells of mice, but not after inhalation exposure or the administration of vinyl acetate in drinking water. A micronucleus test in germ cells (spermatids) yielded negative results. The substance is therefore not classified as a germ cell mutagen.

7 References

BAuA (Bundesanstalt für Arbeitsschutz und Arbeitsmedizin) (German Federal Institute for Occupational Safety and Health) (2002) *Risk assessment vinyl acetate*, draft of 04.02.2002, BAuA, Dortmund

Bogdanffy MS (2002) Vinyl acetate induced intracellular acidification: implications for risk assessment. *Toxicol Sci 66*: 320–326

Bogdanffy MS, Taylor ML (1993) Kinetics of nasal carboxylesterase-mediated metabolism of vinyl acetate. *Drug Metabol Dispos 21*: 1107–1111

Bogdanffy MS, Tyler TR, Vinegar MB, Rickard RW, Carpanini FMB, Cascieri TC (1994a) Chronic toxicity and oncogenicity study with vinyl acetate in the rat: *in utero* exposure in drinking water. *Fundam Appl Toxicol 23*: 206–214

Bogdanffy MS, Dreef-van der Meulen HC, Beems RB, Ferron VJ, Cascieri TC, Tyler TR, Vinegar MB, Rickard RW (1994b) Chronic toxicity and oncogenicity inhalation study with vinyl acetate in the rat and mouse. *Fundam Appl Toxicol 23*: 215–229

Bogdanffy MS, Gladnick NL, Kegelman T, Frame SR (1997) Four-week inhalation cell proliferation study of the effects of vinyl acetate on rat nasal epithelium. *Inhalat Toxicol 9*: 331–350

Bogdanffy MS, Sarangapani R, Plowchalk DR, Jarabek A, Andersen ME (1999) A biologically based risk assessment for vinyl acetate-induced cancer and noncancer toxicity. *Toxicol Sci 51*: 19–35

Bogdanffy MS, Plowchalk DR, Sarangapani R, Starr TB, Andersen ME (2001) Mode of action-based dosimeters for interspecies extrapolation of vinyl acetate inhalation risk. *Inhalat Toxicol 13*: 377–396

Carpenter CP, Smyth HF, Pozzani UC (1949) The assay of acute vapour toxicity, and the grading and interpretation of results on 96 chemical compounds. *J Ind Hyg Toxicol 31*: 343–346

Fedtke N, Wiegand HJ (1990) Hydrolysis of vinyl acetate in human blood. *Arch Toxicol 64*: 428–429

Fiserova-Bergerova V, Pierce JT, Droz PO (1990) Dermal absorption potential of industrial chemicals: criteria for skin notation. *Am J Ind Med 17*: 617–635

Guy RH, Potts RO (1993) Penetration of industrial chemicals across the skin: a predictive model. *Am J Ind Med 23*: 711–719

Halvorson MR, Noffsinger JK, Peterson CM (1993) Studies of whole blood-associated acetaldehyde levels in teetotalers. *Alcohol 10*: 409–413

He SM, Lambert B (1985) Induction and persistence of SCE-inducing damage in human lymphocytes exposed to vinyl acetate and acetaldehyde *in vitro*. *Mutat Res 158*: 201–208

Korte A, Obe G, Ingwersen I, Rueckert G (1981) Influence of chronic ethanol uptake and acute acetaldehyde treatment on the chromosomes of bone-marrow cells and peripheral lymphocytes of Chinese hamsters. *Mutat Res 88*: 389

Kuykendall JR, Bogdanffy MS (1992) Reaction kinetics of DNA-histone crosslinking by vinyl acetate and acetaldehyde. *Carcinogenesis 13*: 2095–2100

Kuykendall JR, Taylor ML, Bogdanffy MS (1993) Cytotoxicity and DNA–protein crosslink formation in rat nasal tissues exposed to vinyl acetate are carboxylesterase-mediated. *Toxicol Appl Pharmacol 123*: 283–292

Laib RJ, Bolt HM (1986) Vinyl acetate, a structural analog of vinyl carbamate, fails to induce enzyme-altered foci in rat liver. *Carcinogenesis 7*: 841–843

Lambert B, Chen Y, He SM, Sten M (1985) DNA cross-links in human leucocytes treated with vinyl acetate and acetaldehyde *in vitro*. *Mutat Res 146*: 301–303

Maltoni C, Ciliberti A, Lefemine G, Soffritti M (1997) Results of a long-term experimental study on the carcinogenicity of vinyl acetate monomer in mice. *Ann NY Acad Sci 837*: 209–238

Ministry of Labour of Japan (1998) *Toxicology and carcinogenesis studies of vinyl acetate in F 344/DuCr Rats and Crj:BDF$_1$ Mice (drinking water study)*, unpublished

Morita T, Takeda K, Okumura K (1990) Evaluation of clastogenicity of formic acid, acetic acid and lactic acid on cultured mammalian cells. *Mutat Res 240*: 195–202

Morita T, Asano N, Awogi T, Sasaki YF, Sato S-I, Shimada H, Sutou S, Suzuki T, Wakata A, Sofuni T, Hayashi M (1997) Evaluation of the rodent micronucleus assay in the screening of IARC carcinogens (Group 1, 2A and 2B). The summary report of the 6th collaborative study by CSGMT/JEMS MMS. *Mutat Res 389*: 3–122

Morris JB, Symanowicz P, Sarangapani R (2002) Regional distribution and kinetics of vinyl acetate hydrolysis in the oral cavity of the rat and mouse. *Toxicol Lett 126*: 31–99

Norppa H, Tursi F, Pfäffli P, Mäki-Paakkanen J, Järventaus H (1985) Chromosome damage induced by vinyl acetate through *in vitro* formation of acetaldehyde in human lymphocytes and Chinese hamster ovary cells. *Cancer Res 45*: 4816–4821

Ott MG, Teta MJ, Greenberg HL (1989) Lymphatic and hematopoietic tissue cancer in a chemical manufacturing environment. *Am J Ind Med 16*: 631–643

Plowchalk DR, Andersen ME, Bogdanffy MS (1997) Physiologically based modelling of vinyl acetate uptake, metabolism, and intracellular pH changes in the rat nasal cavity. *Toxicol Appl Pharmacol 142*: 386–400

Ristow H, Obe G (1975) Acetaldehyde induces cross-links in DNA and causes sister-chromatid exchanges in human cells. *Mutat Res 58*: 115–119

Takeshita T, Iijama S, Higurashi M (1986) Vinyl-acetate induced sister chromatid exchanges in murine bone marrow cells. *Proc Jpn Acad 62*: 239–242

Union Carbide (1969) *Vinyl acetate, range finding toxicity studies*, Report 32–99, unpublished

Valentine R, Bamberger JR, Szostek B, Frame SR, Hansen JF, Bogdanffy MS (2002) Time- and concentration-dependent increases in cell proliferation in rats and mice administered vinyl acetate in drinking water. *Toxicol Sci 67*: 190–197

Vinyl Acetate Toxicology Group (1995) *Delayed contact hypersensitivity study in guinea pigs (Buehler technique) of: vinyl acetate*, Hill Top Biolabs Inc, Cincinnati, OH, USA, report No. 94-8486-21, unpublished

Waxweiler RJ, Smith AH, Falk H, Tyroler HA (1981) Excess lung cancer risk in a synthetic chemicals plant. *Environ Health Perspect 41*: 159–165

WHO (World Health Organization) (1995) *Acetaldehyde*. IPCS – Environmental health criteria 167, WHO, Geneva

Woutersen RA, Appelman LM, Van Garderen-Hoetmer A, Feron VJ (1986) Inhalation toxicity of acetaldehyde in rats. III. Carcinogenicity study. *Toxicology 41*: 213–231

completed 28.02.2002

Authors of Documents in Volume 21

Professor Dr. Dr. H. Bolt, Institut für Arbeitsphysiologie an der Universität Dortmund, Ardeystraße 67, D-44139 Dortmund

Professor Dr. T. Brüning, Berufsgenossenschaftliches Forschungsinstitut für Arbeitsmedizin (BGFA), Bürkle-de-la-Camp-Platz 1, D-44789 Bochum

Professor Dr. med. E. Hallier, Georg-August-Universität Göttingen, Abteilung für Arbeits- und Sozialmedizin, Waldweg 37, D-37073 Göttingen

Dr. H. Lessmann, Informationsverbund Dermatologischer Kliniken an der Universitäts-Hautklinik, von-Siebold-Straße 3, D-37075 Göttingen

Professor Dr.rer.nat. H.-G. Neumann, Peter-Haupt-Straße 27, D-97080 Würzburg

Frau Dr. I. Neumann, Senatskommission der DFG zur Prüfung gesundheitsschädlicher Arbeitsstoffe, Technische Universität München, D-85350 Freising-Weihenstephan

Dr. E. Nies, Berufsgenossenschaftliches Institut für Arbeitssicherheit (BIA), Alte Heerstraße 111, D-53757 St. Augustin

Frau Dr. U. Reuter, Senatskommission der DFG zur Prüfung gesundheitsschädlicher Arbeitsstoffe, Technische Universität München, D-85350 Freising-Weihenstephan

Professor Dr. A. Seeber, Thierschweg 14, D-44141 Dortmund

Professor Dr. med. K. Straif, PhD, Institut für Epidemiologie und Sozialmedizin der Universität Münster, Domagkstraße 3, D-48129 Münster

Professor Dr. M. Wilhelm, Institut für Hygiene, Sozial- und Umweltmedizin, Universitätsstraße 150, D-44801 Bochum

Frau Dr. K. Ziegler-Skylakakis, Senatskommission der DFG zur Prüfung gesundheitsschädlicher Arbeitsstoffe, Technische Universität München, D-85350 Freising-Weihenstephan

Index for Volumes 1–21

6-12® **V** 345
abachi **XIII** 288, **XVIII** 283
absolute ethanol **XII** 129
Abstensil **V** 165
Abstinyl® **V** 165
Acacia melanoxylon **XIII** 285, 291–293
acetaldehyde **III** 1–9
acetene **X** 91
acetic acid 2-butoxyethyl ester **VI** 53
acetic acid butyl ester **XIX** 79
acetic acid *sec*-butyl ester **XIX** 89
acetic acid *tert*-butyl ester **XIX** 93
acetic acid 1,1-dimethylethyl ester **XIX** 93
acetic acid ethenyl ester **V** 229
acetic acid 2-ethoxyethyl ester **VI** 213
acetic acid ethyl ester **XII** 167
acetic acid isobutyl ester **XIX** 211
acetic acid 2-methoxy-1-methylethyl ester **V** 217
acetic acid methyl ester **XVIII** 191
acetic acid 1-methylpropyl ester **XIX** 89
acetic acid 2-methylpropyl ester **XIX** 211
acetic acid pentyl ester **XI** 211
acetic acid 2-propoxyethyl ester **XII** 187
acetic acid vinyl ester **V** 229, **XXI** 271
acetic anhydride **XIII** 43–46
acetic ether **XII** 167
acetic oxide **XIII** 43
acetic peroxide **VII** 229
acetone **VII** 1–8
acetonitrile **XIX** 1–41
1-acetoxyethylene **V** 229
2-acetoxypentane **XI** 211
acetyl hydroperoxide **VII** 229
acetyl oxide **XIII** 43
aclarubicin **I** 5
acquinite **VI** 105
acraldehyde **XVI** 1
acrolein **I** 41–43, 60, **XVI** 1–33
acrylaldehyde **XVI** 1
acrylamide **III** 11–21
acrylic acid *n*-butyl ester **V** 5, **XII** 57, **XVI** 35
acrylic acid ethyl ester **VI** 217, **XVI** 41
acrylic acid 2-ethylhexyl ester **XVI** 47
acrylic acid 2-hydroxyethyl ester **XVI** 89
acrylic acid hydroxypropyl ester **XVI** 95
acrylic acid methyl ester **VI** 253, **XVI** 177
acrylic acid monoester with propanediol **XVI** 95

acrylic acid pentaerythritol triester **XVI** 193
acrylic acid polymer, neutralized, cross-linked **XV** 1–29
acrylic acid 1,1,1-(trihydroxymethyl)propane triester **XVI** 201
acrylic aldehyde **XVI** 1
acrylic amide **III** 11
Acticide® 45 **XVI** 263
actinolite **II** 96, 184, 185
actinomycin D **I** 5
adriamycin **I** 5
AEPD® **IX** 223
aeropur® **I** 125
aerosols **XII** 271–292, **XVI** 289
afara **XIII** 288
African acajou **XVIII** 239
African afzelia **XIII** 285
African black walnut **XIV** 299
African blackwood **XIII** 286, 305
AfricaN'cherry' **XIII** 288
African ebony **XIII** 286, **XIV** 287
African mahogany **XIII** 285, 286, 287, **XVIII** 239
African maple **XIII** 288, **XVIII** 283
African satinwood **XIII** 286, **XIV** 291
African whitewood **XIII** 288, **XVIII** 283
Afrormosia elata **XIII** 287
Afzelia spp. **XIII** 285
AGE **VII** 9
alabaster **II** 117
alcohol **XII** 129
allyl alcohol **XV** 31–40
allyl chloride **XVIII** 1–18
allyl 2,3-epoxypropyl ether **VII** 9
allyl glycidyl ether **VII** 9–16
1-allyloxy-2,3-epoxypropane **VII** 9
allyl trichloride **IX** 171
altretamine **I** 2
aluminium **II** 69–93
aluminium hydroxide **II** 69–93
aluminium oxide **II** 69–93, **VIII** 141–338
Amazon mahogany **XVIII** 253
American arborvitae **XIII** 288
American black walnut **XIII** 286
American mahogany **XIII** 287, **XVIII** 253
American red oak **XVIII** 247
American walnut **XIII** 286
Amerimnum ebenus **XIII** 285, 299
Amine 220® **V** 373

Amine CS-1246 **IX** 275
aminic acid **XIX** 169
2-aminoaniline **XIII** 215
3-aminoaniline **VI** 287
4-aminoaniline **VI** 311
m-aminoaniline **VI** 287
o-aminoaniline **XIII** 215
p-aminoaniline **VI** 311
2-aminoanisole **X** 1
o-aminoanisole **X** 1
aminobenzene **VI** 17
4-(4-aminobenzyl)aniline **VII** 37
4-aminobiphenyl **I** 257–259
6-aminocaproic acid lactam **IV** 65
1-amino-2-chlorobenzene **III** 31
1-amino-3-chlorobenzene **III** 37
1-amino-4-chlorobenzene **III** 45
m-aminochlorobenzene **III** 37
1-amino-3-chloro-6-methylbenzene **VI** 143
2-amino-4-chlorotoluene **VI** 143
2-amino-5-chlorotoluene **VI** 127
3-amino-*p*-cresol methyl ether **IV** 135
m-amino-*p*-cresol methyl ether **IV** 135
aminodiglycol **IX** 215
amino-dimethyl-benzene isomers **XIX** 299
2-aminodimethylethanol **IX** 229
2-aminoethanol **XII** 15–35
2-(2-aminoethoxy)ethanol **IX** 215–222
2-aminoethoxyethanol **IX** 215
2-amino-6-ethoxynaphthalene **VII** 17
6-amino-2-ethoxynaphthalene **VII** 17–19
β-aminoethyl alcohol **XII** 15
3-amino-9-ethylcarbazole **V** 1–3
3-amino-*N*-ethylcarbazole **V** 1
2-amino-2-ethyl-1,3-propanediol **IX** 223–227
Aminoform **V** 355
aminoglutethimide **I** 2
6-aminohexanoic acid cyclic lactam **IV** 65
2-aminoisobutanol **IX** 229
beta-aminoisobutanol **IX** 229
alpha-aminoisopropylalcohol **IX** 237
aminomethane **VII** 145
1-amino-2-methoxybenzene **X** 1
1-amino-2-methoxy-5-methylbenzene **IV** 135
3-amino-4-methoxytoluene **IV** 135
2-amino-4-methylanisole **IV** 135
1-amino-2-methylbenzene **III** 307
1-amino-4-methylbenzene **III** 323
2-amino-1-methylbenzene **III** 307
4-amino-1-methylbenzene **III** 323
1-amino-2-methyl-5-nitrobenzene **VI** 271
2-amino-2-methyl-1-propanol **IX** 229–236
amino-methyl-toluene isomers **XIX** 299
6-aminonaphthol ether **VII** 17

4-amino-2-nitroaniline **IV** 295, **XIII** 207
4-amino-2-nitrophenol **IV** 289
2-amino-4-nitrotoluene **I** 167, **VI** 271, **XXI** 3
p-aminophenyl ether **VI** 277
4-aminophenyl ether **VI** 277
1-amino-2-propanol **IX** 237–244
2-aminotoluene **III** 307
4-aminotoluene **III** 323
o-aminotoluene **III** 307
p-aminotoluene **III** 323
3-amino-*p*-toluidine **VI** 339
5-amino-*o*-toluidine **VI** 339
aminotriazole **XVIII** 19
2-amino-1,3,4-triazole **IV** 1
3-amino-1,2,4-triazole **IV** 1
3-amino-1*H*-1,2,4-triazole **IV** 1
1-amino-2,4,5-trimethylbenzene **IV** 335
aminoxylene isomers **XIX** 299
1,2-aminozophenylene **II** 231
amitrole **IV** 1–10, **XVIII** 19–34
Ammoform **V** 355
ammonia **I** 40, **VI** 1–16, **XIII** 47–48
ammonium molybdate **XVIII** 199
amosite **II** 95–116, 188
AMP® **IX** 229
AMS **XV** 137
1-amyl acetate **XI** 211
n-amyl acetate **XI** 211
sec-amyl acetate **XI** 211
tert-amyl acetate **XI** 211
α-amylase **XI** 1–3
amylcarbinol **IX** 283
anaesthetic ether **XIII** 149
anatase **II** 199
anhydrite **II** 117
anhydrous hydrobromic acid **XIII** 187
aniline **I** 40, **III** 32, 42, 50, 151, 152, 159, 160, **IV** 132, **VI** 17–36, **XXI** 3
anilinomethane **VI** 263
Aningeria spp. **XIII** 285
aningré **XIII** 285
2-anisidine **X** 1
o-anisidine **X** 1–13, **XXI** 3
p-anisidine **XXI** 3
anone **X** 35
anprolene **V** 181
Antabuse® **V** 165
Antadix **V** 165
anthophyllite **II** 96, 182, 183, 187
anthracite dust **XVIII** 107
apyonine auramine base **IV** 12
aqua fortis **III** 233
aquinite **VI** 105
arborvitae **XIII** 288, **XVIII** 263

arsenic acid **XXI** 49
arsenic and its inorganic compounds **XXI** 49–106
arsenic compounds **II** 48, 136
arsenic pentoxide **XXI** 49
arsenic trihydride **XV** 41
arsenic trioxide **XXI** 49
arsenic(lll) acid **XXI** 49
arsenic(lll) oxide **XXI** 49
arsenic(V) acid **XXI** 49
arsenic(V) oxide **XXI** 49
arsenous acid **XXI** 49
arsine **XV** 41–50
artificial almond oil **XVII** 13
artificial oil of ants **XVIII** 189
Arubren CP® **III** 81
asbestos **I** 28, **II** 48, 95–116, 136, 181–198, **VIII** 141–338
ASP 47 **XIII** 237
Aspalatus ebenus **XIII** 285, 299
asphalt **XVII** 37–118
Asulgan® K **V** 289
asymmetrical dichloroethylene **VIII** 109
atmospheric pressure **XIV** 11–16
attapulgite **VIII** 141–338
Aucoumea klaineana **XIII** 285
auramine **I** 250, **IV** 11–25
auramine base **IV** 11–25
Australian blackwood **XIII** 285, 291
Australian silk(y) oak **XIII** 286, **XIV** 295
avodiré **XIII** 288
ayan **XIII** 286, **XIV** 291
ayous **XVIII** 283
1-aza-2-cycloheptanone **IV** 65
5-azacytidine **I** 4
1-aza-3,7-dioxa-5-ethylbicyclo[3.3.0]octane **IX** 275
azathioprine **I** 4
azimethylene **XIII** 141
azimidobenzene **II** 231
aziminobenzene **II** 231
azotic acid **III** 233
BADGE **XIX** 43
barium chromate **III** 101–122, **VII** 33
basic yellow 2 **IV** 11
basic zinc chromate **XV** 289
battery acid **XV** 165
Bayer E393 **XIII** 237
Baywood **XIII** 287, **XVIII** 253
α-BCH **V** 193
β-BCH **V** 193
BCNU **I** 2
beech wood dust **IV** 363, **XVIII** 221
Beisugi **XVIII** 263

Belize mahogany **XVIII** 253
benzal chloride **VI** 79
benzaldehyde **XVII** 13–36
benzenamine **VI** 17
benzene azimide **II** 231
benzenecarbonyl chloride **VI** 80
benzene chloride **XII** 103
1,2-benzenediamine **XIII** 215
1,3-benzenediamine **VI** 287
1,4-benzenediamine **VI** 311, **XIV** 137
m-benzenediamine **VI** 287
o-benzenediamine **XIII** 215
p-benzenediamine **VI** 311
benzene-1,2-dicarboxylic acid **VII** 241
benzene-1,3-dicarboxylic acid **VII** 241
benzene-1,4-dicarboxylic acid **VII** 241
1,2-benzenedicarboxylic acid anhydride **VII** 247
1,2-benzenedicarboxylic acid di-2-propenyl ester **IX** 11
1,3-benzenediol **XX** 239
1,4-benzenediol **X** 113
p-benzenediol **X** 113
α-benzene hexachloride **V** 193
β-benzene hexachloride **V** 193
benzenyl chloride **VI** 80
benzenyl trichloride **VI** 80
benzidine **I** 258–259
1,2-benzisothiazol-3(2*H*)-one **II** 221–230
benzohydroquinone **X** 113
benzoic acid chloride **VI** 80
benzoic aldehyde **XVII** 13
benzoic trichloride **VI** 80
benzoisotriazole **II** 231
benzo[*a*]pyrene **I** 41–43, 58–60, 112, 180
1,2,3-benzotriazole **II** 231
1*H*-benzotriazole **II** 231–238
1*H*-benzotriazole, methyl **II** 295
benzotrichloride **VI** 80
2*H*-3,1-benzoxazine-2,4-[1*H*]-dione **XIII** 97
benzoyl chloride **VI** 80
(benzoyloxy)tributylstannane **I** 315
benzoyl peroxide **III** 249–256
benztriazole **II** 231
benzyl alcohol mono(poly)hemiformal **II** 239–247
benzyl disulfide **II** 283
α-(benzyldithio)toluene **II** 283
benzylene chloride **VI** 79
benzyl hemiformal **II** 239
benzyl hydroxymethyl ether **II** 239
benzylidene chloride **VI** 79
benzylidyne chloride **VI** 80
benzyl oxy methanol **II** 239

benzyl trichloride **VI** 80
bertrandite **XXI** 107
beryl **XXI** 107
beryllium **III** 23–29
beryllium acetate **XXI** 107
beryllium and its inorganic compounds **XXI** 107–160
beryllium carbonate **XXI** 107
beryllium chloride **XXI** 107
beryllium citrate **XXI** 107
beryllium compounds **III** 23–29
beryllium fluoride **XXI** 107
beryllium hydroxide **XXI** 107
beryllium nitrate **XXI** 107
beryllium oxide **XXI** 107
beryllium silicate **XXI** 107
beryllium stearate **XXI** 107
beryllium sulfate **XXI** 107
bété **XIII** 287, **XIV** 299
N,N'-bianiline **IX** 83
bianisidine **V** 153
bicarburetted hydrogen **X** 91
1,2-bichloroethane **III** 137
bicyclopentadiene **V** 125
biethylene **XV** 51
big-leaf mahogany **XIII** 287
binitrobenzene **I** 147
Bioban® CS-1246 **IX** 275
Bioban® P-1487 **IX** 295
[1,1'-biphenyl]-2-ol **II** 299
2-biphenylol **II** 299
o-biphenylol **II** 299
[1,1'-biphenyl]-2-ol sodium salt **II** 299
2-biphenylol sodium salt **II** 299
(2-biphenyloxy)sodium **II** 299
[1,1'-biphenyl]-3,3',4,4'-tetramine **III** 131
3,3',4,4'-biphenyltetramine **III** 131
bis(4-amino-3-chlorophenyl)methane **VII** 193
bis(4-aminophenyl)ether **VI** 277
bis(4-aminophenyl)methane **VII** 37
bis(p-aminophenyl)methane **VII** 37
bis(4-aminophenyl)sulfide **IV** 331
bis(p-aminophenyl)sulfide **IV** 331
2,6-bis($tert$-butyl)phenol **V** 341
bis[(3-carboxyacryloyl)oxy]dioctylstannane diisooctyl ester **VII** 93
bis(β-chloroethyl)methylamine **I** 211
N,N-bis(2-chloroethyl)-N-methylamine **I** 211
bischloroethyl nitrosourea (BCNU) **I** 2
bis(2-chloroethyl)sulfide **IV** 27
bis(β-chloroethyl)sulfide **IV** 27–43
biscyclopentadiene **V** 125
bis(3,5-dichloro-2-hydroxyphenyl) sulfide **IX** 245

bis((diethylamino)thioxomethyl)disulfide **V** 165
bis(diethylthiocarbamoyl) disulfide **V** 165
bis(diethylthiocarbamyl) disulfide **V** 165
bis(N,N-diethylthiocarbamyl) disulfide **V** 165
4,4'-bis(dimethylamino)diphenylmethane **I** 249
p,p'-bis(dimethylamino)diphenylmethane **I** 249
bis(p-(dimethylamino)phenyl)methane **I** 249
bis(p-dimethylaminophenyl)methyleneimine **IV** 11
2,6-bis(1,1-dimethylethyl)phenol **V** 341
bis(dimethylthiocarbamoyl) disulfide **XV** 163
bis(dimethylthiocarbamyl) disulfide **XV** 163
1,3-bis(2,3-epoxypropoxy)benzene **VII** 69
m-bis(2,3-epoxypropoxy)benzene **VII** 69
2,2-bis(4-(2,3-epoxypropoxy)phenyl)propane **XIX** 43
m-bis(glycidyloxy)benzene **VII** 69
bis(4-glycidyloxyphenyl)dimethylmethane **XIX** 43
2,2-bis(p-glycidyloxyphenyl)propane **XIX** 43
bis(2-hydroxy-3,5-dichlorophenyl) sulfide **IX** 245
bis(2-hydroxyethyl) ether **X** 73
bis(2-hydroxyethyl)methylamine **IX** 291
bis(hydroxymethyl)acetylene **XV** 71
1,3-bis(hydroxymethyl)urea **V** 289–293
N,N'-bis(hydroxymethyl)urea **V** 289
bis(4-hydroxyphenyl)dimethylmethane diglycidyl ether **XIX** 43
2,2-bis(4-hydroxyphenyl)propane **XIII** 49
2,2-bis(4-hydroxyphenyl)propane diglycidyl ether **XIX** 43
bis(1,4-isocyanatophenyl)methane **VIII** 65
bis(p-isocyanatophenyl)methane **VIII** 65
bis(isooctyloxymaleoyloxy)dioctylstannane **VII** 93
bis(2-methoxyethyl)ether **IX** 41
bisphenol A **XIII** 49–87
bisphenol A diglycidyl ether **XIX** 43–78
bis(phenylmethyl)disulfide **II** 283
bis(tri-n-butyltin)oxide **I** 315
bithionol **IX** 245–262
bithionol sulfide **IX** 245
4,4'-bi-o-toluidine **V** 153
bitumen **XVII** 37–118
bituminous coal dust **XVIII** 107
bivinyl **XV** 51
black afara **XVIII** 259
black coal dust **XVIII** 107
black ebony **XIII** 286
black sucupira **XIII** 285, 295

black walnut **XIII** 286
black wattle **XIII** 285, 291
Bladafum® **XIII** 237
bleomycin **I** 5
BNPD **II** 249
body burden and hyperbaric pressure **XIV** 11–16
Bombay blackwood **XIII** 285, 301
boracic acid **V** 295
boric acid **V** 295–321
boric anhydride **XIII** 89
boric oxide **XIII** 89
boroethane **XIV** 83
Borofax® **V** 295
boron fluoride **XIII** 93
boron oxide **XIII** 89–91
boron sesquioxide **XIII** 89
boron trifluoride **XIII** 93–95
boron trioxide **XIII** 89
Bowdichia nitida **XIII** 285, 295–297
Brazilian mahogany **XVIII** 253
Brazilian rosewood **XIII** 286, 309
brilliant oil yellow **IV** 11
broadleaf mahogany **XIII** 287, **XVIII** 253
bromic ether **VII** 115
2-bromo-2-(bromomethyl)glutaronitrile **XIV** 87
2-bromo-2-(bromomethyl)pentanedinitrile **XIV** 87
bromoethane **VII** 115
bromofluoroform **VI** 37
bromoform **VII** 319
bromomethane **VII** 155
2-bromo-2-nitro-1,3-propanediol **II** 249–270
β-bromo-β-nitrotrimethylene glycol **II** 249
bromotrifluoromethane **VI** 37–45
Bronopol® **II** 249
brookite **II** 199
brown ebony **XIII** 285, 299
brucite **VIII** 141–338
Brya ebenus **XIII** 285, 299–300
busulfan **I** 2
α,γ-butadiene **XV** 51
1,3-butadiene **XV** 51–70
butadiene diamer **XIV** 185
butane **XX** 1–9
n-butane **XX** 1
butanesulfone **IV** 45
δ-butane sultone **IV** 45
1,4-butane sultone **IV** 45–49
2,4-butane sultone **IV** 45–49
butanethiol **XXI** 161–170
1-butanol **XIX** 99
2-butanol **XIX** 117

2-butanone **XII** 37–56
2-butanone peroxide **III** 250, 252, 254, 255
cis-butenedioic anhydride **IV** 275
1-butene oxide **V** 13
1,2-butene oxide **V** 13
3-buten-2-one **IX** 91
butoxydiethylene glycol **VII** 59
butoxydiglycol **VII** 59
1-*n*-butoxy-2,3-epoxypropane **IV** 51
1-*tert*-butoxy-2,3-epoxypropane **IV** 51
2-butoxyethanol **V** 210, **VI** 47–52
n-butoxyethanol **VI** 47
2-butoxyethanol acetate **VI** 53
2-(2-butoxyethoxy)ethanol **VII** 59
2-[2-(2-butoxyethoxy)ethoxy]ethanol **IX** 315
2-butoxyethyl acetate **VI** 53–55
butoxymethyl oxirane **IV** 51
tert-butoxymethyl oxirane **IV** 51
butoxytriethylene glycol **IX** 315
butoxytriglycol **IX** 315
buttercup yellow **XV** 289
2-butyl acetate **XIX** 89
n-butyl acetate **XIX** 79–88
sec-butyl acetate **XIX** 89–92
tert-butyl acetate **XIX** 93–97
n-butyl acrylate **V** 5–12, **XII** 57–62, **XVI** 35–40
n-butyl alcohol **XIX** 99–115
sec-butyl alcohol **XIX** 117–124
tert-butyl alcohol **XIX** 125–140
butyl carbamic acid 3-iodo-2-propynyl ester **XVI** 247
butyl carbitol **VII** 59
Butyl Cellosolve® **VI** 47
Butyl Cellosolve® acetate **VI** 53
O-butyl diethylene glycol **VII** 59
butyl diglycol **VII** 59
butyl dioxitol **VII** 59
1,2-butylene oxide **V** 13–18
α-butylene oxide **V** 13
1,4-butylene sulfone **IV** 45
O-butyl ethylene glycol **VI** 47
n-butyl glycidyl ether **IV** 51–64
tert-butyl glycidyl ether **IV** 51–64
butyl glycol **VI** 47
butyl glycol acetate **VI** 53
tert-butyl hydroperoxide **III** 250, 253, 254
butyl-3-iodo-2-propynylcarbamate **XVI** 247
N-butyl mercaptan **XXI** 161
tert-butyl methyl ether **XVII** 119–145
tert-butyl peracetate **III** 250
tert-butyl peroxyacetate **III** 250
4-*tert*-butylphenol **XI** 5
p-*tert*-butyl phenol **XI** 5–25

butyl 2-propenoate **V** 5, **XII** 57
n-butyl triethylene glycol **IX** 315
n-butyl triglycol **IX** 315
butynediol **XV** 71–74
2-butyne-1,4-diol **XV** 71
2-butyne-2,4-diol **XVI** 233
cadmium **I** 40, **V** 19–50
cadmium compounds **V** 19–50, **VII** 21
cadmium sulfide **VII** 21–25
calcium carbimide **V** 51
calcium cyanamide **V** 51–64
calcium molybdate **XVIII** 199
calcium sodium metaphosphate **VIII** 141–338
calcium sulfate **II** 117–128, **VIII** 141–338
Californian redwoood **XIII** 287
Calocedrus decurrens **XIII** 285, **XIV** 275–278
CAM **IX** 263
Cameroon ebony **XIV** 287
camphechlor **XIX** 281
camphochlor **XIX** 281
6-caprolactam **IV** 65
ε-caprolactam **IV** 65–78
caproyl alcohol **IX** 283
caprylic alcohol **XX** 227
captan **I** 292, 295
Carbamol® **V** 289
carbanil **XVII** 267
carbazotic acid **XVII** 273
carbinamine **VII** 145
carbinol **XVI** 143
carbomethene **XX** 191
carbon bichloride **III** 271
carbon bisulfide **XII** 63, **XXI** 171
carbon black **XVIII** 35–80
carbon dichloride **III** 271
carbon disulfide **XII** 63–79, **XXI** 171–185
carbon monoxide **I** 40, 42–43, 109, 111, 119, **IV** 79-95, 174, 188
carbon oxide **IV** 79
carbonic oxide **IV** 79
4,4′-carbonimidoylbis(*N*,*N*-dimethylbenzenamine) **IV** 11, 12
4,4′-carbonimidoylbis[*N*,*N*-dimethylbenzenamine] monochloride **IV** 11
carbon sulfide **XII** 63, **XXI** 171
carbon tetrachloride **XVIII** 81–106
Carbowax® **X** 247
N-carboxyanthranilic anhydride **XIII** 97–98
Caribbean mahogany **XIII** 287, **XVIII** 253
carmustine **I** 2
caustic soda **XII** 195
caviuna **XIII** 287, 321

CCNU **I** 2
CDM **VI** 57
cedar of Lebanon **XIII** 285
Cedrus spp. **XIII** 285
Cellosolve **VI** 205
cellosolve acetate **VI** 213
ceramic fibres **VIII** 141–338
cereal flour dusts **XIII** 99–100
Cereclor® **III** 81
Ceylonese ebony **XIII** 286, **XIV** 287
CFC 12 **V** 109
CFC 13 **I** 69
CFC 21 **V** 119
CFC 31 **V** 75
CFC 112 **I** 297, **III** 366
CFC 142b **I** 65
'cherry mahogany' **XIII** 288
chlorallylene **XVIII** 1
chlorambucil **I** 2
chlordimeform **VI** 57–78
chlorfenamidine **VI** 57
chlorinated camphene **XIX** 281
chlorinated hydrocarbon waxes **III** 81
chlorinated naphthalenes **XIII** 101–110
chlorinated paraffins **III** 81, **VII** 27–32
chlorinated paraffin waxes **III** 81
α-chlorinated toluenes **VI** 79–103
chlormethine **I** 211
chloroacetaldehyde **XII** 81–102
2-chloroacetaldehyde **XII** 81
chloroacetaldehyde monomer **XII** 81
chloroacetamide-*N*-methylol **IX** 263–267
chloroacetic acid methyl ester **IX** 1–8
2-chloroacrylonitrile **X** 15–19
4-chloro-2-aminotoluene **VI** 143
5-chloro-2-aminotoluene **VI** 127
2-chloroaniline **III** 31
3-chloroaniline **III** 37
4-chloroaniline **III** 45
m-chloroaniline **III** 37–43, **XXI** 3
o-chloroaniline **III** 31–35, 42, **XXI** 3
p-chloroaniline **III** 42, 45–61, **IV** 132, **XXI** 3
chlorobenzal **VI** 79
α-chlorobenzaldehyde **VI** 80
2-chlorobenzenamine **III** 31
3-chlorobenzenamine **III** 37
4-chlorobenzenamine **III** 45
chlorobenzene **XII** 103–127
chlorobenzol **XII** 103
p-chlorobenzotrichloride **X** 21–30
chlorocamphene **XIX** 281
2-chloro-*N*-(2-chloroethyl)-*N*-methylethanamine **I** 211
1-chloro-2-(β-chloroethylthio)ethane **IV** 27

p-chlorocresol **II** 271
p-chloro-m-cresol **II** 271–282
chlorodeoxyglycerol **V** 83
1-chloro-2,2-dichloroethylene **X** 201
chlorodifluoroethane **I** 65
1-chloro-1,1-difluoroethane **I** 65–67
chlorodifluoromethane **I** 69, **III** 63–71
2-chloro-2-(difluoromethoxy)-1,1,1-trifluoro-
   ethane **VII** 127
2-chloro-1-(difluoromethoxy)-1,1,2-trifluoro-
   ethane **IX** 51
1-chloro-2,3-dihydroxypropane **V** 83
3-chloro-1,2-dihydroxypropane **V** 83
1-chloro-2,4-dinitrobenzene **XIII** 111–115,
   XXI 3
2-chloro-1-ethanal **XII** 81
chloroethane **III** 73–79
2-chloroethanol **V** 65–73
β-chloroethanol **V** 65
chloroethene **V** 66, 241
chloroethene homopolymer **II** 149
2-chloroethyl alcohol **V** 65
1-(2-chloroethyl)-3-cyclohexyl-1-nitrosourea
   I 2
chloroethylene **V** 241
chloroethylidene fluoride **I** 65
chlorofluoromethane **III** 63, 69, 70, **V** 75–77
chloroform **IV** 173, **XIV** 19–58
N-chloroformylmorpholine **V** 79–81
chlorohydric acid **VI** 231
α-chlorohydrin **V** 83–98
2-chloro-N-(hydroxymethyl)acetamide **IX**
   263
6-chloro-3-hydroxytoluene **II** 271
γ-chloroisobutylene **IV** 97
chloromethane **I** 69, **VII** 173
3-chloro-6-methylaniline **VI** 143
4-chloro-2-methylaniline **VI** 127
5-chloro-2-methylaniline **VI** 143
4-chloro-2-methylbenzenamine **VI** 127
5-chloro-2-methylbenzenamine **VI** 143
(chloromethyl)benzene **VI** 79
4-chloro-2-methylbenzeneamine **VI** 127
5-chloro-2-methyl-2,3-dihydroisothiazol-
   3-one **V** 323–339
5-chloro-2-methyl-4-isothiazolin-3(2H)-one
   V 323
5-chloro-2-methyl-3-isothiazolone **V** 323
4-chloro-3-methylphenol **II** 271
N′-(4-chloro-2-methylphenyl)-N,N-dimethyl-
   methanimidamide **VI** 57
3-chloro-2-methylpropene **IV** 97–106
3-chloro-2-methyl-1-propene **IV** 97

1-chloro-2-nitrobenzene **IV** 107
1-chloro-3-nitrobenzene **IV** 115
1-chloro-4-nitrobenzene **IV** 121
2-chloro-1-nitrobenzene **IV** 107
3-chloro-1-nitrobenzene **IV** 115
4-chloro-1-nitrobenzene **IV** 121
m-chloronitrobenzene **IV** 107, 115–120, 121,
   XXI 3
o-chloronitrobenzene **IV** 107–114, 115, 121,
   XXI 3
p-chloronitrobenzene **IV** 107, 115, 121–133,
   XXI 3
1-chloro-1-nitropropane **XI** 27–29
chloroparaffins **III** 81–100, **VII** 27
chlorophenamidine **VI** 57
2-chlorophenylamine **III** 31
3-chlorophenylamine **III** 37
4-chlorophenylamine **III** 45
m-chlorophenylamine **III** 37
o-chlorophenylamine **III** 31
p-chlorophenylamine **III** 45
p-chlorophenyl chloride **IV** 141, **XX** 33
chlorophenylmethane **VI** 79
Chlorophora excelsa **XIV** 279–285
Chlorophora spp. **XIII** 285
chloropicrin **VI** 105–111
chloroprene **III** 205
1-chloro-2,3-propanediol **V** 83
3-chloro-1,2-propanediol **V** 83
3-chloro-1-propene **XVIII** 1
2-chloro-2-propenenitrile **X** 15
3-chloropropylene **XVIII** 1
3-chloropropylene glycol **V** 83
chlorothalonil **VI** 113–126
α-chlorotoluene **VI** 79
ω-chlorotoluene **VI** 79
4-chloro-2-toluidine **VI** 127
4-chloro-o-toluidine **VI** 127–141, 144, **XXI** 3
5-chloro-o-toluidine **VI** 143–144, **XXI** 3
N′-(4-chloro-o-tolyl)-N,N-dimethyl-
   formamidine **VI** 57
1-chloro-4-(trichloromethyl)benzene **X** 21
1-chloro-2,2,2-trifluoroethyldifluoromethyl
   ether **VII** 127
2-chloro-1,1,2-trifluoroethyldifluoromethyl
   ether **IX** 51
chlorotrifluoromethane **I** 69–70
Chlorowax® **III** 81
chlorphenamidine **VI** 57
chlorpromazine **IX** 9–10
chlorvinphos **IV** 201
chromates **III** 101–122
chromic acid zinc salt **XV** 289

chromium compounds **III** 101–122
chromium(VI) compounds **II** 47, 48, **III** 101–122, **VII** 33–35
chromium potassium zinc oxide **XV** 289
chrysotile **I** 28, **II** 95–116, 182, 184, **VIII** 141–338
C.I. 23060 **V** 99
C.I. 24110 **V** 139
C.I. 37105 **VI** 271
C.I. 37230 **V** 153
C.I. 41000 **IV** 11
C.I. 76000 **VI** 17
C.I. 76025 **VI** 287
C.I. 76035 **VI** 339
C.I. 76050 **VI** 145
C.I. 76060 **VI** 311
C.I. 76070 **IV** 295, **XIII** 207
C.I. 76555 **IV** 289
C.I. 77947 **XVIII** 305
C.I. azoic red 83 **IV** 135
cigarette smoke **I** 39–62, **XIII** 3–39
cinnamene **XX** 285
C.I. oxidation base 22 **IV** 295, **XIII** 207
C.I. pigment 4 **XVIII** 305
C.I. pigment yellow 36 **XV** 289
cisplatin **I** 6
citric acid **XVI** 209–230
citron yellow **XV** 289
classification of carcinogenic chemicals **XII** 3–12
Cloparin® **III** 81
coal mine dust **XVIII** 107–156
coastal redwood **XIII** 287
cobalt **III** 123–130, **X** 31–34
cobalt compounds **III** 123–130, **X** 31–34
Cobratec® 99 **II** 231
cocobolo **XIII** 286, 313
cocuswood **XIII** 285, 299
colloidal mercury **XV** 81
colophony **XI** 239
common oak **XVIII** 247
Compound 469 **VII** 127
Contralin **V** 165
coromandel **XIII** 286, **XIV** 287
p-cresidine **IV** 135–139
o-cresol **XIV** 59
m-cresol **XIV** 59
p-cresol **XIV** 59
cresol (all isomers) **XIV** 59–82
o-cresylic acid **XIV** 59
m-cresylic acid **XIV** 59
p-cresylic acid **XIV** 59
cristobalite **II** 182, **XIV** 205
crocidolite **II** 95–116, **VIII** 141–338

Cronetal **V** 165
cross-linked polyacrylates **XV** 1
crystalline silicon dioxide **XIV** 205–271
Cuban mahogany **XIII** 287, **XVIII** 253
cumene **XIII** 117–128
cumene hydroperoxide **III** 250, 253–255
cumol **XIII** 117
cyanamide, calcium salt (1:1) **V** 51
cyanoacrylates **XIII** 201
2-cyanoacrylic acid ethyl ester **I** 233, **XIII** 201
2-cyanoacrylic acid methyl ester **I** 233, **XIII** 201
cyanomethane **XIX** 1
2-cyano-2-propenoic acid ethyl ester **I** 233
2-cyano-2-propenoic acid methyl ester **I** 233
cyclohexane **XIII** 129–140
cyclohexanone **X** 35–51
cyclohexanone iso-oxime **IV** 65
cyclohexenylethylene **XIV** 185
cyclohexyl ketone **X** 35
1,3-cyclopentadiene dimer **V** 125
cyclophosphamide **I** 2
Cystamin **V** 355
Cystogen **V** 355
cytarabine **I** 4
2,4-D **IV** 191, **XI** 61
2,4-DAA **VI** 145
dacarbazine **I** 6
DACPM **VII** 193
dactinomycin **I** 5
Dalbergia latifolia **XIII** 285, 301–303
Dalbergia melanoxylon **XIII** 286, 305–307
Dalbergia nigra **XIII** 286, 309–311
Dalbergia retusa **XIII** 286, 313–316
Dalbergia stevensonii **XIII** 286, 317–319
DAPM **VII** 37
daunomycin **I** 5
daunorubicin hydrochloride **I** 5
dawsonite **VIII** 141–338
DBDCB **XIV** 87
1,2-DCE **III** 137
DDVP **IV** 201
decyl 9-octadecenoate **IX** 269
decyl oleate **IX** 269–274
demeton **XIX** 141–149
demeton-O **XIX** 141
demeton-S **XIX** 141
deodar cedar **XIII** 285
DGA® **IX** 215
DGEBA **XIX** 43
DGEBPA **XIX** 43
diallyl phthalate **IX** 11–19
diamide **I** 171

diamide monohydrate **XIII** 181
diamine **I** 171
2,4-diamineanisole **VI** 145
2,4-diaminoanisole **VI** 145–156, **XXI** 3
m-diaminoanisole **VI** 145
2,4-diaminoanisole base **VI** 145
2,4-diaminoanisole sulfate **VI** 145–156
1,2-diaminobenzene **XIII** 215, **XXI** 3
1,3-diaminobenzene **VI** 287, **XXI** 3
1,4-diaminobenzene **VI** 311, **XXI** 3
m-diaminobenzene **VI** 287
o-diaminobenzene **XIII** 215
p-diaminobenzene **VI** 311
3,3′-diaminobenzidine **III** 131–135
3,3′-diaminobenzidine tetrahydrochloride **III** 131–135
4,4′-diamino-3,3′-bichlorobiphenyl **V** 99
di(4-amino-3-chlorophenyl)methane **VII** 193
4,4′-diamino-3,3′-dichlorodiphenyl **V** 99
4,4′-diamino-3,3′-dichlorodiphenylmethane **VII** 193
4,4′-diamino-3,3′-dimethoxybiphenyl **V** 139
4,4′-diamino-3,3′-dimethylbiphenyl **V** 153
4,4′-diaminodiphenyl ether **VI** 277
4,4′-diaminodiphenylmethane **VII** 37–57
p,p′-diaminodiphenylmethane **VII** 37
4,4′-diaminodiphenyl oxide **VI** 277
4,4′-diaminodiphenylsulfide **IV** 331
p,p′-diaminodiphenylsulfide **IV** 331
diaminoditolyl **V** 153
1,3-diamino-4-methoxybenzene **VI** 145
2,4-diamino-1-methoxybenzene **VI** 145
1,3-diamino-4-methylbenzene **VI** 339
2,4-diamino-1-methylbenzene **VI** 339
1,4-diamino-2-nitrobenzene **IV** 295, **XIII** 207
di(4-aminophenyl)methane **VII** 37
di(*p*-aminophenyl)sulfide **IV** 331
2,4-diaminotoluene **VI** 339
dianilinomethane **VII** 37
dianisidine **V** 139
diatomaceous earth **II** 157
1,4-diazacyclohexane **IX** 303, **XII** 177
diazinon **XI** 31–59
diazomethane **XIII** 141–148
dibenzoyl peroxide **III** 249–256
dibenzyl disulfide **II** 283–286
diborane **XIV** 83–85
diborane(6) **XIV** 83
diboron hexahydride **XIV** 83
[2,7-dibromo-9-(*o*-carboxyphenyl)-6-hydroxy-3-oxo-3*H*-xanthen-4-yl] hydroxymercury disodium salt **XV** 75
1,2-dibromo-2,4-dicyanobutane **XIV** 87–90

(2′,7′-dibromo-3′,6′-dihydroxy-3-oxo-spiro-[isobenzofuran-1(3H),9′-[9H]-xanthen]-4′-yl)hydroxymercury disodium salt **XV** 75
1,2-dibromoethane **III** 144
dibromohydroxymercurifluorescein disodium salt **XV** 75
di-*tert*-butyl hydroperoxide **III** 254
2,6-di-*tert*-butylphenol **V** 341–344
dicarboxybenzene **VII** 241
dichloroacetylene **VI** 157–163
1,2-dichlorobenzene **I** 71–80, **XX** 11–32
1,3-dichlorobenzene **I** 83–89
1,4-dichlorobenzene **I** 71, 73, 83, 196, 347, **IV** 141–171, **XX** 33–92
m-dichlorobenzene **I** 83
o-dichlorobenzene **I** 71, **XX** 11
p-dichlorobenzene **I** 71, 73, 83, 196, 347, **IV** 141, **XX** 33
3,3′-dichlorobenzidine **V** 99–107
o,o′-dichlorobenzidine **V** 99
3,3′-dichloro(1,1′-biphenyl)-4,4′-diamine **V** 99
3,3′-dichlorobiphenyl-4,4′-diamine **V** 99
3,3′-dichloro-4,4′-biphenylenediamine **V** 99
3,3′-dichloro-4,4′-diaminobiphenyl **V** 99
3,3′-dichloro-4,4′-diaminodiphenylmethane **VII** 193
2,2′-dichlorodiethylether **V** 66, 71, 72
2,2′-dichlorodiethyl sulfide **IV** 27
β,β′-dichlorodiethyl sulfide **IV** 27
dichlorodifluoromethane **I** 336, 341, **V** 109–117
dichlorodioctylstannane **VII** 91
1,2-dichloroethane **III** 137–147
α,β-dichloroethane **III** 137
1,1-dichloroethene **V** 66, 72, **VIII** 109
2,2-dichloroethenyl dimethyl phosphate **IV** 201
2,2-dichloroethenyl phosphoric acid dimethyl ester **IV** 201
1,1-dichloroethylene **VIII** 109
di(2-chloroethyl)methylamine **I** 211
1,2-dichloroethylmethyl ether **I** 91
α,β-dichloroethylmethyl ether **I** 91
di-2-chloroethyl sulfide **IV** 27
dichloroethyne **VI** 157
dichlorofluoromethane **I** 69, **V** 119–124
α-dichlorohydrin **I** 95
sym-dichloroisopropyl alcohol **I** 95
dichloromethane **IV** 173–190, **XVII** 147–161
1,2-dichloromethoxyethane **I** 91–93
(dichloromethyl)benzene **VI** 79
2,2′-dichloro-*N*-methyldiethylamine **I** 211

dichloromonofluoromethane **V** 119
dichloronaphthalenes **XIII** 101
2,4-dichlorophenoxyacetic acid **IV** 191–200, **XI** 61–104
2,4-dichlorophenoxyacetic acid salts and esters **XI** 61
dichlorophos **IV** 201
1,2-dichloropropane **IX** 21–39
1,3-dichloro-2-propanol **I** 95–98
α,α-dichlorotoluene **VI** 79
2,2-dichloro-1,1,1-trifluoroethane **X** 53–71
2,2-dichlorovinyl dimethyl phosphate **IV** 201
2,2-dichlorovinyl phosphoric acid dimethyl ester **IV** 201
dichlorovos **IV** 201
dichlorvos **IV** 201–216
dicumyl hydroperoxide **III** 252
dicumyl peroxide **III** 251
1,3-dicyanotetrachlorobenzene **VI** 113
dicyclopentadiene **V** 125–133
diesel engine emissions **I** 101–120, **II** 136
diethanolmethylamine **IX** 291
diethylamine **I** 25–28, 32–34
(diethylamino)ethane **XIII** 267
2-diethylaminoethanol **XIV** 91–100
β-diethylamino ethyl alcohol **XIV** 91
diethylcarbinol acetate **XI** 211
1,4-diethylenediamine **IX** 303, **XII** 177
1,4-diethylene dioxide **XX** 105
diethylene ether **XX** 105
diethylene glycol **X** 73–90
diethylene glycol *n*-butyl ether **VII** 59
diethylene glycol dimethyl ether **IX** 41–50
diethylene glycol monobutyl ether **VII** 59–67
1,4-diethylene oxide **XX** 105
N,*N*-diethylethanamine **XIII** 267
diethyl ether **I** 125, **XIII** 149–160
O,*O*-diethyl-*O*-(2-(ethylthio)ethyl)-phosphorthioate **XIX** 141
O,*O*-diethyl-*S*-(2-(ethylthio)ethyl)-phosphorthioate **XIX** 141
diethylmethylmethane **IV** 269
diethyl monosulfate **XX** 93
diethylnitrosamine **I** 23–35, 41, 60, 261–263, 268, 286
diethyl oxide **XIII** 149
diethyl sulfate **XX** 93–104
Diflamoll® TP **II** 321
difluorochloromethane **III** 63
difluorodichloromethane **V** 109
1,1-difluoroethene **V** 135
1,1-difluoroethylene **V** 135–137
difluoromonochloroethane **I** 65
difluoromonochloromethane **III** 63

1,2-difluoro-1,1,2,2-tetrachloroethane **I** 297
1,3-diformyl propane **VIII** 45, **XVI** 59
diglycidyl ether **IV** 59
diglycidyl ether of bisphenol A **XIX** 43
1,3-diglycidyloxybenzene **VII** 69
diglycidyl resorcinol ether **VII** 69–74
diglycolamine **IX** 215
Diglycolamine® Agent **IX** 215
diglycol monobutyl ether **VII** 59
diglyme **IX** 41
1,3-dihydro-1,3-dioxo-isobenzofuran **VII** 247
dihydrooxirene **V** 181
1,4-dihydroxybenzene **X** 113
m-dihydroxybenzene **XX** 239
4,4′-dihydroxydiphenylpropane **XIII** 49
p,*p*′-dihydroxydiphenylpropane **XIII** 49
1,2-dihydroxyethane **IV** 225
2,2′-dihydroxyethyl ether **X** 73
β,β′-dihydroxyisopropyl chloride **V** 83
2,4-dihydroxy-2-methylpentane **XVI** 233
N,*N*′-dihydroxymethylurea **V** 289
2,3-dihydroxypropyl chloride **V** 83
2,2′-dihydroxy-3,3′,5,5′-tetrachloro-diphenylsulfide **IX** 245
diisobutylketone **XVIII** 157–164
4,4′-diisocyanatodiphenylmethane **VIII** 65
p,*p*′-diisocyanatodiphenylmethane **VIII** 65
2,4-diisocyanato-1-methylbenzene **XX** 291
2,6-diisocyanato-1-methylbenzene **XX** 291
diisooctyl[(dioctylstannylene)dithio]diacetate **VII** 92
diisopropyl ether **XXI** 187–193
diisopropyl peroxydicarbonate **III** 250
dilauroyl peroxide **III** 250–255
dimethanol urea **V** 289
3,3′-dimethoxybenzidine **V** 139–152
3,3′-dimethoxy-[1,1′-biphenyl]-4,4′-diamine **V** 139
3,3′-dimethoxy-4,4′-diaminobiphenyl **V** 139
dimethoxyphosphine oxide **I** 133
dimethylamine **I** 25–28, 32–34, **VII** 75–89
N,*N*-dimethylaminobenzene **III** 149
4,4′-dimethylaminobenzophenonimide **IV** 11
dimethylaminosulfonyl chloride **I** 143
2,4-dimethylaniline **XXI** 3
2,6-dimethylaniline **XXI** 3
N,*N*-dimethylaniline **III** 149–161, **XXI** 3
dimethylaniline isomers **XIX** 299
N,*N*-dimethylbenzenamine **III** 149
dimethylbenzenamine isomers **XIX** 299
dimethylbenzene **V** 263, **XV** 257
3,3′-dimethylbenzidine **V** 153–164
α,α-dimethylbenzyl hydroperoxide **III** 250, 253–255

3,3′-dimethyl-[1,1′-biphenyl]-4,4′-diamine V 153
3,3′-dimethylbiphenyl-4,4′-diamine V 153
2,2-dimethylbutane IV 269
2,3-dimethylbutane IV 269
N,N-dimethylcarbamoyl chloride I 143, 145
dimethyl 2,2-dichlorovinyl phosphate IV 201
O,O-dimethyl-O-(2,2-dichlorovinyl)-phosphate IV 201
3,3′-dimethyldiphenyl-4,4′-diamine V 153
dimethyldithiocarbamic acid iron salt XIX 163
dimethylene oxide V 181
dimethyl ether I 125–131
[(1,1-dimethylethoxy)methyl]oxirane IV 51
4-(1,1-dimethylethyl)phenol XI 5
dimethylformaldehyde VII 1
dimethylformamide VIII 1–44
N,N-dimethylformamide VIII 1
2,6-dimethyl-4-heptanone XVIII 157
1,2-dimethylhydrazine I 176
dimethylketal VII 1
dimethyl ketone VII 1
N,N-dimethylmethanamide VIII 1
dimethyl monosulfate IV 217
dimethylnitromethane III 241
dimethylnitrosamine I 23–35, 41–44, 59, 261–264, 286
1,3-dimethylolurea V 289
N,N′-dimethylolurea V 289
N,N-dimethylphenylamine III 149
dimethylphenylamine isomers XIX 299
dimethyl phosphite I 133
dimethyl phosphonate I 133
1,1-dimethylpropyl acetate XI 211
dimethylpropylmethane IV 269
N,N-dimethylsulfamoyl chloride I 143–145
N,N-dimethylsulfamyl chloride I 143
dimethyl sulfate IV 48, 217–223
dimethyl sulfoxide III 163–171
dinitrobenzene (all isomers) I 147–167, XXI 3
2,4-dinitro-1-chlorobenzene XIII 111
4,6-dinitro-o-cresol XIX 151–161, XXI 3
dinitrogen monoxide IX 115
dinitrophenylmethane VI 165
dinitrotoluenes I 167, 361, VI 165–198
ar,ar-dinitrotoluene VI 165
2,3-dinitrotoluene VI 165
2,4-dinitrotoluene VI 165, XXI 3
2,5-dinitrotoluene VI 165
2,6-dinitrotoluene VI 165, XXI 3
3,4-dinitrotoluene VI 165

3,5-dinitrotoluene VI 165
dioctyldichlorostannane VII 91
2,2-dioctyl-1,3,2-dioxastannepin-4,7-dione VII 93
1,3-dioctyl-1,3-dioxodistannoxane VII 93
dioctyloxostannane VII 91
dioctylstannium dichloride VII 91
4,4′-[(dioctylstannylene)bis(oxy)]bis(4-oxo-2-butenoic acid) diisooctyl ester VII 93
2,2′-[(dioctylstannylene)-bis(thio)]bis(acetic acid)bis(2-ethylhexyl) ester VII 92
2,2′-[(dioctylstannylene)bis(thio)]bis(acetic acid) diisooctyl ester VII 92
dioctylstannylene maleate VII 93
di-n-octyltin bis(2-ethylhexyl mercapto-acetate) VII 92
di-n-octyltin bis(2-ethylhexyl thioglycolate) VII 92
di-n-octyltin bis(isooctyl maleate) VII 93
dioctyltin-S,S′-bis(isooctyl mercaptoacetate) VII 92
di-n-octyltin bis(isooctyl thioglycolate) VII 92
di-n-octyltin compounds VII 91–114
di-n-octyltin dichloride VII 91
dioctyltin di(isooctyl thioglycolate) VII 92
di-n-octyltin dithioglycolic acid 2-ethylhexyl ester VII 92
di-n-octyltin-2-ethylhexyl-dimercapto-ethanoate VII 92
di-n-octyltin maleate VII 93
di-n-octyltin oxide VII 91
Diolane XVI 233
Diospyros spp. XIII 286, XIV 287–290
1,4-dioxacyclohexane XX 105
1,4-dioxane XX 105–133
p-dioxane XX 105
2,5-dioxo-dihydrofuran IV 275
1,3-dioxophthalan VII 247
dioxyethylene ether XX 105
dipentene I 185
1,4-diphenyl-2,3-dithiabutane II 283
1,2-diphenylhydrazine IX 83
N,N′-diphenylhydrazine IX 83
4,4′-diphenylmethanediamine VII 37
diphenylmethane-4,4′-diisocyanate VIII 65
diphenylmethane-p,p′-diisocyanate VIII 65
di(phenylmethyl)disulfide II 283
o-diphenylol II 299
diphenylolpropane XIII 49
dipropylamine I 25
dipropylene glycol monomethyl ether VI 199–204
dipropyl methane XI 165

disodium 2′,7′-dibromo-4′-(hydroxymercury)fluorescein **XV** 75
Distemonanthus benthamianus **XIII** 286, **XIV** 291–293
disulfiram **V** 165–180
disulfuram **V** 165
2,6-ditertiary butyl phenol **V** 341
dithio **XIII** 237
1,1-dithiobis(*N*,*N*-diethylthioformamide) **V** 165
dithiocarbamates **XV** 147–149
dithiocarbonic anhydride **XII** 63, **XXI** 171
dithione **XIII** 237
dithiophos **XIII** 237
4,4′-di-*o*-toluidine **V** 153
divinyl **XV** 51
DMF **VIII** 1
DMFA **VIII** 1
DMSO **III** 163
DNCB **XIII** 111
DNOC **XIX** 151
DNT **VI** 165
dolomite **II** 182
Dominican mahogany **XVIII** 253
DOTC **VII** 91
DOTO **VII** 91
DOTTG **VII** 92
douka **XVIII** 277
doussié **XIII** 285
Dowanol® TBH **IX** 315
Dowicide® **II** 299
doxorubicin **I** 5
dry-zone mahogany **XIII** 287
Dumoria africana **XVIII** 277
Dumoria heckelii **XIII** 288, **XVIII** 277
Durmast oak **XVIII** 247
dusts **II** 3–46, 47–57, **XI** 281–301, **XII** 239, 271, **XVI** 287–315
Dymel A® **I** 125
eastern white cedar **XIII** 288, **XVIII** 263
East Indian ebony **XIV** 287
East Indian rosewood **XIII** 285, 301
EDAO **IX** 275, **XVI** 231
elayl **X** 91
endrin **XVIII** 165–188
enflurane **IX** 51–68
Entandrophragma spp. **XVIII** 229–233
environmental tobacco smoke **XIII** 3
epirubicin **I** 5
epirubicin hydrochloride **I** 5
1,2-epoxy-3-allyoxypropane **VII** 9
1,2-epoxybutane **V** 13
1,2-epoxy-4-(epoxyethyl)cyclohexane **I** 391
1,2-epoxyethane **V** 181

1-(epoxyethyl)-3,4-epoxycyclohexane **I** 391
3-(epoxyethyl)-7-oxabicyclo(4.1.0)heptane **I** 391
4-(epoxyethyl)-7-oxabicyclo(4.1.0)heptane **I** 391
3-(1,2-epoxyethyl)-7-oxabicyclo(4.1.0)-heptane **I** 391
4-(1,2-epoxyethyl)-7-oxabicyclo(4.1.0)-heptane **I** 391
1,2-epoxy-3-isopropoxypropane **VII** 141
1,2-epoxy-3-phenoxypropane **IV** 305
1,2-epoxypropane **V** 221
2,3-epoxypropane **V** 221
2,3-epoxy-1-propanol **XX** 179
2,3-epoxypropyl phenyl ether **IV** 305
(2,3-epoxypropyl)trimethylammonium chloride **IV** 247
erionite **VIII** 141–338
erythrene **XV** 51
Esperal® **V** 165
essence of mirbane **XIX** 227
ESTOL IDCO 3667 **XVI** 257
estramustine **I** 2
Etabus **V** 165
ethanal **III** 1
ethane dichloride **III** 137
1,2-ethanediol **IV** 225
ethanenitrile **XIX** 1
ethaneperoxoic acid **VII** 229
ethanethiol **XXI** 195–203
ethanoic acid ethenyl ester **V** 229
ethanol **XII** 129–165
ethanolamine **XII** 15
Ethanox 701® **V** 341
ethene **X** 91
ethene oxide **V** 181
ethenone **XX** 191
ethenyl acetate **V** 229
ethenylbenzene **XX** 285
9-ethenyl-9*H*-carbazole **XIV** 181
4-ethenyl-1-cyclohexene **XIV** 185
ethenyl ethanoate **V** 229
1-ethenyl-2-pyrrolidinone **V** 249
ether **XIII** 149
ethinyl trichloride **X** 201
ethohexadiol **V** 345
2-ethoxy-6-aminonaphthalene **VII** 17
ethoxyethane **XIII** 149
2-ethoxyethanol **I** 205, 209, **V** 210, **VI** 205–211, **XI** 105–119
2-ethoxyethanol acetate **VI** 213, **XI** 105
2-ethoxyethyl acetate **VI** 213–216, **XI** 105–119
ethyl acetate **XII** 167–176

ethyl acrylate **VI** 217–229, **XVI** 41–46
ethyl alcohol **XII** 129
ethyl aldehyde **III** 1
ethylamine **I** 28, 34
ethyl bromide **VII** 115–120
ethylcellosolve **VI** 205
ethyl chloride **III** 73
ethyl 2-cyanoacrylate **I** 233–245, **XIII** 201–206
ethyl α-cyanoacrylate **I** 233, **XIII** 201
7α-ethyldihydro-1H,3H,5H-oxazolo-[3,4-c]oxazole **IX** 275, **XVI** 231
10-ethyl-4,4-dioctyl-7-oxo-8-oxa-3,5-dithia-4-stannatetradecanoic acid 2-ethylhexyl ester **VII** 92
5-ethyl-3,7-dioxa-1-azabicyclo[3.3.0]octane **IX** 275–281, **XVI** 231–232
ethyldithiurame **V** 165
ethylene **X** 91–107
ethylene carboxamide **III** 11
ethylene chloride **III** 137
ethylene chlorohydrin **V** 65
ethylene dichloride **III** 137
ethylene glycol **IV** 225–245
ethylene glycol n-butyl ether **VI** 47
ethylene glycol chlorohydrin **V** 65
ethylene glycol dimethacrylate **XIII** 161–163
ethylene glycol ethyl ether **VI** 205
ethylene glycol ethyl ether acetate **VI** 213
ethylene glycol isopropyl ether **V** 207
ethylene glycol methacrylate **XIII** 193
ethylene glycol methyl ether **VI** 239
ethylene glycol methyl ether acetate **VI** 249
ethylene glycol monoacrylate **XVI** 89
ethylene glycol monobutyl ether **VI** 47
ethylene glycol monobutyl ether acetate **VI** 53
ethylene glycol monoethyl ether **VI** 205
ethylene glycol monoethyl ether acetate **VI** 213
ethylene glycol monoisopropyl ether **V** 207
ethylene glycol monomethacrylate **XIII** 193
ethylene glycol monomethyl ether **VI** 239
ethylene glycol monomethyl ether acetate **VI** 249
ethylene glycol mono-n-propyl ether **XII** 179
ethylene glycol monopropyl ether acetate **XII** 187
ethylene monochloride **V** 241
ethylene oxide **V** 181–192, **XIII** 165–168
1-ethyleneoxy-3,4-epoxycyclohexane **I** 391
ethylene tetrachloride **III** 271
ethylene thiourea **XI** 121–163
ethylene trichloride **X** 201

ethyl ether **XIII** 149
ethyl-ethylene oxide **V** 13
ethyl formate **XIX** 181
ethyl glycol acetate **VI** 213
2-ethylhexane-1,3-diol **V** 345
2-ethyl-1,3-hexanediol **V** 345–353
2-ethylhexanol **XX** 135–178
2-ethyl-1-hexanol **XX** 135
2-ethylhexyl acrylate **XVI** 47–50
ethyl hexylene glycol **V** 345
2-ethylhexyl propenoate **XVI** 47
2-ethyl-2-(hydroxymethyl)-1,3-propanediol triacrylate **XVI** 201
ethyl mercaptan **XXI** 195
ethyl(2-mercaptobenzoato-S)mercury sodium salt **XV** 249
ethyl methacrylate **XVI** 51–58
ethyl α-methacrylate **XVI** 51
ethyl methyl ketone **XII** 37
ethyl 2-methyl-2-propenoate **XVI** 51
4,4′-(2-ethyl-2-nitro-1,3-propanediyl)-bismorpholine **IX** 295–301
4,4′-(2-ethyl-2-nitro-trimethylene)-dimorpholine **IX** 295
ethyl orthosilicate **III** 299
ethyl oxide **XIII** 149
ethyl oxirane **V** 13
ethyl propenate **VI** 217
ethyl-2-propenoate **VI** 217
2-ethyl-3-propyl-1,3-propanediol **V** 345
ethyl silicate **III** 299
ethyl sulfate **XX** 93
ethyl sulfhydrate **XXI** 195
ethyl thiram **V** 165
Ethyl Thiurad **V** 165
ethyltrimethylmethane **IV** 269
etoposide **I** 6
ETS **XIII** 3
European walnut **XIII** 286
excursion limits **I** 15–22
exhaust fumes **I** 101–120
Exhoran® **V** 165
exposure peaks **I** 15–22
fast dark blue base R **V** 153
F-13B1 **VI** 37
F 134a **XIII** 251
FC 11 **I** 335, **III** 366
FC 12 **I** 336, 341, **III** 366, **V** 109
FC 13 **I** 69
FC 21 **V** 119
FC 22 **III** 63
FC 31 **V** 75
FC 112 **I** 297
FC 113 **III** 365, 366

FC 123 **X** 53
FC 142b **I** 65
ferbam **XIX** 163–168
fermentation butyl alcohol **XIX** 217
ferric dimethyldithiocarbamate **XIX** 163
ferric oxide **II** 135
ferrous oxide **II** 135
fibrous dust **VIII** 141–338
fine dust **II** 53
flint **II** 157
flowers of zinc **XVIII** 305
Fluorocarbon 134a **XIII** 251
fluorochloroform **I** 335
fluorodichloromethane **V** 119
fluorotrichloromethane **I** 335
5-fluorouracil **I** 4
flutamide **I** 6
FLX-0012® **III** 81
formaldehyde **I** 41–42, 44, 60, **II** 240, 250, 287, 331, **III** 1–8, 173–189, **IV** 136, **XVII** 163–201
formalin **III** 173
formic acid **XIX** 169–180
formic acid dimethylamide **VIII** 1
formic acid ethyl ester **XIX** 181–186
Formin **V** 355
N-formyldimethylamine **VIII** 1
2-formylfuran **IX** 69
formylic acid **XIX** 169
fraké **XIII** 288
framiré **XVIII** 259
Freon® 13B1 **VI** 37
Freon® FE 1301 **VI** 37
fumes **II** 59–67, **XII** 271
Fundal® **VI** 57
Fungicide E® **XVI** 263
2-furaldehyde **XVIII** 189
2-furanaldehyde **IX** 69
2-furancarbinol **VII** 121, **XIX** 187
2-furancarbonal **IX** 69
2-furancarboxaldehyde **IX** 69, **XVIII** 189
2,5-furandione **IV** 275
2-furanmethanol **VII** 121, **XIX** 187
furfural **IX** 69–79, **XVIII** 189–190
furfural alcohol **VII** 121
furfuralcohol **XIX** 187
furfuraldehyde **IX** 69
furfuryl alcohol **VII** 121–125, **XIX** 187–188
furyl alcohol **VII** 121
2-furylaldehyde **IX** 69
2-furylcarbinol **VII** 121, **XIX** 187
2-furylmethanal **IX** 69
2-furylmethanol **VII** 121
fused silica **II** 157

Gaboon ebony **XIV** 287
Gaboon mahogany **XIII** 285
Galecron® **VI** 57
gasoline engine emissions **I** 109–120
gedu nohor **XVIII** 229
general threshold limit value for dust **II** 3–46, **XII** 239–270
germ cell mutagens **I** 9–13, **XVII** 3–9
giant arborvitae **XIII** 288, **XVIII** 263
giant cedar **XIII** 288, **XVIII** 263
glass fibres **VIII** 141–338
glass wool **VIII** 141–338
glauramine **IV** 12
glutaral **VIII** 45, **XVI** 59
glutaraldehyde **VIII** 45–64, **XVI** 59–64
glutardialdehyde **VIII** 45, **XVI** 59
glutaric dialdehyde **VIII** 45, **XVI** 59
glycerin α-monochlorohydrin **V** 83
glycerol chlorohydrin **V** 83
sym-glycerol dichlorohydrin **I** 95
glycerol-α,γ-dichlorohydrin **I** 95
glycerol trichlorohydrin **IX** 171
glyceryl monothioglycolate **IX** 81–82
glyceryl trichlorohydrin **IX** 171
glycidol **XX** 179–190
glycidyl allyl ether **VII** 9
glycidyl butyl ether **IV** 51
glycidyl isopropyl ether **VII** 141
1-n-glycidyloxybutane **IV** 51
glycidyl phenyl ether **IV** 305
glycidyl trimethylammonium chloride **IV** 247–255
glycol **IV** 225
glycol butyl ether **VI** 47
glycol chlorohydrin **V** 65
glycol dichloride **III** 137
glycol ether DB **VII** 59
glycol ethyl ether **VI** 205
glycol methacrylate **XIII** 193
glycol methyl ether **VI** 239
glycol methyl ether acetate **VI** 249
glycol monobutyl ether acetate **VI** 53
glycolmonochlorohydrin **V** 65
glycol monoethyl ether **VI** 205
glycol monoethyl ether acetate **VI** 213
glycol monomethyl ether **VI** 239
glycol monomethyl ether acetate **VI** 249
G-MAC® **IV** 247
gold 'teak' **XIII** 287
Gonystylus bancanus **XIII** 286, **XVIII** 235–237
Grand Bassam mahogany **XIII** 287, **XVIII** 239
graphite **II** 129–134

green ebony **XIII** 285, 299
Grevillea robusta **XIII** 286, **XIV** 295–297
Grotan® BK **II** 331
Grotan® HD **IX** 263
gypsum **II** 117, **VIII** 141–338
haematite **II** 136
halloysite **VIII** 141–338
Halon® 1301 **VI** 37
hard coal dust **XVIII** 107
hard metal **III** 123, 124
HCB **XVI** 65
α-HCH **V** 193
β-HCH **V** 193
γ-HCH **XVI** 113
2-(8-heptadecenyl)-4,5-dihydro-1-*H*-imidazole-1-ethanol **V** 373
2-(8-heptadecenyl)-2-imidazoline-1-ethanol **V** 373
2-heptadecenyl-2-imidazoline-1-ethanol **V** 373
n-heptane **XI** 165–176
Heritiera utilis **XIII** 288
hexabutyldistannoxane **I** 315
hexachlorobenzene **XVI** 65–88
α-1,2,3,4,5,6-hexachlorocyclohexane **V** 193
β-1,2,3,4,5,6-hexachlorocyclohexane **V** 193
1α,2α,3β,4α,5α,6β-hexachlorocyclohexane **XVI** 113
1α,2α,3β,4α,5β,6β-hexachlorocyclohexane **V** 193
1α,2β,3α,4β,5α,6β-hexachlorocyclohexane **V** 193
α-hexachlorocyclohexane **V** 193–206
β-hexachlorocyclohexane **V** 193–206
γ-1,2,3,4,5,6-hexachlorocyclohexane **XVI** 113
1,2,3,4,10,10-hexachloro-6,7-epoxy-1,4,4a,5,6,7,8,8a-octahydro-*endo*,*endo*-1,4:5,8-dimethanonaphthalene **XVIII** 165
hexachloronaphthalenes **XIII** 101
(1aα2β,2aβ,3α,6α,6aβ,7β,7aα)-3,4,5,6,9,9-hexachloro-1a,2,2a,3,6,6a,7,7a-octahydro-2,7:3,6-dimethanonaphth[2,3-*b*]oxirene **XVIII** 165
hexahydro-2*H*-azepin-2-one **IV** 65
hexahydro-2*H*-azepin-7-one **IV** 65
hexahydrobenzene **XIII** 129
hexahydro-1,4-diazine **IX** 303, **XII** 177
hexahydrophthalic anhydride **X** 109–111
hexahydropyrazine **IX** 303, **XII** 177
hexahydro-1,3,5-tris(2-hydroxyethyl)-s-triazine **II** 331
hexamethylene **XIII** 129
hexamethyleneamine **V** 355

hexamethylenetetramine **V** 355–372
hexamethylmelamine **I** 2
hexamine **V** 355
hexanaphthene **XIII** 129
n-hexane **IV** 257–268, **XIV** 101–126
hexane (all isomers except *n*-hexane) **IV** 269–273
6-hexanelactam **IV** 65
1-hexanol **IX** 283–290
n-hexanol **IX** 283
hexone **XIII** 169–180
n-hexyl alcohol **IX** 283
hexylene glycol **XVI** 233–246
Heyderia decurrens **XIII** 285, **XIV** 275
HFC-134a **XIII** 251
HMT **V** 355
HMTA **V** 355
HN2 **I** 211
Honduras mahogany **XIII** 287, **XVIII** 253
Honduras rosewood **XIII** 286, 317
Hordaflex® **III** 81
hydrazine **I** 40, 171–182, **XIII** 181–186
hydrazine anhydrous **XIII** 181
hydrazine hydrate **XIII** 181–186
hydrazine monohydrate **XIII** 181
hydrazine salts **XIII** 181–186
hydrazinobenzene **XI** 225
hydrazobenzene **IX** 83–89
hydrobromic acid **XIII** 187
hydrobromic ether **VII** 115
hydrochloric acid **VI** 231
hydrochloric ether **III** 73
hydrocyanic acid **I** 42, **XIX** 189
hydrogen arsenide **XV** 41
hydrogen bromide **XIII** 187–191
hydrogen carboxylic acid **XIX** 169
hydrogen chloride **VI** 231–238
hydrogen chloride gas **VI** 231
hydrogen cyanide **XIX** 189–210
hydrogen nitrate **III** 233
hydrogen peroxide **III** 249, 252, 253
hydrogen sulfate **XV** 165
α-hydro-ω-hydroxypoly(oxy-1,2-ethanediyl) **X** 247
α-hydro-ω-hydroxypoly[oxy(methyl-1,2-ethandiyl)] **X** 271
hydroquinone **III** 195, **X** 113–145
2-hydroxybiphenyl **II** 299
o-hydroxybiphenyl **II** 299
2-hydroxybutane **XIX** 117
4-hydroxy-2-butane sulfonic acid **IV** 45
1-hydroxy-4-*tert*-butylbenzene **XI** 5
hydroxycarbamide **I** 7
2-hydroxyethyl acrylate **XVI** 89–93

2-hydroxyethyl methacrylate **XIII** 193–199
2-hydroxy-1,3-di-*tert*-butylbenzene **V** 341
2-hydroxy-3,5-dichlorophenyl sulfide **IX** 245
2-hydroxydiphenyl **II** 299
hydroxy ether **VI** 205
2-(2-hydroxyethoxy)ethylamine **IX** 215
2,2′-hydroxyethoxyethylamine **IX** 215
2-hydroxyethylamine **XII** 15
1-hydroxyethyl-2-heptadecenylglyoxalidine **V** 373
1-(2-hydroxyethyl)-2-heptadecenylglyoxalidine **V** 373
1-(2-hydroxyethyl)-2-(8-heptadecenyl)-2-imidazoline **V** 373
1-(2-hydroxyethyl)-2-*n*-heptadecenyl-2-imidazoline **V** 373
1-hydroxyethyl-2-heptadecenyl-imidazoline **V** 373–375
β-hydroxyethyl isopropyl ether **V** 207
2-(*N*-2-hydroxyethyl-*N*-methylamino)ethanol **IX** 291
N-(2-hydroxyethyl)-3-methyl-2-quinoxalinecarboxamide 1,4-dioxide **IX** 151
1-hydroxy hexane **IX** 283
1-hydroxyisopropyl-(2-methoxypropyl)ether **VI** 200
1-hydroxy-1′-methoxydiisopropylether **VI** 200
2-hydroxy-2′-methoxydi-*n*-propylether **VI** 200
1-hydroxy-2-methylbenzene **XIV** 59
1-hydroxy-3-methylbenzene **XIV** 59
1-hydroxy-4-methylbenzene **XIV** 59
hydroxymethylene benzyl ether **II** 239
hydroxymethylene mono(poly)oxymethylene benzyl ether **II** 239
2-hydroxymethylfuran **VII** 121, **XIX** 187
2-hydroxymethyl-*n*-heptan-4-ol **V** 345
2-(hydroxymethyl)-2-nitro-1,3-propanediol **II** 287–293
1-hydroxymethylpropane **XIX** 217
4-hydroxy-3-nitroaniline **IV** 289
hydroxy octyl oxostannane **VII** 93
2-hydroxypropanamine **IX** 237
3-hydroxy-1-propanesulfonic acid sulfone **IV** 313
3-hydroxy-1-propanesulfonic acid sultone **IV** 313
α-hydroxypropanetricarboxylic acid **XVI** 209
2-hydroxy-1,2,3-propanetricarboxylic acid **XVI** 209
hydroxypropyl acrylate (all isomers) **XVI** 95–104
1-hydroxy-2-propyl acrylate **XVI** 95

1-hydroxy-3-propyl acrylate **XVI** 95
2-hydroxy-1-propyl acrylate **XVI** 95
2-hydroxypropylamine **IX** 237
3-hydroxypropylene oxide **XX** 179
2-hydroxypropyl methacrylate **XVI** 105–111
2-hydroxypropyl-(1-methoxyisopropyl)ether **VI** 200
2-hydroxypropyl 2-methyl-2-propenoate **XVI** 105
1-hydroxy-2-(1*H*)-pyridinethione sodium salt **X** 287
2-hydroxytoluene **XIV** 59
3-hydroxytoluene **XIV** 59
4-hydroxytoluene **XIV** 59
1-hydroxy-2,4,5-trichlorobenzene **XII** 209
2-hydroxytriethylamine **XIV** 91
hydroxyurea **I** 7
hyperbaric pressure and body burden **XIV** 11–16
hyponitrous acid anhydride **IX** 115
hytrol O **X** 35
idigbo **XVIII** 259
ifosfamide **I** 2
IGE **VII** 141
2-imidazolidinethione **XI** 121
4,4′-imidocarbonyl-bis(*N*,*N*-dimethylaniline) **IV** 11
4,4′-imidocarbonyl-bis(*N*,*N*-dimethylbenzenamine) **IV** 11
inactive limonene **I** 185
incense cedar **XIII** 285, **XIV** 275
Indian ebony **XIII** 286, **XIV** 287
Indian rosewood **XIII** 285, 301
inhalable dust **XII** 239
inhalation kinetics **XIV** 3–10
inorganic fibres **VIII** 141–338
inorganic lead compounds **XVII** 203
inorganic tin compounds **XIV** 149–165
iodomethane **VII** 219
3-iodo-2-propynylbutylcarbamate **XVI** 247–256
3-iodo-2-propynyl-*N*-butylcarbamate **XVI** 247
3-iodo-2-propynylcarbamic acid butyl ester **XVI** 247
Ionox 99® **V** 341
ipe **XIII** 288, **XIV** 311
ipé peroba **XIII** 287, **XIV** 307
iroko **XIII** 285, **XIV** 279
iron oxides **II** 135–144
isatoic acid anhydride **XIII** 97
isatoic anhydride **XIII** 97
isoamyl acetate **XI** 211
1,3-isobenzofurandione **VII** 247

isobutane **XX** 1
isobutanol-2-amine **IX** 229
isobutenyl chloride **IV** 97
isobutyl acetate **XIX** 211–215
isobutyl alcohol **XIX** 217–226
isobutyltrimethylmethane **I** 347
isocyanatobenzene **XVII** 267
isocyanic acid methylene di-*p*-phenylene ester **VIII** 65
isocyanic acid polymethylenepolyphenylene ester **VIII** 65
isodecyl oleate **XVI** 257–261
isoflurane **VII** 127–140
isohexane **IV** 269
Isol **XVI** 233
isonitropropane **III** 241
isooctane **I** 347
isoparaffins **IV** 269–273
isopentyl acetate **XI** 211
isophorone **I** 196
isophosphamid **I** 2
isophthalic acid **VII** 241
isopropanolamine **IX** 237
isopropenylbenzene **XV** 137
4-isopropyl-1-methyl-1-cyclohexene **I** 185
2-isopropoxyethanol **V** 207-211
(isopropoxymethyl)oxirane **VII** 141
2-isopropoxypropane **XXI** 187
isopropylacetone **XIII** 169
isopropylamine **I** 34
isopropylbenzene **XIII** 117
isopropylcarbinol **XIX** 217
isopropyl cellosolve **V** 207
isopropyl ether **XXI** 187
isopropyl ethylene glycol ether **V** 207
isopropyl glycidyl ether **VII** 141–144
isopropyl glycol **V** 207
4,4′-isopropylidenediphenol **XIII** 49
4,4′-isopropylidenediphenol diglycidyl ether **XIX** 43
3-isopropyloxypropylene oxide **VII** 141
isosystox **XIX** 141
isothiourea **I** 301
'jacaranda' **XIII** 287, 321
Jamaica mahogany **XVIII** 253
jeweler's rouge **II** 135
Juglans spp. **XIII** 286
kambala **XIII** 285, **XIV** 279
Kathon® **V** 323
Kathon® 893 MW **XVI** 263
Kathon® biocide **V** 323
Kaurit® S **V** 289
ketene **XX** 191–196
ketohexamethylene **X** 35

ketone propane **VII** 1
β-ketopropane **VII** 1
Khaya mahogany **XIII** 286–287, **XVIII** 239
Khaya spp. **XIII** 286–287, **XVIII** 239–246
kieselguhr **II** 157
kosipo **XVIII** 229
lapacho **XIII** 288, **XIV** 311
Larvacide 100 **VI** 105
laughing gas **IX** 115
lead **XVII** 203–244
lead chromate **III** 101–122, **VII** 33
lead compounds **XVII** 203
lead molybdate **XVIII** 199
lead tetraethyl **XV** 223
lead tetramethyl **XV** 237
Libocedrus decurrens **XIII** 285, **XIV** 275
limba **XIII** 288
limitation of exposure peaks **I** 15–22
limonene **I** 185–200, 347
lindane **XVI** 113–141
α-lindane **V** 193
β-lindane **V** 193
lithium hydride **III** 191–193
loadstone **II** 135
lodestone **II** 135
longleaf pine **XIII** 287
long-term threshold values **XI** 281
Lorol C 6 **IX** 283
low cristobalite **XIV** 205
low quartz **XIV** 205
low tridymite **XIV** 205
Lubrimet® P600, P900 **X** 271
lusamba **XIII** 288
lye **XII** 195
Macassar ebony **XIII** 286, **XIV** 287
Machaerium scleroxylon **XIII** 287, 321–323
macore **XVIII** 277
Mafu® **IV** 201
magnesia **II** 145
magnesium oxide **II** 145–148
magnesium oxide sulfate **VIII** 141–338
magnetite **II** 136
mahogany **XIII** 287
mainstream smoke **XIII** 3
makoré **XIII** 288, **XVIII** 277
maleic acid anhydride **IV** 275, **XI** 177
maleic anhydride **IV** 275–287, **XI** 177–178
manganese **XII** 293–328
manganese compounds **XII** 293
man-made mineral fibres **VIII** 141–338
Mansonia altissima **XIII** 287, **XIV** 299–305
maritime pine **XIII** 287
MBOCA **VII** 193
MDEA **IX** 291

MDI **VIII** 65
mechlorethamine **I** 211
mecrylate **I** 233
MEK **XII** 37
melphalan **I** 3
p-mentha-1,8-diene **I** 185
1,8(9)-*p*-menthadiene **I** 185
merbromin **XV** 75–79
mercaptoethane **XXI** 195
2-mercaptoimidazoline **XI** 121
mercaptomethane **XX** 217
mercaptophos **XIX** 141
6-mercaptopurine **I** 4
mercurialin **VII** 145
mercurochrome **XV** 75
Mercurothiolate **XV** 249
mercury and inorganic mercury compounds **XV** 81–122
mercury, organic compounds **XV** 123–136
mesitylene **IV** 341
metallic mercury **XV** 81
metal-working fluids **II** 207–220, **IX** 195–213, **XVI** 207–286, **XX** 197–215
metaphenylenediamine **VI** 287
methacrylic acid ethyl ester **XVI** 51
methacrylic acid 2-hydroxyethyl ester **XIII** 193
methacrylic acid 2-hydroxypropyl ester **XVI** 105
methacrylic acid methyl ester **XVI** 181
2-methallyl chloride **IV** 97
β-methallyl chloride **IV** 97
methanal **III** 173, **XVII** 163
methanamine **VII** 145
methane base **I** 249
methane dichloride **IV** 173, **XVII** 147
methanethiol **XX** 217
methanoic acid **XIX** 169
methanol **XVI** 143–175
methenamine **V** 355
methotrexate **I** 3
4-methoxy-2-aminoanisole **IV** 135
2-methoxyaniline **X** 1, **XXI** 3
4-methoxyaniline **XXI** 3
o-methoxyaniline **X** 1
2-methoxybenzenamine **X** 1
4-methoxy-1,3-benzenediamine **VI** 145
2-methoxyethanol **V** 210, **VI** 239–248
2-methoxyethanol acetate **VI** 249
2-methoxyethyl acetate **VI** 249–251
methoxyhydroxyethane **VI** 239
2-methoxy-5-methylaniline **IV** 135
2-methoxy-5-methylbenzenamine **IV** 135
(2-methoxymethylethoxy)propanol **VI** 199

2-(2-methoxy-1-methylethoxy)-1-propanol **VI** 200
1-(2-methoxy-1-methylethoxy)-2-propanol **VI** 200
2-(2-methoxy-2-methylethoxy)-1-propanol **VI** 200
1-(2-methoxy-2-methylethoxy)-2-propanol **VI** 200
2-methoxy-2-methylpropane **XVII** 119
1-methoxy-2-nitrobenzene **IX** 103
o-methoxyphenylamine **X** 1
p-methoxy-*m*-phenylenediamine **VI** 145
4-methoxy-*m*-phenylenediamine **VI** 145
1-methoxy-2-propanol **V** 213–216, **XIV** 127–129
1-methoxypropan-2-ol **I** 204, **V** 213
2-methoxy-1-propanol **I** 203, 207
1-methoxy-2-propanol acetate **V** 217
2-methoxy-1-propanol acetate **I** 203, 207
1-(1-methoxy-2-propoxy)-2-propanol **VI** 200
1-(2-methoxy-1-propoxy)-2-propanol **VI** 200
2-(1-methoxy-2-propoxy)-1-propanol **VI** 200
2-(2-methoxy-1-propoxy)-1-propanol **VI** 200
1-methoxypropyl-2-acetate **V** 217–220
2-methoxypropyl-1-acetate **I** 207–209
4-methoxy-*m*-toluidine **IV** 135
methural **V** 289
methyl acetate **XVIII** 191–196
methyl acrylate **VI** 253–262, **XVI** 177–180
2-methylacrylic acid methyl ester **III** 195
methyl alcohol **XVI** 143
methyl aldehyde **III** 173, **XVII** 163
methyl allyl chloride **IV** 97
methylamine **I** 28, 34, **VII** 145–153
1-methyl-2-aminobenzene **III** 307
2-methyl-1-aminobenzene **III** 307
(methylamino)benzene **VI** 263
N-methylaminobenzene **VI** 263
N-methylaminodiglycol **IX** 291
1-methyl-2-aminoethanol **V** 237
N-methylaniline **I** 25, **VI** 263–270, **XXI** 3
2-methylaniline **III** 307
4-methylaniline **III** 323
p-methylaniline **III** 323
5-methyl-*o*-anisidine **IV** 135
p-methylanisidine **IV** 135
1-methylazacyclopentan-2-one **X** 147
2-methylbenzenamine **III** 307
4-methylbenzenamine **III** 323
4-methyl-1,3-benzenamine **VI** 339
N-methylbenzenamine **VI** 263
methylbenzene **VII** 257
p-methylbenzeneamine **III** 323
methylbenzotriazole **II** 295

methyl-1H-benzotriazole **II** 295–298
N-methyl-bis(2-chloroethyl)amine **I** 211–230
methylbis(2-hydroxyethyl)amine **IX** 291
methyl bromide **VII** 155–172
2-methyl-1-butanol acetate **XI** 211
2-methyl-2-butanol acetate **XI** 211
3-methyl-1-butanol acetate **XI** 211
1-methylbutyl acetate **XI** 211
2-methylbutyl acetate **XI** 211
3-methylbutyl acetate **XI** 211
β-methylbutyl acetate **XI** 211
methyl-tert-butyl ether **XVII** 119
methyl carbinol **XII** 129
Methyl Cellosolve® **VI** 239
Methyl Cellosolve® acetate **VI** 249
methyl chloride **VII** 173–191
methyl chloroacetate **IX** 1
2-methyl-4-chloroaniline **VI** 127
3-methyl-4-chlorophenol **II** 271
methyl cyanide **XIX** 1
methyl 2-cyanoacrylate **I** 233–245, **XIII** 201–206
methyl α-cyanoacrylate **I** 233, **XIII** 201
methyldibromoglutaronitrile **XIV** 87
N-methyl-2,2′-dichlorodiethylamine **I** 211
methyldiethanolamine **IX** 291–294
N-methyldiethanolamine **IX** 291
N-methyldiethanolimine **IX** 291
2-methyl-2,3-dihydroisothiazol-3-one **V** 323–339
methyldinitrobenzene **VI** 165
1-methyl-2,3-dinitrobenzene **VI** 165
1-methyl-2,4-dinitrobenzene **VI** 165
1-methyl-3,4-dinitrobenzene **VI** 165
1-methyl-3,5-dinitrobenzene **VI** 165
2-methyl-1,3-dinitrobenzene **VI** 165
2-methyl-1,4-dinitrobenzene **VI** 165
4-methyl-1,2-dinitrobenzene **VI** 165
2-methyl-4,6-dinitrophenol **XIX** 151
methylene acetone **IX** 91
methylene base **I** 249
methylene bichloride **IV** 173, **XVII** 147
4,4′-methylenebis(aniline) **VII** 37
4,4′-methylenebis(benzenamine) **VII** 37
4,4′-methylene(bis)chloroaniline **VII** 193
4,4′-methylenebis(2-chloroaniline) **VII** 193–218
p,p′-methylenebis(alpha-chloroaniline) **VII** 193
4,4′-methylenebis(2-chlorobenzenamine) **VII** 193
4,4′-methylene-bis(N,N-dimethylaniline) **I** 249-256, **IV** 15
4,4′-methylene-bis(N,N-dimethyl)benzenamine **I** 249
1,1′-methylenebis(4-isocyanatobenzene) **VIII** 65
4,4′-methylene-bis(2-methylaniline) **XVIII** 197–198
methylenebis(4-phenylene isocyanate) **VIII** 65
methylenebis(p-phenylene isocyanate) **VIII** 65
methylene bisphenyl isocyanate **VIII** 65
methylene chloride **IV** 173, **XVII** 147
4,4′-methylenedianiline **VII** 37
p,p′-methylenedianiline **VII** 37
methylene dichloride **IV** 173, **XVII** 147
methylenedi-4-phenylene diisocyanate **VIII** 65
methylenedi-p-phenylene diisocyanate **VIII** 65
methylene di(phenylene isocyanate) **VIII** 65
4,4′-methylenediphenylene isocyanate **VIII** 65
4,4′-methylenediphenyl diisocyanate **VIII** 65
4,4′-methylene diphenyl isocyanate **VIII** 65, **XIV** 131–133
methylene glycol **III** 173
methylene oxide **III** 173, **XVII** 163
(1-methylethenyl)-benzene **XV** 137
methyl ether **I** 125
methyl ethoxol **VI** 239
2-(1-methylethoxy)ethanol **V** 207
[(1-methylethoxy)methyl]oxirane **VII** 141
(1-methylethyl)benzene **XIII** 117
methyl ethyl carbinol **XIX** 117
methylethylene oxide **V** 221
4,4′-(1-methylethylidene)bisphenol **XIII** 49
2,2′-[(1-methylethylidene)bis(4,1-phenyleneoxymethylene)]bis-oxirane **XIX** 43
methyl ethyl ketone **XII** 37
methyl ethyl ketone peroxide **III** 250, 252, 254, 255
methyl glycol **VI** 239
methyl glycol acetate **VI** 249
methyl glycol monoacetate **VI** 249
2,2′-(methylimino)bisethanol **IX** 291
2,2′-(methylimino)diethanol **IX** 291
N-methyliminodiethanol **IX** 291
N-methyl-2,2′-iminodiethanol **IX** 291
methyl iodide **VII** 219–228
methyl isobutyl ketone **XIII** 169
1-methyl-4-isopropenyl-1-cyclohexene **I** 185
2-methyl-4-isothiazolin-3(2H)-one **V** 323
2-methyl-3(2H)-isothiazolone **V** 323

methyl ketone **VII** 1
N-methyl-2-ketopyrrolidine **X** 147
N-methyl lost **I** 211
methyl mercaptan **XX** 217–226
methyl methacrylate **III** 195–231, **XVI** 181–191
methyl α-methacrylate **III** 195
N-methylmethanamine **VII** 75
methyl methylacrylate **III** 195, **XVI** 181
1-methyl-4-(1-methylethenyl)cyclohexene **I** 185
methyl 2-methyl-2-propenoate **III** 195, **XVI** 181
methyl monochloroacetate **IX** 1
2-methyl-5-nitroaniline **VI** 271
6-methyl-3-nitroaniline **VI** 271
2-methyl-5-nitrobenzenamine **VI** 271
1-methyl-2-nitrobenzene **VIII** 97
2-methyl-1-nitrobenzene **VIII** 97
o-methylnitrobenzene **VIII** 97
2-methyl-5-nitrobenzeneamine **VI** 271
(*Z*)-(2-methylnonyl)-9-octadecenoate **XVI** 257
N-methyl-*N*-oleoyl-aminoacetic acid **XVI** 281
4-methyl-3-oxapentan-1-ol **V** 207
3-methyl-1,2-oxathiolane 2,2-dioxide **IV** 45
methyloxirane **V** 221
N-methyl-*N*-(1-oxo-9-octadecenyl)glycine **XVI** 281
N-methyl-2-oxypyrrolidine **X** 147
2-methylpentane **IV** 269
3-methylpentane **IV** 269
2-methyl-2,4-pentanediol **XVI** 233
4-methyl-2-pentanone **XIII** 169
2-methylphenol **XIV** 59
3-methylphenol **XIV** 59
4-methylphenol **XIV** 59
N-methylphenylamine **VI** 263
4-methyl-*m*-phenylenediamine **VI** 339
1-methyl-1-phenylethylene **XV** 137
2-methylpropane **XX** 1
1-methyl-1,3-propane sultone **IV** 45
methyl propenate **VI** 253
methyl-2-propenoate **VI** 253
2-methyl-2-propenoic acid ethyl ester **XVI** 51
2-methyl-2-propenoic acid 2-hydroxyethyl ester **XIII** 193
2-methyl-2-propenoic acid 2-hydroxypropyl ester **XVI** 105
2-methyl-2-propenoic acid methyl ester **III** 195, **XVI** 181
1-methyl-1-propanol **XIX** 117
2-methyl-1-propanol **XIX** 217

2-methyl-2-propanol **XIX** 125
2-methyl-1-propyl acetate **XIX** 211
1-methyl propyl alcohol **XIX** 117
beta-methylpropyl ethanoate **XIX** 211
N-methyl-2-pyrrolidinone **X** 147
1-methyl-2-pyrrolidinone **X** 147
1-methyl-5-pyrrolidinone **X** 147
N-methylpyrrolidone **X** 147
1-methyl-2-pyrrolidone **X** 147
N-methyl-2-pyrrolidone **X** 147–170
α-methyl styrene **XV** 137–140
methyl sulfhydrate **XX** 217
methylsulfinylmethane **III** 163
methyl sulfoxide **III** 163
methyltetrahydrophthalic anhydride **X** 109–111
N-methyl-*N*,2,4,6-tetranitroaniline **XI** 179–185, **XXI** 3
N-methyl-*N*,2,4,6-tetranitrobenzenamine **XI** 179
methyl toluene **V** 263
methyltoluidine isomers **XIX** 299
methyl tribromide **VII** 319
1-methyl-2,4,6-trinitrobenzene **I** 359
2-methyl-1,3,5-trinitrobenzene **I** 359
methyl vinyl ketone **IX** 91–101
Michler's base **I** 249
Michler's hydride **I** 249
Michler's ketone **IV** 12, 13, 20, 21
Michler's methane **I** 249
Microberlinia spp **XIII** 287
Mimusops africana **XVIII** 277
Mimusops heckelii **XIII** 288, **XVIII** 277
mineral fibres **II** 185
MIPA **IX** 237
mists **XII** 271
mithramycin **I** 5
mitobronitol **I** 7
mitomycin **I** 5
mitoxantrone **I** 5
MMA **III** 195, **XVI** 181
MME **III** 195
4-MMPD **VI** 145
MOCA **VII** 193
molybdenite **XVIII** 199
molybdenum **XVIII** 199–219
molybdenum compounds **XVIII** 199–219
molybdenum dichloride **XVIII** 199
molybdenum dioxide **XVIII** 199
molybdenum dioxide **XVIII** 199
molybdenum disulfide **XVIII** 199
molybdenum pentachloride **XVIII** 199
molybdenum trichloride **XVIII** 199
molybdenum trioxide **XVIII** 199

molybdic anhydride **XVIII** 199
molybdophosphoric acid **XVIII** 199
monobromoethane **VII** 115
monobromomethane **VII** 155
monobutyl glycol ether **VI** 47
monochloroacetaldehyde **XII** 81
monochloroacetic acid methyl ester **IX** 1
monochlorobenzene **XII** 103
monochlorodifluoromethane **III** 63
monochloroethane **III** 73
2-monochloroethanol **V** 65
monochloroethene **V** 241
monochloroethylene **V** 241
monochlorofluoromethane **V** 75
α-monochlorohydrin **V** 83
monochloromethane **VII** 173
monochloromonofluoroethane **V** 75
monochloronaphthalenes **XIII** 101
monochloropropanediol **V** 83
monochlorotrifluoromethane **I** 69
monocyclic aromatic amino and nitro compounds **XXI** 3–45
monoethanolamine **XII** 15
monoethyleneglycol **IV** 225
mono-isopropanolamine **IX** 237
monomethylamine **VII** 145
monomethylaniline **VI** 263
monomethyl glycol acetate **VI** 249
mononitroethane **XIX** 245
mononitromethane **XIX** 251
mono-*n*-octyltin compounds **VII** 91–114
monooctyltin hydroxide **VII** 93
mono-*n*-octyltin oxide **VII** 93
monooctyltin thioglycolate **VII** 94
mono-*n*-octyltin trichloride **VII** 93
mono-*n*-octyltin tris(2-ethylhexyl thioglycolate) **VII** 94
mono-*n*-octyltin tris(2-ethylhexylmercaptoacetate) **VII** 94
mono-*n*-octyltin tris(isooctylthioglycolate) **VII** 94
mopidamol **I** 4
morpholine **I** 25, 28, 33–34
4-morpholinecarbonyl chloride **V** 79
morpholinylcarbamoyl chloride **V** 79
morpholinylcarbonyl chloride **V** 79
MOTC **VII** 93
MOTO **VII** 93
MOTTG **VII** 94
Morus excelsa **XIV** 279
muriatic acid **VI** 231
mustard gas **IV** 27
mustargen **I** 211
myleran **I** 2–3

Mystox WFA® **II** 299
nadone **X** 35
naphthalene **XI** 187–210
2-naphthylamine **I** 258–259
Natriphene® **II** 299
Natrium-Pyrion® **X** 287
natural isobutyl acetate **XIX** 211
natural rubber latex (*Hevea* species) **XV** 141–143
NCI-C56417 **V** 295
2-NDB **XIII** 207
nemalite **VIII** 141–338
neohexane **IV** 269
niangon **XIII** 288
Nicaragua mahogany **XVIII** 253
nickel compounds **I** 40, **II** 47, 48, 136
nicotine **I** 28, 40, 42–44, 59
Nigerian ebony **XIV** 287
Nigerian satinwood **XIII** 286, **XIV** 291
Nigerian walnut **XIV** 299
nimustine **I** 3
nitramine **XI** 179
nitric acid **III** 233–240
nitrite **I** 27–34, 59
2-nitro-4-aminophenol **I** 167, **IV** 289–293, **XXI** 3
o-nitro-*p*-aminophenol **IV** 289
4-nitro-2-aminotoluene **VI** 271
nitroaniline **I** 149–150, 166
4-nitroaniline **XXI** 3
2-nitroanisole **IX** 103–114, **XXI** 3
o-nitroanisole **IX** 103
nitrobenzene **XIX** 227–243, **XXI** 3
2-nitro-1,4-benzenediamine **IV** 295, **XIII** 207
nitrobenzol **XIX** 227
4-nitrobiphenyl **I** 257–259
4-nitro-1,1′-biphenyl **I** 257
p-nitrobiphenyl **I** 257
4-(2-nitrobutyl)morpholine **IX** 295–301
N-(2-nitrobutyl)morpholine **IX** 295
nitrocarbol **XIX** 251
o-nitrochlorobenzene **IV** 107
m-nitrochlorobenzene **IV** 115
p-nitrochlorobenzene **IV** 121
nitrochloroform **VI** 105
2-nitro-1,4-diaminobenzene **IV** 295, **XIII** 207
4-nitrodiphenyl **I** 257
p-nitrodiphenyl **I** 257
nitroethane **XIX** 245–250
nitrogen dioxide **XXI** 205–260
nitrogen mustard **I** 211
nitrogen oxide **IX** 115
nitrogen oxides **I** 28, 40, 42–44, 58

nitrogen peroxide **XXI** 205
nitro-isobutylglycerol **II** 287
nitroisopropane **III** 241
nitromethane **XIX** 251–260
3-nitro-6-methylaniline **VI** 271
1-nitro-2-methylbenzene **VIII** 97
2-nitro-1,4-phenylenediamine **IV** 295, **XIII** 207
2-nitro-*p*-phenylenediamine **IV** 295–303, **XIII** 207–210, **XXI** 3
o-nitro-*p*-phenylenediamine **IV** 295, **XIII** 207
o-nitrophenyl methyl ether **IX** 103
1-nitropropane **XIII** 211–213
2-nitropropane **III** 241–247
nitropyrenes **I** 112, 118, 258
N-nitrosamines **I** 23–35, 42, 58–59, 261–286, **II** 250
nitrosation of amines **I** 23–35
N-nitrosodi-*n*-butylamine **I** 24, 29, 261–262, 274–276, 286
N-nitrosodiethanolamine **I** 27, 29, 261–262, 277–278, 286
N-nitrosodiethylamine **I** 23–35, 41, 60, 261–262, 268–271, 286
N-nitrosodiisopropylamine **I** 261–262, 274, 286
N-nitrosodimethylamine **I** 3–35, 41–44, 59, 261–267, 286
N-nitrosodiphenylamine **I** 27
N-nitrosodi-*n*-propylamine **I** 261–262, 271–273, 286
N-nitrosoethylphenylamine **I** 261–262, 280, 286
N-nitrosomethylethylamine **I** 261–262, 267, 286
N-nitrosomethylphenylamine **I** 29, 261–262, 279, 286
N-nitrosomorpholine **I** 29, 34, 261–262, 280–282, 286
p-nitrosophenol **XIV** 135
4-nitrosophenol **XIV** 135–136
N-nitrosopiperidine **I** 261–262, 282–284, 286
N-nitrosopyrrolidine **I** 40, 261–262, 284–285, 286
2-nitrotoluene **VIII** 97–108, **XXI** 3
3-nitrotoluene **XXI** 3
4-nitrotoluene **XXI** 3
o-nitrotoluene **VIII** 97
5-nitro-*o*-toluidine **VI** 271–275, **XXI** 3
nitrotrichloromethane **VI** 105
nitrous oxide **IX** 115–150
nitroxanthic acid **XVII** 273
NO_x **I** 28

normal butyl thioalcohol **XXI** 161
northern red oak **XVIII** 247
northern white cedar **XIII** 288, **XVIII** 263
Noxal **V** 165
1-NP **XIII** 211
2-NP **XIII** 207
2-NPPD **XIII** 207
Nuran® **IV** 201
oak wood dust **IV** 363, **XIII** 287, **XVIII** 221
obeche **XIII** 288, **XVIII** 283
octachlorocamphene **XIX** 281
octachloronaphthalene **XIII** 101
(*Z*)-9-octadecenoic acid **XVII** 245
9-octadecenoic acid (*Z*) decyl ester **IX** 269
(*Z*)-9-octadecenoic acid isodecyl ester **XVI** 257
1-octanol **XX** 227–237
Octhilinone **XVI** 263
n-octyl alcohol **XX** 227
octylene glycol **V** 345
2-octyl-4-isothiazolin-3-one **XVI** 263–280
N-*n*-octyl-3-isothiazolone **XVI** 263
2-*n*-octyl-3-isothiazolone **XVI** 263
2-octyl-3(2*H*)-isothiazolone **XVI** 263
2,2′,2″-[(octylstannylidyne)tris(thio)]tris-(acetic acid)tri-isooctyl ester **VII** 95
2,2′,2″-[(octylstannylidyne)tris(thio)]tris-(acetic acid)tris(2-ethylhexyl) ester **VII** 95
[(octylstannylidyne)trithio]tri(acetic acid)-tris(2-ethylhexyl ester) **VII** 94
octyltin trichloride **VII** 93
octyltin tris(isooctyl thioglycolate) **VII** 94
octyltrichlorostannane **VII** 93
octyltris(2-ethylhexyloxycarbonylmethylthio)-stannane **VII** 94
oil of mirbane **XIX** 227
oil of turpentine **XIV** 173, **XVII** 315
oil of vitriol **XV** 165
OKO® **IV** 201
okoumé **XIII** 285
olaquinox **IX** 151–169
olefiant gas **X** 91
oleic acid **XVII** 245–266
oleic acid decyl ester **IX** 269
oleic acid isodecyl ester **XVI** 257
oleic sarcosine **XVI** 281
oleoyl *N*-methylaminoacetic acid **XVI** 281
oleoyl *N*-methylglycine **XVI** 281
oleoyl sarcosine **XVI** 281–286
N-oleoyl sarcosine **XVI** 281
Onyxide® 500 **II** 249
opal **II** 157
organic fibres **VIII** 205

orthoboric acid **V** 295
oxacyclopropane **V** 181
3-oxa-1-hepatanol **VI** 47
oxane **V** 181
3-oxapentane-1,5-diol **X** 73
1,2-oxathiane 2,2-dioxide **IV** 45
1,2-oxathiolane 2,2-dioxide **IV** 313
Oxazolidine E **IX** 275
oxidation base 25 **IV** 289
α,β-oxidoethane **V** 181
oxiran **V** 181
oxirane **V** 181, **XIII** 165
oxirane methane ammonium-*N*,*N*,*N*-trimethyl chloride **IV** 247
oxiranemethanol **XX** 179
3-oxiranyl-7-oxabicyclo[4.1.0]heptane **I** 391
Oxitol **VI** 205
3-oxo-1,2-benzisothiazoline **II** 221
2-oxobutane **XII** 37
oxodioctylstannane **VII** 91
2-oxo-hexamethylenimine **IV** 65
oxomethane **III** 173, **XVII** 163
oxybis(4-aminobenzene) **VI** 277
4,4′-oxybisaniline **VI** 277
4,4′-oxybisbenzenamine **VI** 277
1,1′-oxybisethane **XIII** 149
oxybismethane **I** 125
1,1′-oxybis(2-methoxyethane) **IX** 41
2,2′-oxybispropane **XXI** 187
4,4′-oxydianiline **VI** 277–286
2,2-oxydiethanol **X** 73
4,4′-oxydiphenylamine **VI** 277
oxyfume **V** 181
oxymethurea **V** 289
oxymethylene **III** 173, **XVII** 163
ozone **X** 171–199
Pacific red cedar **XIII** 288, **XVIII** 263
PAH **I** 42, 58–59, 111, 112
palygorskite **VIII** 141–338
pâo ferro **XIII** 287, 321
paraphenylenediamine **VI** 311
Paratecoma peroba **XIII** 287, **XIV** 307–309
Parmetol® F85 **XVI** 263
Paroil® **III** 81
passive smoking at work **I** 39–62, **XIII** 3–39
PCM **I** 291
PCMC **II** 271
pedunculate oak **XVIII** 247
pencil cedar **XIII** 285, **XIV** 275
pentachloronaphthalenes **XIII** 101
pentachlorophenol **III** 261–270
pentaerythritol triacrylate **XVI** 193–199
1,5-pentanedial **VIII** 45, **XVI** 59
1,5-pentanedione **VIII** 45

3-pentanol acetate **XI** 211
2-pentanol acetate **XI** 211
pentyl acetate **XI** 211–223
1-pentyl acetate **XI** 211
3-pentyl acetate **XI** 211
pentylcarbinol **IX** 283
per **III** 271
peracetic acid **VII** 229–239
perchlorobenzene **XVI** 65
perchloroethylene **III** 271
perchloromercaptan **I** 291
perchloromethane **XVIII** 81
perchloromethylmercaptan **I** 291–295
periclase **II** 145
Pericopsis elata **XIII** 287
peroba do(s) campo(s) **XIII** 287, **XIV** 307
peroxides **III** 249–260
peroxoacetic acid **VII** 229
peroxyacetic acid **III** 250, 253, 254, **VII** 229
petrol engine emissions **I** 109–120
3-phenoxy-1,2-epoxypropane **IV** 305
(phenoxymethyl)oxirane **IV** 305
phenoxypropene oxide **IV** 305
phenoxypropylene oxide **IV** 305
phenylamine **VI** 17
phenyl carbimide **XVII** 267
phenyl chloroform **VI** 80
2,2′-[1,3-phenylenebis(oxymethylene)]-bisoxirane **VII** 69
1,2-phenylenediamine **XIII** 215
1,3-phenylenediamine **VI** 287
1,4-phenylenediamine **VI** 311
m-phenylenediamine **VI** 287–299, **XXI** 3
o-phenylenediamine **XIII** 215–235, **XXI** 3
p-phenylenediamine **VI** 311–337, **XIV** 137–141, **XXI** 3
p-phenylenediamine compounds **XV** 151–153
phenyl 2,3-epoxypropyl ether **IV** 305
phenylethylene **XX** 285
phenyl glycidyl ether **IV** 59, 305–311
phenylhydrazine **XI** 225–234
phenyl isocyanate **XVII** 267–272
phenyl methanal **XVII** 13
phenylmethane **VII** 257
(phenylmethoxy)methanol **II** 239
[(phenylmethoxy)methoxy]methanol **II** 239
[[(phenylmethoxy)methoxy]methoxy]-methanol **II** 239
N-phenylmethylamine **VI** 263
1-phenyl-1-methylethylene **XV** 137
phenyl mono(poly)methoxy methanol **II** 239
4-phenylnitrobenzene **I** 257
p-phenylnitrobenzene **I** 257

2-phenylphenol **II** 299
o-phenylphenol **II** 299–316
o-phenylphenol sodium salt **II** 299
2-phenylpropane **XIII** 117
2-phenylpropene **XV** 137
phenyl trichloromethane **VI** 80
philosopher's wool **XVIII** 305
phosphomolybdic acid **XVIII** 199
phosphonic acid dimethyl ester **I** 133
phosphoric acid 2,2-dichloroethenyl dimethyl ester **IV** 201
phosphoric acid 2,2-dichlorovinyl dimethyl ester **IV** 201
phosphoric acid tributyl ester **XVII** 285
phosphoric acid triphenyl ester **II** 321
phosphorothioic acid *O,O*-diethyl *O*-[2-(ethylthio)ethyl] ester **XIX** 141
phosphorothioic acid *O,O*-diethyl *O*-[6-methyl-2-(1-methylethyl)-4-pyrimidinyl] ester **XI** 31
phosphorous acid dimethyl ester **I** 133
1,3-phthalandione **VII** 247
phthalic acid **VII** 241–246
phthalic acid anhydride **VII** 247, **XI** 235
phthalic acid diallyl ester **IX** 11
phthalic anhydride **VII** 247–255, **XI** 235–237
physical activity and inhalation kinetics **XIV** 3–10
Pic-clor **VI** 105
Picfume® **VI** 105
picric acid **XVII** 273–284
Picride **VI** 105
picronitric acid **XVII** 273
picrylmethylnitramine **XI** 179
picrylnitromethylamine **XI** 179
pimelic ketone **X** 35
Pinus spp. **XIII** 287
piperazidine **IX** 303, **XII** 177
piperazine **IX** 303–313, **XII** 177–178
piperidine **I** 25
pitch pine **XIII** 287
plaster **II** 117
plaster of Paris **II** 117
plicamycin **I** 5
Pluracol® P410, P1010, P2010, P4010 **X** 271
Pluriol® P600, P2000 **X** 271
PMDI **VIII** 65–95, **XIV** 131
PNB **I** 257
polychlorinated camphenes **XIX** 281
polychlorocamphene **XIX** 281
polyethylene glycol **X** 247–270
polyethylene oxide **X** 247
polyglycol **X** 247

Polyglycol P425, P1200, P2000, P4000 **X** 271
"polymeric MDI" **VIII** 65–95, **XIV** 131
polyoxyethylene **X** 247
poly(propane-1,2-diol) **X** 271
polypropylene glycol **X** 271–285
polypropylene oxide **X** 271
Poly-Solv® TB **IX** 315
polyvinyl chloride **II** 149–156
portland cement **XI** 303–333
potassium citrate **XVI** 209
potassium cyanide **XIX** 189
potassium titanates **VIII** 141–338
potassium zinc chromate **XV** 289
PPG **X** 271
Preventol® CMK **II** 271
Preventol® D 2 **II** 239
Preventol® O Extra **II** 299
Preventol® ON Extra **II** 299
procarbazine **I** 7
1,2-propanediol-1-acrylate **XVI** 95
1,3-propanediol-1-acrylate **XVI** 95
1,2-propanediol-2-acrylate **XVI** 95
1,3-propane sultone **IV** 46, 313–321
2-propanone **VII** 1
propargyl alcohol **XXI** 261–269
2-propenal **XVI** 1
2-propenamide **III** 11
propene oxide **V** 221
2-propenoic acid butyl ester **V** 5, **XII** 57, **XVI** 35
2-propenoic acid ethyl ester **VI** 217, **XVI** 41
2-propenoic acid 2-ethylhexyl ester **XVI** 47
2-propenoic acid 2-ethyl-2-[[(1-oxo-2-propenyl)oxy]methyl]-1,3-propane-diyl ester **XVI** 201
2-propenoic acid homopolymer, sodium salt **XV** 1
2-propenoic acid 2-hydroxyethyl ester **XVI** 89
2-propenoic acid 2-hydroxy-1-methylethyl ester **XVI** 95
2-propenoic acid 2-(hydroxymethyl)-2-[[(1-oxo-2-propenyl)oxy]methyl]-1,3-propanediyl ester **XVI** 193
2-propenoic acid 2-hydroxypropyl ester **XVI** 95
2-propenoic acid 3-hydroxypropyl ester **XVI** 95
2-propenoic acid hydroxypropyl ester **XVI** 95
2-propenoic acid methyl ester **VI** 253, **XVI** 177
1-propenol-3 **XV** 31

2-propen-1-ol **XV** 31
[(2-propenyloxy)methyl]oxirane **VII** 9
2-propoxyethanol **XII** 179–186
2-propoxyethanol acetate **XII** 187
2-propoxyethyl acetate **XII** 187–193
propyl carbinol **XIX** 99
propyl cellosolve **XII** 179
propylene chloride **IX** 21
propylene dichloride **IX** 21
propylene epoxide **V** 221
propylene glycol 1-methyl ether **V** 213, **XIV** 127
propylene glycol 2-methyl ether **I** 203
propylene glycol 1-methyl ether 2-acetate **V** 217
propylene glycol 2-methyl ether 1-acetate **I** 207
propylene glycol monoacrylate **XVI** 95
propylene glycol monomethyl ether **V** 213
1,2-propylene oxide **V** 221–228
n-propyl glycol **XII** 179
Protectol® DMU **V** 289
Proxel® **II** 221
pruno **XIII** 287, **XIV** 299
pseudothiourea **I** 301
PVC **II** 149–156
pyrazine hexahydride **IX** 303, **XII** 177
2-pyridinethiol 1-oxide sodium salt **X** 287
pyroacetic acid **VII** 1
pyroacetic ether **VII** 1
pyrrolylene **XV** 51
QUAB 151® **IV** 247
quartz **I** 113, **II** 129–134, 182–195, **XIV** 205–271
quartz glass **II** 157
Quercus spp **XIII** 287, **XVIII** 247–252
quinone monoxime **XIV** 135
quinone oxime **XIV** 135
R 11 **I** 335, **III** 366
R 12 **V** 109
R 13 **I** 69
R 21 **V** 119
R 22 **III** 63
R 31 **V** 75
R 112 **I** 297
R 113 **III** 365, 366
R 142b **I** 65
radon **II** 136
ramin **XIII** 286, **XVIII** 235
RDGE **VII** 69
red cedar **XIII** 288, **XVIII** 263
red oak **XVIII** 247
red oxide of zinc **XVIII** 305
red pine **XIII** 287

reduced Michler's ketone **I** 249
Remol TRF® **II** 299
Reomet® SBT 75 **II** 317
resorcin **XX** 239
resorcinol **XX** 239–273
resorcinol bis(2,3-epoxypropyl) ether **VII** 69
resorcinol diglycidyl ether **VII** 69
resorcinyl diglycidyl ether **VII** 69
respirable dust **XII** 239
Rewopon® IM OA **V** 373
RH 893 HQ Technical **XVI** 263
RH 893 T **XVI** 263
Ribeclor® **III** 81
rock wool **VIII** 141–338
rosewood **XIII** 286, 313
rosin **XI** 239–242
Ro-sulfiram® **V** 165
rotenone **XIX** 261–271
rubber components **XV** 145–164
Rutgers 612® **V** 345
rutile **II** 199
rye flour dust **XIII** 99–100
samba **XVIII** 283
Samba scleroxylon **XIII** 288
Santos rosewood **XIII** 287, 321
sapele **XVIII** 229
sapeli **XVIII** 229
sapupira (da mata) **XIII** 285, 295
Sarkosyl® O **XVI** 281
Sarkosyl® OT **XVI** 281
Scots pine **XIII** 287
selenite **II** 117
sepiolite **VIII** 141–338
Sequoia sempervirens **XIII** 287
serpentine **II** 182, 183
sessile oak **XVIII** 247
shinglewood **XIII** 288, **XVIII** 263
short-term exposures **I** 15–22
sidestream smoke **XIII** 3
silica, amorphous **II** 157–179
silica, crystalline **XIV** 205–271
silica fume **II** 160, 165
silica gel **II** 157
silica glass **II** 157
silicic acid tetraethyl ester **III** 299
silicon carbide **VIII** 141–338
silicon dioxide, crystalline **XIV** 205–271
silk(y) oak **XIII** 286
silver oak **XIII** 286, **XIV** 295
sipo **XVIII** 229
Skane® M-8 **XVI** 263
Skane® M-8 HQ **XVI** 263
slag wool **VIII** 141–338
soapstone **II** 181

soda lye **XII** 195
sodium arsenate **XXI** 49
sodium arsenite **XXI** 49
sodium azide **XX** 275–284
sodium (1,1'-biphenyl)-2-olate **II** 299
sodium 2-biphenylolate **II** 299
sodium citrate **XVI** 209
sodium cyanide **XIX** 189
sodium ethylmercurithiosalicylate **XV** 249
sodium hydrate **XII** 195
sodium hydroxide **XII** 195–207
sodium 2-hydroxybiphenyl **II** 299
sodium 1-hydroxy-2-(1H)-pyridinethione **X** 287
sodium molybdate **XVIII** 199
Sodium Omadine® **X** 287
sodium o-phenylphenate **II** 299
sodium o-phenylphenol **II** 299–316
sodium o-phenylphenolate **II** 299
sodium o-phenylphenoxide **II** 299
sodium pyridinethione **X** 287
sodium pyrithione **X** 287–311
solvent yellow 34 **IV** 11
South American mahogany **XIII** 287
Southern silk(y) oak **XIII** 286, **XIV** 295
spanon **VI** 57
spartalite **XVIII** 305
spinel **III** 128
spirit of hartshorn **VI** 1
spirits of salt **VI** 231
spirits of turpentine **XIV** 173, **XVII** 315
spirits of wine **XII** 129
steatite **II** 181
sterlingite **XVIII** 305
Stopetyl **V** 165
Stopmold B® **II** 299
strychnidin-10-one **XIX** 273
strychnine **XIX** 273–279
styrene **XX** 285–290
styrol **XX** 285
succinic acid peroxide **III** 254
sucupira **XIII** 285, 295
sulfate of lime **II** 117
sulfate turpentine **XIV** 173, **XVII** 315
sulfinylbismethane **III** 163
sulfotep(p) **XIII** 237
sulfuric acid **XV** 165–222
sulfuric acid calcium salt **II** 117
sulfuric acid diethyl ester **XX** 93
sulfuric acid dimethyl ester **IV** 217
sulfuric ether **XIII** 149
sulfur mustard **IV** 27
sulourea **I** 301

γ-sultone **IV** 45
Swietenia spp. **XIII** 287, **XVIII** 253–258
systox **XIX** 141
2,4,5-T **XI** 243
Tabebuia spp. **XIII** 288, **XIV** 311–315
talc **II** 181–198
tantalum **XVI** 293–296
Tarrietia utilis **XIII** 288
Tasmanian blackwood **XIII** 285, 291
TBP **IX** 245
2,4-TDI **XX** 291
2,6-TDI **XX** 291
teak **XIII** 288, **XIV** 317
Tecoma peroba **XIII** 287
Tectona grandis **XIII** 288, **XIV** 317–323
TEDP **XIII** 237–249
tegafur **I** 4
TEL **XV** 223
teniposide **I** 6
terephthalic acid **VII** 241
Terminalia ivorensis **XVIII** 259–262
Terminalia superba **XIII** 288, **XVIII** 259–262
3,3',4,4'-tetraaminobiphenyl **III** 131
1,3,5,7-tetraazaadamantane **V** 355
1,3,5,7-tetraazatricyclo[3.3.1.1^{3,7}]decane **V** 355
2,4,5,6-tetrachloro-1,3-benzene-dicarbonitrile **VI** 113
2,4,5,6-tetrachloro-3-cyanobenzonitrile **VI** 113
1,1,2,2-tetrachloro-1,2-difluoroethane **I** 297–299
tetrachloroethene **III** 271
tetrachloroethylene **III** 271–297
2,4,5,6-tetrachloroisophthalonitrile **VI** 113
tetrachloromethane **XVIII** 81
tetrachloronaphthalenes **XIII** 101
m-tetrachlorophthalodinitrile **VI** 113
m-tetrachlorophthalonitrile **VI** 113
α,α,α,4-tetrachlorotoluene **X** 21
Tetradine **V** 165
tetraethoxysilane **III** 299
tetraethyl dithiopyrophosphate **XIII** 237
tetraethyllead **XV** 223–235
tetraethyl orthosilicate **III** 299
tetraethylplumbane **XV** 223
tetraethyl silicate **III** 299–305
tetraethylthioperoxydicarbonic diamide **V** 165
tetraethylthiram disulfide **V** 165
N,N,N',N'-tetraethylthiuram disulfide **V** 165
Tetraetil **V** 165

1,1,1,2-tetrafluoroethane **XIII** 251–266
4,5,6,7-tetrahydro-1*H*-1,2,3-benzotriazole **II** 317
4,5,6,7-tetrahydro-1*H*-benzotriazole **II** 317–319
[2*R*-(2α,6aα,12aα)]-1,2,12,12a-tetrahydro-8,9-dimethoxy-2-(1-methylethenyl)-[1]-benzopyrano[3,4-*b*]furo[2,3-*h*][1]benzopyran-6(6a*H*)-one **XIX** 261
tetrahydro-*p*-dioxin **XX** 105
3α,4,7,7α-tetrahydro-4,7-methano-1*H*-indene **V** 125
1,2,3,4-tetrahydrostyrene **XIV** 185
tetralit **XI** 179
tetralite **XI** 179
tetramethyldiaminodiphenylacetimine **IV** 11
N,N,N',N'-tetramethyl-*p,p'*-diaminodiphenylmethane **I** 249
tetramethyl-*p*-diamino-imido-benzophenone **IV** 11
tetramethyllead **XV** 237–248
tetramethylplumbane **XV** 237
tetramethylthioperoxydicarbonic diamide **XV** 163, **XIX** 141
tetramethylthiuram disulfide **XV** 163–164
tetranitromethane **IV** 323–330
Tetrosin OE® **II** 299
tetryl **XI** 179
teturamin **V** 165
thiazols **XV** 155–157
thimerosal **XV** 249–255
thioaniline **IV** 331
4,4'-thiobis(aniline) **IV** 331
4,4'-thiobisbenzenamine **IV** 331
1,1'-thiobis(2-chloroethane) **IV** 27
2,2'-thiobis(4,6-dichlorophenol) **IX** 245
thiobutyl alcohol **XXI** 161
thiocarbamide **I** 301, **XIV** 143
thiocarbonyl tetrachloride **I** 291
4,4'-thiodianiline **IV** 331–334
p,p'-thiodianiline **IV** 331
thiodi-*p*-phenylenediamine **IV** 331
thiodiphosphoric acid tetraethyl ester **XIII** 237
thioethyl alcohol **XXI** 195
thioguanine **I** 4
Thiomersal **XV** 249
Thiomersalate **XV** 249
thiomethyl alcohol **XX** 217
thiopyrophosphoric acid tetraethyl ester **XIII** 237
thiotepa **I** 3
thiotepp **XIII** 237
thiourea **I** 301–312, **XIV** 143–148

thiram **XV** 163
thiurams **XV** 159–164
Thiuranide **V** 165
THT **IX** 323
Thuja spp. **XIII** 288, **XVIII** 263–276
Tieghemella africana **XVIII** 277–281
Tieghemella heckelii **XIII** 288, **XVIII** 277–281
tin **XIV** 149–165
tin compounds, inorganic **XIV** 149–165
tin compounds, organic **I** 317
titanium dioxide **II** 199–204
titanium oxide **II** 199
TMTD **XV** 163
TNT **I** 359
tobacco smoke **I** 28, 39–62, **XIII** 3
2-tolidine **V** 153
3,3'-tolidine **V** 153
o-tolidine **V** 153
tolite **I** 359
toluene **VII** 257–318
2,4-toluenediamine **VI** 339
toluene-2,4-diamine **VI** 339–352, **XXI** 3
m-toluenediamine **VI** 339
toluene diisocyanate **XX** 291–338
toluene-2,4-diisocyanate **XX** 291
toluene-2,6-diisocyanate **XX** 291
toluene trichloride **VI** 80
2-toluidine **III** 307
4-toluidine **III** 323
o-toluidine **III** 307–322, 329, **VI** 144, **XXI** 3
p-toluidine **III** 323–330, **XXI** 3
toluol **VII** 257
2,4-toluylenediamine **VI** 339
m-toluylenediamine **VI** 339
o-tolylamine **III** 307
p-tolylamine **III** 323
tolyl chloride **VI** 79
tolyltriazole **II** 295
Topane® **II** 299
total dust **II** 52
toxaphene **XIX** 281–297
tremolite **II** 96, 182–187
treosulfan **I** 3
tretamine **I** 3
tri **X** 201
triatomic oxygen **X** 171
1,2,3-triazaindene **II** 231
1,3,5-triazin-1,3,5(2*H*,4*H*,6*H*)-triethanol **II** 331
1*H*-1,2,4-triazol-3-amine **IV** 1, **XVIII** 19
tribromomethane **VII** 319–332
tributylchlorostannane **I** 315
tributylfluorostannane **I** 315

tributyl-(2-methyl-1-oxo-2-propenyl)oxy-
  stannane **I** 316
tributylmono(naphthenoyl-oxy)stannane
  derivatives **I** 316
tributyl-(1-oxo-9,12-octadecadienyl)oxy-
  (Z,Z)-stannane **I** 316
tributyl phosphate **XVII** 285–314
tri-*n*-butyl phosphate **XVII** 285
tri-*n*-butyltin benzoate **I** 315
tri-*n*-butyltin chloride **I** 315
tri-*n*-butyltin compounds **I** 315–331
tri-*n*-butyltin fluoride **I** 315
tri-*n*-butyltin linoleate **I** 316
tri-*n*-butyltin methacrylate **I** 316
tri-*n*-butyltin naphthenate **I** 316
tributyltin oxide **I** 315
1,2,3-trichlorobenzene **III** 331–363
1,2,4-trichlorobenzene **I** 72, **III** 331–363,
  XIV 167–172
1,3,5-trichlorobenzene **III** 331–363
1,1,2-trichloroethane **V** 66, 72
trichloroethene **X** 201
trichloroethylene **III** 273, 369, **X** 201–244
trichlorofluoromethane **I** 69, 335–344
trichlorohydrin **IX** 171
trichloromethane **XIV** 19
trichloromethanesulfenyl chloride **I** 291
(trichloromethyl)benzene **VI** 80
trichloromethylsulfenyl chloride **I** 291
trichloromonofluoromethane **I** 335
trichloronaphthalenes **XIII** 101
trichloronitromethane **VI** 105
trichlorooctylstannane **VII** 93
2,4,5-trichlorophenol **XII** 209–221
2,4,5-trichlorophenoxyacetic acid **XI** 243–
  278
2,4,5-trichlorophenoxyacetic acid salts and
  esters **XI** 243
trichlorophenylmethane **VI** 80
1,2,3-trichloropropane **IX** 171–192
α,α,α-trichlorotoluene **VI** 80
ω,ω,ω-trichlorotoluene **VI** 80
1,1,2-trichloro-1,2,2-trifluoroethane **III** 365–
  370
tridymite **XIV** 205
triethanolamine **I** 27
triethoxybutanol **IX** 315
triethylamine **I** 26, 33, **XIII** 267–280
triethylene glycol *n*-butyl ether **IX** 315–321
triethylene glycol monobutyl ether **IX** 315
1,3,5-triethylol-hexahydro-s-triazine **II** 331
trifluoroborane **XIII** 93
trifluorobromomethane **VI** 37
trifluorochloromethane **I** 69

1,1,1-trifluoro-2,2-dichloroethane **X** 53
trifluoromethyl chloride **I** 69
trifluoromonochlorocarbon **I** 69
trifluorotrichloroethane **III** 365
triglycol monobutyl ether **IX** 315
trihydroxymethylnitromethane **II** 287
trimethylamine **I** 26, 28
1,2,4-trimethyl-5-aminobenzene **IV** 335
2,4,5-trimethylaniline **IV** 335–340, **XXI** 3
2,4,5-trimethylbenzenamine **IV** 335
sym-trimethylbenzene **IV** 341
1,3,5-trimethylbenzene **IV** 341–348
2,4,5-trimethylbenzeneamine **IV** 335
trimethyl carbinol **XIX** 125
trimethylolmelamine **I** 7
trimethylolnitromethane **II** 287
trimethylolpropane triacrylate **XVI** 201–206
N,N,N-trimethyl-oxiranemethanaminium
  chloride **IV** 247
2,2,4-trimethylpentane **I** 196, 347–356
2,4,6-trinitrophenol **XVII** 273
2,4,6-trinitrophenylmethylnitramine **XI** 179
2,4,6-trinitrophenyl-*N*-methylnitramine **XI**
  179
trinitrotoluene (all isomers) **I** 148, 359–360
2,4,6-trinitrotoluene **XXI** 3
3,6,9-trioxa-1-tridecanol **IX** 315
triphenyl phosphate **II** 321–330
Triplochiton scleroxylon **XIII** 288, **XVIII**
  283–289
tris(dimethyldithiocarbamato)iron **XIX** 163
tris(2-hydroxyethyl)-hexahydro-1,3,5-triazine
  II 331
N,N',N''-tris(β-hydroxyethyl)-hexahydro-
  1,3,5-triazine **II** 331–339, **IX** 323–324
tris(ω-hydroxyethyl)-hexahydro-1,3,5-triazine
  II 331
tris(hydroxymethyl)nitromethane **II** 287
Tris Nitro® **II** 287
tris(oxymethyl)nitromethane **II** 287
tritol **I** 359
triton **I** 359
trofosfamide **I** 3
trotyl **I** 359
Troysan® KK-108A **XVI** 247
Troysan® Polyphase® Anti-Mildew **XVI** 247
Troysan® Polyphase® P100 **XVI** 247
true teak **XIII** 288
TTD **V** 165
Tumescal OPE® **II** 299
turpentine **XIV** 173–179, **XVII** 315–332
turpentine oil **XIV** 173, **XVII** 315
Turraeanthus africanus **XIII** 288
ultrafine aerosol particles **XVI** 289–292

Uniclor® **III** 81
Uritone® **V** 355
Urotovet® **V** 355
Urotropin® **V** 355
Ursol D® **VI** 311
utile **XVIII** 229
δ-valerosultone **IV** 45
vanadic anhydride **IV** 349
vanadium compounds **IV** 349–361
vanadium(V) oxide **IV** 349
vanadium pentoxide **IV** 349–361
Vapona® **IV** 201
Varine® O **V** 373
VDC **VIII** 109
vinblastine sulfate **I** 6
vincristine sulfate **I** 6
vindesine **I** 6
vinegar naphtha **XII** 167
vinyl acetate **V** 229–239, **XXI** 271–294
vinyl alcohol 2,2-dichloro dimethyl phosphate **IV** 201
vinylbenzene **XX** 285
vinylbutyrolactam **V** 249
vinylcarbazole **XIV** 181–183
N-vinylcarbazole **XIV** 181
vinyl carbinol **XV** 31
vinyl chloride **I** 40, **II** 149–156, **V** 66, 71, 241–248
1-vinylcyclohex-3-ene **XIV** 185
4-vinylcyclohexene **XIV** 185–202
vinyl cyclohexene diepoxide **I** 391
4-vinyl-1,2-cyclohexene diepoxide **I** 391
4-vinyl-1-cyclohexene dioxide **I** 391–395
1-vinyl-3-cyclohexene dioxide **I** 391
vinyl ethanoate **V** 229
vinylethylene **XV** 51
vinylidene chloride **V** 66, 72, **VIII** 109–139
vinylidene dichloride **VIII** 109
vinylidene fluoride **V** 135
1-vinyl-2-pyrrolidinone **V** 249
N-vinylpyrrolidinone **V** 249
1-vinyl-2-pyrrolidone **V** 249

N-vinyl-2-pyrrolidone **V** 249–261
vitreous silica **II** 157
walnut **XIII** 286
wawa **XIII** 288, **XVIII** 283
western red cedar **XIII** 288, **XVIII** 263
wheat flour dust **XIII** 99–100
white acajou **XIII** 287
white afara **XIII** 288
white cedar **XIII** 288, **XVIII** 263
white peroba **XIII** 287, **XIV** 307
wishmore **XIII** 288
Witaclor® **III** 81
wollastonite **VIII** 141–338, **XVI** 297–315
wood dust **IV** 363–373, **XIII** 283–323, **XIV** 273–323, **XVIII** 221–227
wood ether **I** 125
woods **IV** 363–373, **XIII** 283–323, **XIV** 273–323, **XVIII** 229–289
wood turpentine **XIV** 173, **XVII** 315
xylene **V** 263–285, **XV** 257–288
xylidine isomers **XIX** 299–311, **XXI** 3
xylol **V** 263
yellow pine **XIII** 287
zebrano **XIII** 287
zebrawood **XIII** 287
zeolites **VIII** 141–338
zinc chloride **XVIII** 291–303
zinc chloride fume **XVIII** 291–303
zinc chromate **III** 101–122, **XV** 289–294
zinc chromate hydroxide **XV** 289
zinc dichloride **XVIII** 291
zincite **XVIII** 305
zinc molybdate **XVIII** 199
zinc oxide **XVIII** 305–323
zinc oxide fume **XVIII** 305–323
zinc potassium chromate **XV** 289
zinc white **XVIII** 305
zinc yellow **XV** 289
zirconium **XII** 223–236
zirconium alloys and compounds **XII** 223
Zoldine® ZE **IX**